New ICMI Study Series

Volume 16

For further volumes:
http://www.springer.com/series/6351

Published under the auspices of the International Commission on Mathematical Instruction under the general editorship of

Abraham Arcavi, President   Ferdinando Arzarello, Secretary-General

Information on the ICMI Study programme and on the resulting publications can be obtained at the ICMI website http://www.mathunion.org/ICMI/ or by contacting the ICMI Secretary-General, whose email address is available on that website.

Alain Damlamian · José Francisco Rodrigues
Rudolf Sträßer
Editors

Educational Interfaces between Mathematics and Industry

Report on an ICMI-ICIAM-Study

International Commission on
Mathematical Instruction

Springer

Editors
Alain Damlamian
Université Paris-Est Créteil Val de Marne
Créteil Cedex, Val-de-Marne
France

José Francisco Rodrigues
Universidade de Lisboa/CMAF-FCUL
Lisboa
Portugal

Rudolf Sträßer
FB 12 Mathe, Inst. Didaktik der Mathematik
Universität Gießen
Gießen
Germany

ISSN 1387-6872
ISBN 978-3-319-02269-7 ISBN 978-3-319-02270-3 (eBook)
DOI 10.1007/978-3-319-02270-3
Springer Cham Heidelberg New York Dordrecht London

Library of Congress Control Number: 2013953250

Printed on acid-free paper

Springer is part of Springer Science+Business Media (www.springer.com)

Foreword I

It is a great pleasure to introduce to the community the fruits of this novel study, the first to combine the expertise of ICMI on education in the mathematical sciences with knowledge coming from the ICIAM community of the ways mathematics is used in industry. The idea of starting such a cooperation between ICMI and ICIAM arose while Rolf Jeltsch was President of ICIAM, and Rolf has in addition contributed to the project through his involvement with one of the Working Groups. Let me use this foreword to acknowledge his leadership.

The dedication and the persistence of the many people who worked on this project deserve the thanks of all of us. Their names are repeated in the prologue and conclusions, but I would like to single out for my special thanks the editorial team of Alain Damlamian, José Francisco Rodrigues, and Rudolf Sträßer. They are the ones responsible for getting the job done, despite obstacles ranging from the usual procrastination of authors to the unusual eruption of a volcano. Despite the difference in cultures between ICMI and ICIAM, and the fact that neither community had previously focused on this precise question, the study has come together in rather short order to produce a work that outlines the challenges of teaching mathematics in such a way that students are ready to use it outside of mathematics. The study reaches some important conclusions, and lays the groundwork for further studies.

As any good scientist knows, one needs data in order for knowledge to advance, and this book provides, along with its analysis, a library of data in the form of case studies, examples of existing programs, and working group reports. The data include examples from industries where one expects to see mathematics used extensively, and in sophisticated ways, such as aerospace and finance, and from industries where the importance of mathematical understanding might come as a surprise, such as nursing or retail sales. Incidentally, one will also find an international perspective, as our two societies, both international, recruited contributors from industrialized and developing countries alike, from economic leaders, and from those attempting to catch up. Whether browsing the volume or reading it systematically, the reader will be introduced to the many dimensions of "mathematical education for industry," as different contributors present varied experiences, on how mathematics is used and on how mathematics is taught, and on the relation between these two things.

The readers for whom this book is intended—applied mathematicians of all stripes, mathematics educators, and public officials interested in the educational and economic development of their countries—will all find much to learn from its pages. They are not likely to find the answers to their questions neatly spelled out, nor do the authors claim to have produced conclusive results, organized and summarized in some optimal way. That is a project for the future. But a future that improves the education of young people who will work as mathematicians in industry, or who will work as "analysts," "scientists," "specialists," "engineers," or who will work at a job whose title has no relation to mathematics or to science, has been brought measurably closer by the completion of this volume.

Columbus, Ohio, May 30, 2013

Barbara Lee Keyfitz
President, ICIAM

Foreword II

Let me begin with congratulations to the Co-chairs of this ICMI-study. Not only did they overcome active volcanoes and geographic challenges, but also they have managed to bring together two distinct communities and produced a coherent volume that integrates their views, their research, and their aspirations.

Thanks must also go to those who had the vision to initiate this study. ICMI welcomed the opportunity to engage professionally with the Industrial Mathematics community. The result is both rich and productive.

Readers of this volume will quickly realize that a repeating theme is diversity: diversity of communities, diversity of mathematics, diversity of goals, and diversity of actions. It would be easy to quickly conclude that there is little common ground in this theme and abandon the project. Fortunately, the participants of Study 20 did not take this path. Using "boundary objects" as a starting point, mutual interests were identified and their relevance to each of the two communities explored. The key idea of modeling is a prime example of a concept with a powerful role in both industry and mathematics education.

But there is a wide issue being brought into place in this process. As an international community, a central concern is bringing together different cultures, be they ethnic, social, academic, linguistic, or otherwise. The ICMI Study 20 is an exemplification of this process.

Thus, it is with pleasure that I commend the present volume to the research communities of ICIAM and ICMI. In doing so, ICMI willingly takes up the challenge of supporting further research into educational issues in industrial mathematics.

Ferdinando Arzarello

Prologue

At the ICIAM Congress in Zurich (July 17–20, 2007), the President of the Portuguese National Committee of Mathematics (CNM)[1] made a formal proposal to the ICIAM Board for a new ICMI Study entitled "Educational Interfaces between Mathematics and Industry." He had already approached ICMI regarding it and hoped for a joint support. The ICIAM Board accepted the proposal enthusiastically. This was how the ICMI-Study 20 started.

Several other people and institutions felt the need for such a Study, in particular, as a consequence of a recent OECD report on Mathematics and Industry. So, ICMI nominated Rudolf Sträßer as Co-chair while ICIAM proposed Alain Damlamian. José Francisco Rodrigues joined the team as organizer of the Study Conference.

This was the beginning of a very interesting adventure which none of us had expected would last that long. First were the meetings of the International Programme Committee in Óbidos, sponsored by the Centro Internacional de Matemática (CIM) (in October 2008, to draft the Discussion Document for the Study) and in Paris (in November 2009, to plan the Study Conference). The Study Conference itself, originally planned in Lisbon in late April 2010, had to be postponed to October 2010 because of the serious travel disruptions provoked by the Icelandic volcano eruption. There was also a follow-up conference in Macau (in November 2011).

We hope that the results of this study (the Lisbon Conference Proceedings and the present volume) made the journey worth the while.

This Study would not have been possible without the dedication and support of many individuals and organizations. We would like to acknowledge all participants of the conferences in Lisbon and Macau, the contributors to the Lisbon Proceedings, and to this book. Our special thanks go in particular to the members of the IPC (A. Jofré, B. R. Hodgson, B. Lutz-Wetphal, F. Santosa, G. Kaiser, G. Fitzsimons, H. Aslaksen, H. van der Kooij, J. Carvalho e Silva, J. M. Gambi, T-T. Li, N. Nigam, R. Jeltsch, S. Garfunkel and T. Mitsui), to the Co-editors of the Proceedings and Co-organizers of the Lisbon conference (A. Araújo, A. Azevedo

[1] As it turned out, he is one of the Co-editors of this Study.

and A. Fernandes), and to the Co-organizers of the Macau meeting (R. Martins, T. Qian and R. Cheng).

Among the institutions which supported and contributed to this study, in addition to ICMI and ICIAM, we wish to acknowledge the Portuguese CIM and FCT (Fundação para a Ciência e Tecnologia), as well as the Universities of Lisbon and of Macau.

We apologize for any omissions and mistakes still present in this prologue and in the book as a whole.

Paris, 1 March 2013 — Alain Damlamian
Lisboa — José Francisco Rodrigues
Gießen/Münster — Rudolf Sträßer

Contents

Part III WG Education/Training with Industry Participation

Part IV WG University and Academic Technical/Vocational Education

Part I
Discussion Document and Study Report

Discussion Document

The International Programme Committee

1 Introduction

The ICMI/ICIAM-Study on ′Educational Interfaces between Mathematics and Industry′ (EIMI-Study) starts from two assumptions, namely:

1. There are intimate connections between innovation, science, mathematics and the production and distribution of goods and services in society. In short: there are intimate connections between mathematics and industry;
2. In view of these connections, there is a need for a fundamental analysis and reflection on strategies for the education and training of students and maybe the development of new ones.

The EIMI-Study, organised jointly by the International Commission on Mathematical Instruction (ICMI) and the International Council for Industrial and Applied Mathematics (ICIAM), seeks to better understand these connections and to offer ideas and suggestions on how education and training can contribute to enhancing both individual and societal developments.

1.1 Tentative Description of the Field

Historically, there have been productive interactions between mathematics and industry in generating and solving problems associated with the development of humankind, economically and socially. In a modern, technological world, mathematics is said to be used almost everywhere. However, these uses are not gen-

This document is reprinted with authorization from "L'Enseignement Mathématique" (the official organ of ICMI) and the ZDM Mathematics Education (2009) 41:525–533 (DOI: 10.1007/s11858-009-0194–4)

A. Damlamian et al. (eds.), *Educational Interfaces between Mathematics and Industry*, New ICMI Study Series 16, DOI: 10.1007/978-3-319-02270-3_1,

erally visible except to specialists. Even people using mathematics in their workplaces may not recognise its presence.

There have been many studies of the mathematics used in the workplace—ranging from descriptive lists of traditional school-based topics to sociological studies of workplace activities set in context. There have been many collections of applications of problem solving and modelling based on or informed by practical industrial problems, especially at higher levels of mathematics, in fields such as the natural and physical sciences, engineering and finance.

Internationally, there are frequent articles and debates in the popular media citing employer dissatisfaction with the perceived quality of mathematics education. Graduates from schools, vocational colleges and universities often appear unable to draw upon and use mathematics in work situations as opposed to classroom or examination contexts. At all educational levels, students typically have been taught the tools of mathematics with little or no mention of authentic real world applications, and with little or no contact with what is done in the workplace (be it the classical engineering situations or other more recent activities like biotechnology, biomedicine, the financial, insurance and risk sector or consulting engineering companies).

Nowadays, highly complex problems need to be solved and, hence, some training to solve such problems—in particular, real life problems—is necessary. Increasingly, powerful computers make it possible to treat such complex problems and this is achieved not only using off-the-shelf software but with innovation, often mathematical innovation requiring insight and analysis.

In order to better understand these phenomena, the Study starts from a broad definition of **Industry** (from the Organisation for Economic Co-operation and Development) "… broadly interpreted as any activity of economic or social value, including the service industry, regardless of whether it is in the public or private sector" (OECD 2008, p. 4). The term "industry" obviously refers to a diverse range of activities, producing goods and services. Under constraints, such as time and money, these activities generally attempt to optimise limited—sometimes scarce—resources, both material and intellectual. The overarching goal is to maximise benefits for certain groups of people while, ideally, minimising harm to other groups and the natural environment.

"**Mathematics** (or the mathematical sciences, here the two terms are used interchangeably) comprises any activity in the mathematical sciences, including mathematical statistics" (OECD 2008, p. 4). Workers at all levels utilise mathematical ideas and techniques, consciously or unconsciously, in the process of achieving the desired workplace outcome. In other words, mathematics is just one part of a repertoire of tools and strategies of a practical nature. However, as a major factor in decision-making and communication processes, it is crucial that mathematics be used appropriately, accurately, and with confidence. For this Study, we start from the assumption that professional mathematicians are located in the academia, in industry, sometimes in both. The discourse of mathematics in all its various specialisations involves certain ways of thinking and acting. Traditionally, mathematicians consider axioms and definitions and make logical deductions. In mathematical modelling, one formulates problems in mathematical terms. However, the mathematical solution needs to take into account the industrial context.

This Study will examine the implications for education at the intersection of two communities of practice—industrialists and mathematicians or industry and mathematics. We wish to emphasise that there should be a balance between the perceived needs of industry for relevant mathematics education and the needs of learners for lifelong and broad education in a globalised environment. In other words, learners should be equipped for flexibility in an ever-changing work and life environment, globally and locally.

1.2 Rationale for the Study

Who are the intended beneficiaries of this Study? They include, among others:

- students enrolled in formal education systems across all sectors, including vocational, secondary, tertiary and even primary,
- pre-service teachers [teacher students] and practising teachers involved in continuing education or professional development programs,
- teacher educators for the above categories,
- learners undertaking workplace education, from low-skilled workers through to management [and their workplace teachers/trainers],
- industry decision makers,
- mathematicians working in industry,
- policy makers.

What are the aims of the Study? The aims of the Study are:

- to broaden the public awareness of the integral role that mathematics plays in society with respect to low- and high-technology industries,
- to broaden the awareness of industry with respect to what mathematics can and cannot realistically achieve under current circumstances,
- to broaden the awareness of industry with respect to what school and university graduates can and cannot do realistically in terms of mathematics,
- to broaden the awareness of mathematics teachers and educators with regard to industrial practices and needs with respect to education,
- to enhance the appropriate usage of mathematics in society and industry (e.g., by presenting examples of good practice),
- to attract and retain more students, encouraging them to continue their mathematical studies at all levels of education through meaningful and relevant contextualised examples,
- to improve mathematics curricula at all levels of education.

Why is there a need for this Study? This Study is needed:

- to create new and innovative educational practices and support existing good practices,

- to ensure that, when used as an employment selection tool, Mathematics is used appropriately,
- to develop in learners the mathematical reasoning and logical thinking needed in industry,
- to enhance the dialogue and understanding between the communities of mathematicians, workers and industry decision makers, politicians and educators.

2 The Role of Mathematics: Visibility and Black Boxes

We all use mathematics every day; to predict weather, to tell the time, to handle money. Mathematics is more than formulas or equations; its logic, its rationality have for a long time gone beyond just numbers. However, people are often not aware of the importance of the role of mathematics in modern technologies. Many people have a restricted view of what mathematics is and does. We need to make the use of mathematics in modern society more visible.

If young people are not aware of the importance of mathematics and have not personally experienced its applicability, they may not want to study mathematics in school. This may limit their career and educational opportunities later on. Many societies have had some kind of selection process, such as Classics (be it studying Chinese, Greek or Latin). Today, mathematics serves this purpose in many countries. This can be progressive, in that it gives children an opportunity for upward social mobility through studying mathematics. However, it can also be repressive in that it can limit the opportunities of students with problems in mathematics. Some people will manage to compensate for gaps in their mathematical training, but others will not. The consequences, political, cultural and educational, are important.

The role of mathematics is twofold. It can give people highly developed skills in abstraction, analysis of underlying structures and logical thinking. It can also give them experience with the best tools for formulating and solving problems. We will refer to this as analysis. This is in comparison with applying "black boxes", which refers to the packaging mathematics with other conceptual and material tools into (hopefully) automatic solutions to problems, with the consequence of hiding the mathematics from the immediate view of the users. This packaging can be anything from a fast food cash register, where the keys show only pictures of the items instead of numbers, to the search algorithm in *Google*™. We believe strongly that most people will need a combination of the skills listed above. Just knowing how to apply black boxes has many shortcomings:

- It limits innovation, critical analysis and adjustments to the techniques.
- It does not allow analysis in case of failure of the black box.
- It makes it harder for people to judge the appropriateness of various techniques and the validity of the output.

The exact balance of emphasis between analysis and black box techniques and the various levels of description of the inner workings of the black boxes will depend on the nature of the application.

Questions

1. How can mathematics, especially industrial mathematics, be made more visible to the public at large?
2. How can mathematics be made more appealing and exciting to students and the professionals in industry?
3. How can mathematics serve a progressive rather than a restrictive role in education and training for the workplace?
4. What is the best way to teach analytical skills to various groups of students?
5. To what extent is it necessary or desirable to describe the inner workings of black boxes?
6. What are the social implications of not explaining the inner workings of black boxes?

3 Examples of Use of Technology and Mathematics

Modern workplaces are characterised by the use of very different types of technology. 'Technology' is understood in the broadest sense, including traditional machinery, modern information technology and workplace organisation. Examples include a turning lathe, paper forms for reporting on production, technical drawing packages such as Computer Aided/Assisted Design/Drafting (CAD) and the assembly line as a means of organising production in contrast to other forms of workplace organisation.

The conceptual basis of most "modern" information technology is obviously some sort of (often discrete) mathematics. Reports like *Mathematics in Industry* (OECD 2008) mention university-level examples of uses in the chemical industry, oil exploration, medical imaging, micro- and nano-electronics, logistics and transportation, finance, information security and communications and entertainment as areas of industrial use of mathematics. It lists mathematical themes such as complexity, uncertainty, multiple scales, large-scale simulations and data and information (see p. 10). As a consequence, one could think that the growth in technology use implies an increasing presence of mathematics in the workplace. Research findings and anecdotal evidence seem to point to the contrary: Mathematics is said to disappear from the workplace. At least, mathematics is less visible in modern workplaces than in traditional ones. Why?

One way to understand this paradox is the following interpretation of the situation: As discussed above, technology is a means of packaging mathematics with other conceptual and material tools into ideally automatic solutions of problems, thus hiding mathematics from the immediate view of the users. One side-effect of these black boxes is to secure a certain distribution of technological power, giving the control of a whole range of situations to those few who understand the inner working and intricacies of the boxes.

Questions

1. How is it possible to describe and analyse the role of Mathematics within technology? What are insightful examples of the role of technology in showing and/or hiding mathematics in the workplace?
2. Does the existence of special types of technology hiding mathematics from the view of the user imply a change in the mathematical demands on the user? How?
3. To be more precise and to give one example for a more detailed question: With the idea of exact measurement furthered by modern technology, do old competencies like estimation of results and reading of different scales become obsolete when using modern technology? Or, do they become more important?
4. What are the social and political consequences of the 'crystallising' and 'hiding' of mathematics in black boxes—this question is pertinent not only to modern technology like computers, but also to software and hybrid technology.

4 Communication and Collaboration

In the workplace, mathematics is seldom undertaken as an individual activity. Mathematical work, mostly on modelling and problem solving, is almost always a group activity and frequently the groups involved are made up of individuals with diverse expertise and expectations. Communication between and among such groups is crucial. By communication we include listening, writing, speaking and the use of communication technologies as essential skills. A societal or industrial problem may be concerned with making a process faster, cheaper, more robust or, in a general sense, more efficient. But when is the problem amenable to mathematical analysis and solution? Communication at this level is often difficult because managers, mathematical and non-mathematical team members may come with different training, different goals and use different languages. They may have to cope with highly complex situations, often marked by uncertainty, the use of multiple scales, sometimes relying on large-scale simulations. Nonlinearity, data and information are important aspects of modern industry (OECD 2008).

In industries of all sizes, small, medium and large, good communication is extremely important in understanding the nature of a problem and its mathematical components. Communication is clearly essential between team members in "solving" a problem. However, translating a mathematical solution into a workable solution can be challenging as there may be confounding social and political considerations.

It is also necessary to understand that in some industrial settings, intellectual property rights and secrecy may be an issue restricting open communications.

Questions

1. How to identify which societal and/or industrial problems should be worked on?
2. How to better communicate within multi-disciplinary working groups?
3. How to communicate the underlying mathematics to the problem owners and/or general public?
4. How to achieve greater quantitative literacy among school leavers, workers and the general population?

5 Teaching and Learning of Industrial Mathematics: Making Industrial Mathematics More Visible

In Sect. 1, we described the somewhat paradoxical situation of Mathematics being used more and more extensively in modern society, while it is progressively disappearing from societal perception. Let us consider some examples:

- To many primary school students the long division algorithm is a black box.
- Most students have never thought about how calculators compute transcendental functions such as the exponential.
- Google's Page Rank algorithm is a powerful application of eigenvectors in very high-dimensional linear algebra.
- GPS, the Global Positioning System, involves diverse technologies, and its understanding requires geometry, linear algebra and coding theory among others.
- Cryptography is fundamental to modern society. Despite its high sophistication, it involves mathematical ideas that can be presented in a simplified way to a broad audience.
- Weather forecasting shows the power of combining mathematical modelling and high-speed computers.
- Statistics is a useful tool, but many people do not understand how to interpret statistical results properly.
- Computer animation involves many mathematical issues, such as using quaternion's to parameterize rotations.

Logistics and decision making are broader fields, which also imply the use of simple and/or sophisticated mathematical technology. In all fields, most of the mathematical technologies are used without realising that mathematics is used; mathematics is normally not even mentioned in relation to them. Teaching and learning of mathematics has to cope with this fact by either hiding mathematics from the view of the learner or deliberately showing the use of mathematics even if this is not obvious. The (in)visibility of mathematics, especially industrial mathematics implies a major problem in education and training. Nevertheless, describing the inner workings of these procedures can be a great way to motivate teaching and learning of various topics.

Questions

1. Who decides what will be explained and to whom?
2. How to decide the level of explanation for various groups?
3. How to organise teaching and learning in order to make industrial mathematics visible—if this is wanted/necessary?
4. How much is it appropriate to explain for educational purposes in order to generate interest and excitement without overwhelming the learner?

6 Using Technology and Learning with Technology: Modelling and Simulation

For a better understanding of the role of technology in the educational interfaces between industry and mathematics, it may be helpful to distinguish between technology created and used to help industry do its job—often called "indutech"—and technology created especially to foster teaching and learning mathematics and its use for instance in industry—called "edutech". Using either kind of technology, especially new technology, usually requires special efforts to become acquainted with it, to develop routines and practice. To put it in workplace words, a special effort (in terms of time set aside) must be made to teach and learn using a specific technology. This can be an obstacle to switching to a more modern technology as long as the older one still "does the job". On the other hand, change and innovation are necessary in industry (especially in times of globalisation).

There are a number of issues related to using technology in industry (*indutech*) and teaching and learning with technology (*edutech*). We name just a few:

- Technology may be a reason to make obsolete certain competencies by means of routines packed into black boxes—e.g., simple arithmetic and more advanced handheld calculators may change the role of traditional arithmetic performed with paper and pencil or in the head.
- Technology taken as an unquestionable "god" may call for a critical evaluation of the results, both in industry and in education.
- Technology can be a means to enhance learning by simulating the workplace situation (virtual workplace simulation). This can be technology especially created for teaching/learning purposes as well as the usual workplace technology used in a typical educational way. Simulation—especially of unusual, even dangerous situations—for educational purposes can reveal the inner workings of black boxes. Modelling and simulation may be seen as a way to create opportunities to better understand an input–output system, to see the consequences of certain input variables, and even, perhaps, to understand the mechanisms inside the system.

Questions

1. How should one decide on the level of detailed mathematics expected to be taught/learned in a given vocational black box situation?
2. How can mathematics help the transfer of technological procedures and/or solutions between different fields of industry?
3. What criteria should be used to judge the appropriateness of simulation in the teaching and learning of industry-related practice?
4. How can one compensate for the "standardising effects" of any technology that is in widespread use?

7 Teaching and Learning for Communication and Collaboration

As mentioned in Sect. 3, communication and collaboration form an integral and important part of the industrial use of mathematics. Because of the importance of communication skills for work in industry, is it desirable to have these skills taught and learned in all parts of education and training. But there are additional justifications for including teaching and learning communication skills within mathematics education. To cite a few:

- Modelling industrial problem settings at each level of education can be used to teach mathematics in the context of its contemporary uses. This modelling relies on effective communications throughout the process.
- Teaching communication skills is a natural part of the mathematics learning process (listening, writing, speaking and using communication technologies).
- Assessing communication skills (listening, writing, speaking and using communication technologies) is a natural part of the mathematics assessment process.
- Collaborative learning should begin in primary education and be continued as part of a life-long learning process. Examples include: planning and making healthy soft drinks in primary school, analysing strategic games or simulating running a small business in secondary education, and working on real industrial problems in tertiary education.

Questions

1. What communication skills are specific to mathematics?
2. Are there specific skills for use in relation to industrial mathematics?
3. How do we teach mathematics as a "second language"?
4. What is the role of mathematical contests and competitions in developing and assessing communications skills in mathematics?

8 Curriculum and Syllabus Issues

A partnership between mathematics and industry requires adjustments of the mathematics curriculum in order to prepare students for both the needs of mathematics and the requirements of industry. This can also support the teaching of mathematics in general. Students are often taught as if mathematics were a dead science and a finished product. Giving students an opportunity to experience the excitement of realistic applications may enliven their view of the subject.

Questions

1. What are the advantages and disadvantages of identifying a core curriculum of mathematics for industry within the general mathematical curriculum at various levels and for various professions?
2. What are useful ways to introduce mathematics for industry into vocational education?
3. What are the advantages and disadvantages of creating specific courses on mathematics for industry versus including the topic in the standard mathematical courses at various levels?
4. What are the advantages and disadvantages of treating mathematics for industry as an interdisciplinary activity or as part of the traditional mathematics syllabus?

9 Teacher Training

Teachers must be trained in new mathematical content, pedagogy and assessment and to recognise the presence of mathematics in society and industry. This should take place in schools of education for mathematics teachers of all levels, in-service programs, and industrial training programs.

There may be a special need to act on teacher training in areas like statistics, discrete math (recursion, graph theory, matrices), operations research and mathematical modelling in the context of real applications. These curricular changes must be complemented by changes and innovation in the pedagogy offered to teachers, such as stressing collaborative learning, making decisions on appropriate use of educational technology and fostering communication skills. Assessment modes and practices should include informal assessment techniques, portfolio and group assessment.

These changes may be fostered by industry visits or longer-term placements of students who want to become teachers and by practising teachers. Understanding the educational interface between mathematics and industry can be enhanced by current and future teachers actually going into the workplace to observe and talk to workplace personnel as they experience real industrial problems.

Questions

1. What level of understanding of this new content in relation to EIMI is appropriate for each grade level?
2. What are good practices that support this new direction in teacher training?
3. How to implement these changes in an efficient way?

10 Good Practices and Lessons to be Learned

In all sectors of education there are examples of good practice in relation to this Study. To give an example, we mention engineering programs that seem to foster creativity and innovation by final-year projects in a better way than school and university courses in the mathematical sciences. One important aspect of the Study is to communicate and exchange examples of good practices in order to make concrete the ideas already discussed in this document. We would like to collect outstanding examples of innovative practices that are suitable for use and adaptation by teachers at the various levels of education.

Possible examples may include:

- Creating *optimal routes* for garbage collection (appropriate at different levels of sophistication starting with 4-year-olds); graph/network theory.
- Analysing the *probability concepts* which underpin different games of cards; probabilistic decision making.
- *Analysing data* related to the environment and patterns of consumption such as energy use, clothing sales (appropriate at different levels of sophistication starting with 10-year-olds); statistical decision making.
- *Project management* for an event of relevance to the particular students; total project management skills.
- *Game theory* (decision making).
- *Continuous optimisation*—e.g., finding an optimal path on a downhill ski slope.

We are also looking for good examples of how to integrate industry into the educational process. For example, through:

- Study groups which address problems brought into the classroom from industry;
- Internships (for students and teachers into industry; for industrialists into education);
- Summer/winter schools, where industry representatives work together with students on problems of relevance to industry rather than the school curriculum;
- Competitions involving mathematics used to solve industry-related problems under realistic conditions;
- The documentation of existing case studies.

Lessons to be learned from failures are of the same interest as those from successes. The examples should include the intended outcomes, the pedagogical practices involved and a critical evaluation of the process.

11 Research and Documentation

Launching this Study, we start from the assumption that, from a global perspective, there is a need to research existing practices; no coverage of the whole field currently exists. Pertinent topics are diverse—with diverse methodologies being appropriate to the field of educational interfaces between mathematics and industry.

Within this research field, mathematical interfaces can include mathematics "proper", but are usually interdisciplinary in practice. Mathematics is part of the method of coping with industrial situations, but can also be the object of study (especially in studies on societal and industry needs and educational practice and innovation). For example, there is a need for studies on the curricular consequences for vocational, secondary, tertiary and primary education.

Compared to other topics, it seems obvious that national and trans-national documentation is widely missing in the field of mathematics and industry (even with CEDEFOP and other institutional databases). Suggestions and contributions describing existing and future research and documentation of activities in the field of Educational Interfaces between Mathematics and Industry will be most welcome.

International Programme Committee

Alain Damlamian, Faculté de Sciences et de Technologie, Université Paris-Est (France), co-chair. *Rudolf Sträßer*, Institut für Didaktik der Mathematik, Justus-Liebig- Universität Gießen (Germany), co-chair. *José Francisco Rodrigues*, Faculdade de Ciências, Universidade de Lisboa (Portugal), local organiser. *Helmer Aslaksen*, Department of Mathematics, National University of Singapore (Singapore). *Gail FitzSimons*, Faculty of Education, Monash University (Australia). *José M. Gambi*, Escuela Politecnica Superior, Universidad Carlos III de Madrid (Spain). *Solomon Garfunkel*, Comap (USA). *Alejandro Jofré*, Departamento de Ingeniria Matematica, Universidad de Chile (Chile). *Gabriele Kaiser*, Faculty of Education, University of Hamburg (Germany). *Henk Van der Kooij*, Freudenthal Institute, Universiteit Utrecht (Netherlands). *Li Ta-tsien*, Department of Mathematics, Fudan University (China). *Brigitte Lutz-Westphal*, Institut für Didaktik der Naturwissenschaften, der Mathematik und des Sachunterrichts, Universität Vechta (Germany). *Taketomo Mitsui*, Department of Mathematical Sciences, Doshisha University (Japan). *Nilima Nigam*, Department of Mathematics, Simon Fraser University (Canada). *Fadil Santosa*, Institute for Mathematics and its Applications, University of Minnesota (USA). *Bernard R. Hodgson*, Département de mathématiques et de statistique, Université Laval (Canada), *ex officio, Secretary-*

General of ICMI. Rolf Jeltsch, Seminar for Applied Mathematics, ETH Zürich (Switzerland), *ex officio, President of ICIAM.*

The official website for the joint ICMI/ICIAM Study is http://www.cim.pt/eimi/.

Reference

Organisation for Economic Co-operation and Development, Global Science Forum. (2008). *Report on Mathematics in Industry*. http://www.oecd.org/dataoecd/47/1/41019441.pdf.

Report on the Study

Alain Damlamian, José Francisco Rodrigues and Rudolf Sträßer

1 Aim and Scope of the Study

The Study on "Educational Interfaces between Mathematics and Industry (EIMI-Study)" was organized jointly by the International Commission on Mathematical Instruction (ICMI) and the International Council for Industrial and Applied Mathematics (ICIAM). Both institutions were interested in the topic of the study, which was stimulated by a recent OECD Report. ICMI and ICIAM were also keen on cooperating to better understand the connections between the teaching and learning of mathematics on all levels as well as the role of mathematics in industry. As such, it was the 20th ICMI-study and the second developed in cooperation with another association. As usual with ICMI-studies, the EIMI-study wants to offer ideas and suggestions on how education and training can contribute to enhancing both individual and societal developments—in this case also building on the experience and competence of mathematicians from industry.

Following the framework of ICMI-studies, the EIMI-study was begun by appointing an International Program Committee (IPC) with two co-chairs (Alain Damlamian appointed by ICIAM and Rudolf Sträßer from ICMI) and members proposed by both organizations: Helmer Aslaksen (Singapore), Gail Fitzsimons (Australia), José Gambi (Spain), Solomon Garfunkel (USA), Alejandro Jofré (Chile), Gabriele Kaiser (Germany), Henk van der Kooij (Netherlands), Li Ta-tsien

A. Damlamian (✉)
Université Paris-Est Créteil Val de Marne, Créteil Cedex, Val-de-Marne, France
e-mail: damla@u-pec.fr

J. F. Rodrigues
Universidade de Lisboa/CMAF-FCUL, Lisboa, Portugal
e-mail: rodrigue@fc.ul.pt

R. Sträßer
FB 12 Mathe, Inst. Didaktik der Mathematik, Universität Gießen, Gießen, Germany
e-mail: rudolf.straesser@uni-giessen.de

A. Damlamian et al. (eds.), *Educational Interfaces between Mathematics and Industry*, New ICMI Study Series 16, DOI: 10.1007/978-3-319-02270-3_2,

(China), Brigitte Lutz-Westphal (Germany), Taketomo Mitsui (Japan), Nilima Nigam (Canada), José Francisco Rodrigues (Portugal, host country), Fadil Santosa (USA), Bernard Hodgson (ex-officio, ICMI), and Rolf Jeltsch (ex-officio, ICIAM).

As can be read in the Discussion Document, the EIMI-Study starts from two assumptions, namely:

There are intimate connections between innovation, science, mathematics, and the production and distribution of goods and services in society. In short: there are intimate connections between mathematics and industry;
In view of these connections, there is a need for a fundamental analysis and reflection on strategies for the education and training of students and maybe the development of new ones.

Historically, there have been productive interactions between mathematics and industry in generating and solving problems associated with the economic and social development of humankind. In a modern, technological world, mathematics is said to be used almost everywhere. However, these uses are not generally visible except to specialists. Even people using mathematics in their workplaces may not recognize its presence. There have been many studies of the mathematics used in the workplace—ranging from descriptive lists of traditional school-based topics to sociological studies of workplace activities set in context. There have been many collections of applications of problem solving and modeling based on or informed by practical industrial problems, especially at higher levels of mathematics, in fields such as the natural and physical sciences, engineering, and finance.

Internationally, there are frequent articles and debates in the popular media citing employer dissatisfaction with the perceived quality of mathematics education. Graduates from schools, vocational colleges, and universities often appear unable to draw upon and use mathematics in work situations as opposed to classroom or examination contexts. At all educational levels, students typically have been taught the tools of mathematics with little or no mention of authentic real world applications, and with little or no contact with what is done in the workplace (be it the classical engineering situations or other more recent activities like biotechnology, biomedicine, the financial, insurance and risk sectors, or consulting engineering companies). Nowadays, highly complex problems need to be solved and, hence, some training to solve such problems—in particular, real life problems—is necessary. Increasingly, powerful computers make it possible to treat such complex problems and this is achieved not only using off-the-shelf software but with innovation, often mathematical innovation requiring insight and analysis.

In order to better understand these phenomena, the Study starts from a broad definition of Industry (from the Organisation for Economic Co-operation and Development) "… broadly interpreted as any activity of economic or social value, including the service industry, regardless of whether it is in the public or private sector" (OECD 2008, p. 4). The term "industry" obviously refers to a diverse range of activities, producing goods, and services. Under constraints such as time and money, these activities generally attempt to optimize limited -sometimes scarce- resources, both material and intellectual. The overarching goal is to maximize benefits for certain groups of people while, ideally, minimizing harm to other groups and the natural environment. “Mathematics (or the mathematical sciences, here the two terms are used interchangeably) comprises any activity in the mathematical sciences, including mathematical statistics” (OECD 2008, p. 4). Workers at all levels utilize mathematical ideas and techniques, consciously or unconsciously, in the process of achieving the desired workplace outcome. In other words, mathematics is just one part of a repertoire of tools and strategies of a practical nature. However, as a major factor in decision-making and communication processes, it is crucial that mathematics be used appropriately, accurately and with confidence. For this Study, we start from the assumption that professional mathematicians are located in academia, in industry, sometimes in both. The discourse of mathematics in all its various specializations involves certain ways of thinking and acting. Traditionally, mathematicians consider axioms and definitions and make logical deductions. In mathematical modeling, one formulates problems in mathematical terms. However, the mathematical solution needs to take into account the industrial context. This Study examines the implications for education at the intersection of two communities of practice industrialists and mathematicians—or industry and mathematics. We emphasize that there should be a balance between the perceived needs of industry for relevant mathematics education and the needs of learners for lifelong and broad education in a globalized environment. In other words, learners should be equipped for flexibility in an ever-changing work and life environment, globally and locally.

Intended beneficiaries of this study include, among others:

students enrolled in formal education systems across all sectors, including vocational, secondary, tertiary, and even primary,
preservice teachers (teacher students) and practicing teachers involved in continuing education or professional development programs,
teacher educators for the above categories,
learners undertaking workplace education, from low-skilled workers through to management (and their workplace teachers/trainers),
industry decision makers,
mathematicians working in industry (or having an interest in the applications of their field).
and eventually, all policy makers.

The aims of the Study are:

to broaden the public awareness of the integral role that mathematics plays in society with respect to low and high-technology industries,
to broaden the awareness of industry with respect to what mathematics can and cannot realistically achieve under current circumstances,
to broaden the awareness of industry with respect to what school and university graduates can and cannot do realistically in terms of mathematics,
to broaden the awareness of mathematics teachers and educators with regard to industrial practices and needs with respect to education,

to enhance the appropriate usage of mathematics in society and industry (e.g., by presenting examples of good practice),
to attract and retain more students, encouraging them to continue their mathematical studies at all levels of education through meaningful and relevant contextualized examples,
and to improve mathematics curricula at all levels of education.

In the views of the International Program Committee, there is a need for this Study in order.

to create new and innovative educational practices and support existing good practices,
to ensure that, when used as an employment selection tool, mathematics is used appropriately,
to develop in learners the mathematical reasoning and logical thinking needed in industry,
and to enhance the dialog and understanding between the communities of mathematicians, workers and industry decision makers, politicians, and educators.

2 The Discussion Document

In a meeting in Óbidos, Portugal, hosted by the Centro Internacional de Matemática in October 2008, the International Program Committee prepared the Discussion Document (DD) for the EIMI-Study. After presenting the basic definitions and rationale for the Study as described above, this document consists of eight chapters which detail the situation in the field and raise questions to be considered in order to meet the expectations for the study. Below, we give the titles of the chapters of the Discussion Document.

- The role of mathematics—visibility and black boxes.
- Examples of use of technology and mathematics.
- Communication and collaboration.
- Teaching and learning of industrial mathematics—making industrial mathematics more visible.
- Using technology and learning with technology: modeling and simulation.
- Teaching and learning for communication and collaboration.
- Curriculum and syllabus issues.
- Teacher training.

In addition, the DD asks for good practices and lessons to be learned as well as comments and suggestions for research and documentation.

The DD was distributed in the widest way possible—including the publication in the official ICMI-journal “L’enseignement mathématique” and “ZDM The International Journal on Mathematics Education” (see the text in this volume or Damlamian and Sträßer 2009 or http://eimi.glocos.org/?page_id=203). A website was created even before the Óbidos meeting (see http://eimi.glocos.org/).

3 The Contributions Offered

After the publication of the Discussion Document, the International Program Committee received 70 texts from all five continents to be considered for inclusion in the Study. Authors proposed their texts by the end of October 2009. These were evaluated in view of a possible invitation to the Study Conference. To this end, the IPC met in Paris in the beginning of November 2009. Beyond deciding on possible contributions to the Study Conference, the major challenge of this meeting was to find an efficient way to organize the conference and to structure the discussions. The IPC finally agreed to have six plenary talks during the conference, while the major work was to be done in six “Working Groups” with topics to be discussed during the Study Conference. Each Working Group would be given six hours for presentation of texts and discussion of their topic with an opportunity of presenting the results in a plenary session of the conference (for details see the section below).

After revisions, it was decided to gather 54 papers and the Discussion Document in a proceedings book to be published before the conference (planned for April 2010). The volume was made available early in April 2010 (see Araújo et al. 2010). The proceedings were offered to every participant of the Study Conference as a printed book. This collection of papers on “Educational Interfaces between Mathematics and Industry” can be freely downloaded from http://www.cim.pt/files/proceedings_eimi_2010.pdf.

4 The Study Conference

The second major step of every ICMI-study is a Study Conference to bring together experts from all over the world and make them share ideas on the topic of the study at hand. The ICMI-ICIAM-study on “Educational Interfaces between Mathematics and Industry (EIMI-study)” exactly followed this road by asking for contributions and comments to the DD and then by inviting about one hundred people to the Study Conference. This conference was held in Portugal, as a tribute to the fact that its National Committee of Mathematicians made the original suggestion for this Study. It was organized by the Centro Internacional de Matemática (CIM) and scheduled to be held in Lisbon on April 19–23, 2010.

Colleagues from all over the world registered for the conference, but the ashes of the Icelandic volcano Eyjafjallajökull simply made it impossible to run the conference in April 2010. At the planned beginning of the conference, there were no flights over most of Europe for several days, so the conference organizers had to cancel the conference—and soon decided to postpone it to October 11–15, 2010. In October 2010, the postponed conference took place at the University of Lisbon, with the participation of the majority but not all registered participants and discussed the 54 papers published in April in the Proceedings. These were grouped in order to link them with the six Working Groups of the conference, which were planned to act as the places of discussion and preparation of the Study Volume. The working Groups were entitled:

WG 1: The mathematics-industry interface (mainly looking at Sects.2 and 6 of the Discussion Document).
WG 2: Technology issues (mainly looking at Sects.3 and 6 of the Discussion Document).
WG 3: Mathematics-Industry Communication (mainly looking at Sects.4 and 11 of the Discussion Document).
WG 4: Education in Schools (mainly looking at Sects.5, 7, 8, and 9 of the Discussion Document).
WG 5: University and academic technical/vocational education (mainly looking at Sects.5, 7, 8, and 9 of the Discussion Document).
WG 6: Education/training with industry participation (mainly looking at Sect.6 and pertinent parts of Sects.5, 7, 8, and 9).

The Working Groups were bundled into two sets scheduled to meet in the first half and second half of the Study Conference. Consequently, each participant could participate in one of WGs 1, 3, or 6 and later during the week in one of WGs 2, 4, or 5. Each Working Group had 330 min working time and the respective reports were presented during a plenary session.

In addition to the conference opening and closing session, the following six plenary talks were given with the purpose to present expert views on mathematics in industry and on workplace related mathematics in educational settings (i.e., workplace related Didactics of mathematics):

Thomas A. Grandine (Applied Mathematics, The Boeing Company, Seattle, USA): The use of Mathematics in Industry.
Arvind Gupta (Director/CEO of MITACS, University of British Columbia, Canada): Industrial Mathematics in Academia (plenary presented by Nilima Nigam).
Celia Hoyles (Institute of Education, University of London, UK): Research on Mathematics in the Workplace (plenary presentation held via Skype).
Henk van der Kooij (The Freudenthal Institute, Utrecht University, The Netherlands): How to integrate work-related math in mathematics education?
Helmut Neunzert (Fraunhofer ITWM, University of Kaiserslautern, Germany): Models for Industrial Problems: How to find and how to solve them — in industry and in education (plenary presentation held via Skype).

Masato Wakayama (Faculty of Mathematics, Kyushu University, Japan): A Trial for PhD Education and Research in Mathematics for Industrial Technology.

Papers in the proceedings which could not easily be linked to exactly one of the Working Groups were presented in paper presentation sessions (two or three presentations in parallel) (for detailed information see the program of the Study Conference at http://eimi.glocos.org/?page_id=319; there is also a list of the participants of the Study Conference at http://eimi.glocos.org/?page_id=787).

5 Work After the Study Conference

With the purpose of contributing to this on-going study, to complement the Study Conference and, to enhance its recommendations and conclusions for the EIMI-Study Book, a 2-day Workshop was hosted by the University of Macau, China, on November 3–4, 2011. In view of the increasing connections between innovation, science and mathematics, the global need for an analysis and a reflection on strategies for the education and training of new generations was reaffirmed in this Asian meeting that had the participation of six members of the EIMI International Program Committee. The contribution of Li Tatsien in this book also originated from this workshop.

In order to develop the Study Book, the Working Group coordinators were asked to give a written report on their Working Group and suggest papers for inclusion into the Study Book. With this information and in consultation with the coordinators (using sometimes difficult communication channels), the editors finally condensed the Study Book to its present form.

References

Araújo, A., Fernandes, A., Azevedo, A., & Rodrigues, J.F. (Eds.) (2010). Educational interfaces between mathematics and industry. *Proceedings of EIMI 2010 Lisbon conference*. Centro International de Matemática, Portugal Available online at: http://www.cim.pt/files/proceedings_eimi_2010.pdf.

Damlamian, A., & Sträßer, R. (2009). ICMI study 20: Educational interfaces between mathematics and industry (Discussion Document of the EIMI Study 20) in L'Enseignement mathématique. 55, 197–209, also published in Zentralblatt für Didaktik der Mathematik. *41*(4), 525–533 (on behalf of the International Program Committee).

Organisation for Economic Co-operation and Development, Global Science Forum (2008). Report on Mathematics in Industry. http://www.oecd.org/dataoecd/47/1/41019441.pdf.

Part II
Plenary and Invited Lectures

Getting Math off the Ground: Applied Mathematics at Boeing

Stephen P. Keeler and Thomas A. Grandine

1 Introduction

Years ago a director of our group had a visitor, an executive from another Boeing organization. After meeting some mathematicians and reviewing our work, he said, "You people are having too much fun." The aerospace business is a lot of fun for mathematicians, and it provides perpetual challenges. It's challenging because it's a business, and it's fun because it's a business. The challenges arise in creating products that beat the competition, by whatever performance metrics matter to our customers, and math drives the design processes that predict and optimize performance. The fun and satisfaction come from harnessing the math to achieve business success.

Even this article has a business motivation. We hope more prospective mathematicians will consider industrial careers and will be encouraged to find out about industrial jobs and how to prepare for them. We also hope more teachers at all levels will be able to guide their students as they consider these careers. Eventually more mathematicians in the industrial job market will mean a higher level of performance for the companies that offer the most attractive career opportunities.

We hope to address some of the aims of the ICMI/ICIAM Study on "Educational Interfaces between Mathematics and Industry" (Araujo et al. 2010): broadening awareness of the role that mathematics plays in society with respect to high-technology industries; broadening the awareness of mathematics teachers and educators with regard to industrial practices and needs; helping to attract students to mathematics; and providing insights which may help educators improve the

S. P. Keeler (✉)
Applied Mathematics, The Boeing Company, Seattle, USA
e-mail: stephen.p.keeler@boeing.com

T. A. Grandine
The Boeing Company, Seattle, USA

A. Damlamian et al. (eds.), *Educational Interfaces between Mathematics and Industry*, New ICMI Study Series 16, DOI: 10.1007/978-3-319-02270-3_3,

mathematics curriculum. In the process we'll describe what mathematicians in our group do and what it takes to succeed.

Our observations are based on more than 50 combined years of technical and management experience in the Boeing Applied Mathematics organization, and on what we've learned from our colleagues in other companies, government labs, and universities. Our experience certainly doesn't represent every industrial math job, and we've tried to avoid generalizing too broadly. However, we do believe that many industrial jobs have much in common, if not in specific technical requirements then in the general approaches and attitudes that are effective.

2 The Boeing Company

Boeing is the world's largest aerospace company and a leading manufacturer of commercial jetliners and defense, space and security systems. Boeing products and services include commercial and military aircraft, satellites, weapons, electronic and defense systems, launch systems, advanced information and communication systems, and logistics and training.

Boeing employs more than 159,000 people across the United States and in 70 other countries, and supports airlines and US and allied government customers in more than 90 countries. About 77 % of the workforce hold college degrees, including nearly 32,000 advanced degrees, in virtually every business and technical field, from approximately 2,700 colleges and universities worldwide. Boeing is organized into two business units: Boeing Commercial Airplanes (BCA) and Boeing Defense, Space and Security (BDS). Other important components include Boeing Capital Corporation and Boeing Engineering, Operations and Technology, which helps develop, acquire, apply and protect innovative technologies and processes. EO&T includes Boeing Research and Technology, where the Applied Math group is located organizationally.

Boeing Commercial Airplanes' main products are the 737, 747, 767 and 777 families of airplanes and the Boeing Business Jet. New product development efforts are focused on the Boeing 787 Dreamliner and the 747–8 family of new, high-capacity 747 s. The company has nearly 12,000 commercial jetliners in service worldwide, which is roughly 75 % of the world fleet. Competition in the commercial aviation industry is largely based on the life-cycle cost of airplane acquisition and operation. A central factor is of course the design of airplanes, where very small relative improvements in performance are tremendously important over the life of the product. In addition, our airline customers are deeply concerned with fleet mix planning and fleet operations, air traffic management, and many other factors. Commercial transport production is very much a global operation, leading to complex supplier-management considerations. Efficiency in all these areas is heavily dependent on mathematical techniques.

Boeing Defense, Space and Security provides products and services for global military, government and commercial customers. In addition to designing, producing, modifying and supporting fighters, bombers, transports, rotorcraft, aerial refuelers, missiles, munitions and spacecraft, BDS provides network-enabled solutions, communications and intelligence, surveillance and reconnaissance technologies. On the defense and space side life-cycle cost of operation is a major design driver. In addition, these products frequently demand the validation and incorporation of new technologies, which in turn calls for innovative mathematical methods to model and predict performance. A major difference from the commercial transport business is the frequent need to represent and evaluate new designs and operational concepts in their early stages, where high accuracy may not be necessary but rapid turnaround is critical.

3 The Boeing Applied Mathematics Group

Our organization comprises about 60 scientists and managers, nearly all with advanced degrees and about two thirds with doctorates. We work on roughly 200 different projects in the course of a year, many in support of programs and projects in BCA and BDS. Another substantial part of the work is applied research and development of algorithms and software. We do some external contracting, typically with government customers.

Applied Math is organized by technical specialty. The areas of specialization have evolved over the years, driven by the company's needs. For example, at one time we had a very strong program in numerical linear algebra. It was so successful that eventually the demand for further improvements in our capabilities faded, and we reduced our emphasis in this area. We believe that Boeing derives enormous benefit from the centralization of its math organization. It allows us to respond efficiently when a combination of skills is needed, to standardize company-wide on the best mathematical approaches, to manage the math workforce effectively, and to present one face to the international mathematics community.

The Computational Mathematics group specializes in linear algebra, differential and integral equations, and advanced engineering computing. They work closely with engineers in computational fluid dynamics, structures, electromagnetics and other areas where the analytical software primarily solves differential or integral equations. They also deal with algorithms for parallel computing and other advanced computer architectures.

Operations Research deals with resource allocation, planning and scheduling, as well as stochastic, discrete and network optimization. They collaborate with engineers working on air traffic management systems, logistics problems such as inventory management and forecasting, surveillance and battlefield management, and assignment problems involving multiple military aircraft, weapons and targets. A representative project involved "fusing" target tracking data from multiple sensors and scans to obtain improved tracking performance. A critical need was an algorithm for determining which measurements are caused by the same object. Objects might be closely spaced, and real-time solutions were needed. The OR group developed an efficient and near-optimal algorithm for solving this data association problem, formulating it as a set partitioning problem, a type of combinatorial optimization problem, and solving it with a specialized branch and bound algorithm and a nonsmooth optimization algorithm. The method was awarded a US patent.

The Mathematical Modeling group works on a range of problems which expand the traditional role of the Applied Math group. Recent examples include the design and placement of sensor arrays for border security, new approaches for designing antennas which are integral to aircraft structures, and efforts to develop complexity metrics which can be used in product design.

Applied Statistics provides expert guidance and support to statisticians across the company, particularly in the areas of reliability, design of experiments, and statistical quality assurance. Typical projects involve experimental design to improve the efficiency of wind tunnel tests and improved methods of acceptance testing for aircraft parts delivered by suppliers. A project in support of the 787 Dreamliner involved cabin pressurization. Cabin pressure affects passenger comfort and is also a factor in the structural design of the fuselage. On all previous Boeing commercial jet transports the passenger compartment pressure is equivalent to an altitude of 8,000 ft (2,439 m). The all-composite design of the 787 made it possible to consider increasing the cabin pressure. Our statisticians took part in a study published in the *New England Journal of Medicine* (Muhm et al. 2007). Analysis of data from clinical trials indicated that while cabin pressure altitude of 8,000 ft is not harmful to healthy people, passengers will be more comfortable, less tired, and less prone to illness at a cabin pressure of 6,000 ft (1,829 m). On the basis of these findings, Boeing made a decision to set the cabin pressure altitude of the 787 at 6,000 ft.

The Geometry and Optimization group specializes in computer-aided geometric design and numerical optimization. Much of our work over the last 10 years has been in developing new processes and software tools for optimal design of products. These depend on advanced methods of representing and manipulating product geometry as well as advanced optimization methods, which we describe in more detail below.

In Applied Math we often think of our work in terms of projects and funding sources. We start every fiscal year with less funding than we need to keep everyone "billable." That's part of our business model, since we know that new demands will arise in the course of the year and we want to be able to respond to them. But it means that our staff must operate with something less than complete security and must be constantly aware of emerging opportunities to contribute—also part of our business model.

What qualifications do we expect of our staff? Obviously we want to hire good mathematicians, and we sometimes have the opportunity to hire a mathematician whose career is established and whose record tells us all we need to know. But when we hire more junior people, typically new PhDs, we rely heavily on the reputation of the professor and graduate program they came from; we don't give quizzes. This method has served us well, and the cases where a newly hired mathematician wasn't competent are so few as to escape memory.

We tend to focus instead on two more difficult questions: Do you understand what the job entails, and do you want to do the job? The new PhD has spent a lot of years in academia, and it may well be that the only professional mathematicians she knows are the professors she's studied and worked with. Particularly in math, as opposed to engineering, it's very likely that those professors themselves know little about industrial applied math jobs. The distinction, oversimplified, is this: A research mathematician goes where the mathematics leads; an industrial mathematician goes where the business leads. Certainly our staff members are deeply committed to their fields of study and interest, and one of the challenges is to maintain the continuity of their research and its applications over multiple projects and over a period of years. But a colleague who is *primarily* motivated by his or her field of mathematics eventually becomes a burden on the organization, with rare exceptions. For our staff the reward comes from harnessing their expertise to the solution of business problems.

How is a mathematician from academia going to find the answers to those questions? In our organization we welcome opportunities to talk with students about what we do, and the same is true of many of our colleagues in other companies and government labs. We support the SIAM Visiting Lecturer Program (2011), which helps schools arrange for visits from industrial mathematicians. SIAM published a report on industrial mathematics, very carefully researched and very consistent with our experience (SIAM Report on Mathematics in Industry 2012), and the US Bureau of Labor Statistics produces an excellent overview of jobs in math (and other fields) (US Bureau 2011). The very best way to understand what industrial jobs are like is to get one, and the best opportunities are internships and fellowships. At Boeing the process of recruiting and hiring technical interns is heavily focused on undergraduate engineers, but we hire summer interns in Applied Math whenever our workload allows. Our experience is that internships are much more likely to benefit the student and the company when the intern is a graduate student. In addition to internships, we sometimes host visits by graduate students who are supported by their schools and continue their research while they are with us. Such arrangements may arise as part of a collaboration between our staff and their academic supervisors, interactions which we value highly. We have also participated in the industrial postdoctoral fellowship program at the Institute for Mathematics and Its Applications (University of Minnesota Industrial postdoctoral program 2011), and we were happy to hire the postdoctoral fellow at the completion of the program.

You have to understand what the job is, you have to want to do the job, and you have to be a good mathematician. We don't rely entirely on education and training in applied as opposed to pure math (whatever the distinction may be), but the work in our group does call for numerical analysis, approximation theory and numerical optimization. In many cases good mathematicians have successfully expanded their academic education on the job and developed new areas of expertise. Our technical results are almost always delivered in the form of software, and we need people with an adequate level of programming experience, preferably an object-oriented language like C++ or Java. MATLAB and Mathematica are sometimes useful, but often our engineering collaborators need to incorporate our algorithms and code into their existing systems, which aren't supported by MATLAB or Mathematica. We rarely send people to programming classes.

We need mathematicians who can work effectively with other mathematicians and who can learn to work closely with engineers, meeting them much more than halfway in the collaboration. The success of projects—and careers—frequently depends on the mathematician's willingness to learn to communicate technical ideas to non-mathematicians. In the words of the mathematician Charles S. Peirce, "The very first lesson that we have a right to demand that logic shall teach us is, how to make our ideas clear; and a most important one it is, depreciated only by minds who stand in need of it."

4 Big and Little Examples

We provide a service for Boeing called the Math Hotline. A caller said, "I need 20 digits of pi" and the mathematician read them off from a reference. The caller read them back, said "thanks" and hung up.

At the other end of the scale of scope and complexity is the program in design optimization we've been pursuing for many years. Boeing is heavily engaged in advancing its engineering design methods in order to support innovation, shorten the time to market, and improve the performance of its products. The fundamental and familiar strategy is to use computer models to predict and optimize performance, validating these models with laboratory testing and analysis of existing products. In the aerospace industry the most promising approach is "multidisciplinary design optimization," or MDO. At one time our work in optimization dealt primarily with gradient methods for solving very large problems. An example is computing the minimum-fuel trajectory to move a satellite from one stable orbit to another, a problem which might have 250,000 variables and 400,000 constraints. We develop and maintain a very powerful software suite for such problems. However, this paradigm is of limited value to a designer. First of all, high-fidelity engineering computer models are computationally expensive; a typical aerodynamics code might require an hour on a large cluster, and a computational electromagnetics code can run for days. It is therefore critical to limit the number of evaluations of the objective and constraint functions. Moreover, the design problem can rarely be formulated so simply. The design team may spend months in a process of "design space exploration," understanding the technological and physical constraints, weighing competing objectives against each other, and frequently reformulating the problem and reworking the computer codes. Our group has spent years developing methods and computing tools to support this process.

A highly simplified version of our usual procedure is as follows. In all but the simplest design efforts, the process includes various feedback loops and iterations of these three steps.

(1) Develop the "multidisciplinary analysis" or MDA, a single code which combines the relevant engineering disciplines. As a typical aerospace example, a flow code can predict the pressure distribution over the surface of a wing under specified flight conditions, and a structural analysis code can predict how the wing will deform in response to that pressure distribution. But the deformation changes the aerodynamic properties, and therefore the pressure distribution, which in turn changes the structural response, and so on. What's needed is an "aeroelastic" code that models both disciplines simultaneously. The high-fidelity engineering codes used in design not only require long run times, they may produce outputs which are discontinuous or non-differentiable functions of their inputs, and may be known to fail in certain regions of design space. In addition, experience shows that good optimization software has a knack for finding and exploiting previously unknown errors in the engineering codes. Besides the engineering content, the MDA must have the ability to generate vehicle or component geometry as a function of the design parameters. That innocent-sounding requirement has been a major focus of our work for years, described in more detail below. The MDA encodes all the engineering knowledge that will be used in the computational analysis, and creating it is a major undertaking.
(2) Survey design space. We use a method borrowed from statistical design of experiments to identify a computationally affordable number of points in design space where the MDA is evaluated initially, which we call a survey. Analysis of the survey results provides information for the design team, and may lead to a further small number of MDA evaluations. Then the data are used to build "response surface models," in our case kriging models, of those outputs of the MDA which are of interest. These models are used as surrogates for the expensive engineering code.
(3) Optimize the design. We use a specialized sequential quadratic programming method to solve an optimization problem in which the objective and constraints are surrogate models. Depending on progress in this step, additional runs of the MDA are used in a pattern search step and to update and improve the models. Unless a limit is reached on the number of iterations or the number of function evaluations, the pattern search is provably convergent to a local optimum of the engineering code (rather than the surrogate model).

Steps (2) and (3) of the process are implemented in our Design Explorer software suite (Audet et al. 2000). Design Explorer provides a framework to manage the models and the overall process. In addition to the modeling and optimization functions outlined above, it provides capabilities for finding the global optimizer, estimating uncertainty in the results, dealing with multiple objectives (finding the pareto boundary), handling mixed-integer nonlinear programming problems, and solving problems with categorical variables. We license

Design Explorer to Phoenix Integration, Inc. for use in their ModelCenter® product (Phoenix Integration 2011).

There are several useful ways to deal with vehicle geometry in the design process. The simplest is to use some form of "sizing code" which computes estimates of performance measures without ever creating a geometric representation. The method is fast and can be valuable in exploratory and conceptual design. A very powerful method is grid perturbation. The engineer creates geometry in a computer-aided design system (CAD system) and develops a computational mesh around the geometry as input to an aerodynamics code. Then the optimization code modifies the mesh to achieve higher performance. Finally an engineer constructs a new surface from the optimized mesh. Because only small adjustments to the surfaces are possible, the method has been called "computational sandpaper." It is used to produce critical performance improvements in Boeing designs, especially commercial transports.

For several years we have been pursuing a very general approach in which the parameters of a geometric representation are used as optimization variables. The result is a new geometry system currently deployed in both major business units. An early achievement of the development team was to create the name "General Environment for Optimization and Design Using a Common Kernel," or Geoduck. A geoduck (pronounced GOO-ee-duck) is a large clam indigenous to the US Pacific Northwest, and the name is a Native American word meaning "dig deep." Geoduck is fundamentally a scripting language which makes calls to the Spline Tool Kit, a library of some 300 C-language routines. We will set aside the many organizational and architectural complexities that are addressed by Geoduck, and focus instead on how it is used for design optimization. An early Boeing example involved the design of a hypersonic vehicle.

The hypersonic region of flight is generally considered to begin at Mach 5, when aerodynamic heating becomes an important consideration. The earliest hypersonic vehicle was the German V-2 rocket in World War II. Weight and cost considerations have led to the development of scramjet propulsion as a potential alternative to rockets. One of the earliest efforts involving scramjets was the US Defense Advanced Research Projects Agency (DARPA) X-30 program, conceived as a replacement for the space shuttle fleet but cancelled in 1993. A subsequent effort, the NASA X-43A, was the first program to prove the viability of scramjet propulsion on an actual flight vehicle. This vehicle sustained scramjet propulsion at Mach 10 for 10 s, during which it traveled 19 miles (30.6 km). Following this successful demonstration, Boeing decided to take a fresh look at a vehicle like the X-30, drawing on 10 years of technological advances (Bowcutt et al. 2008). The design test case was the second stage of a two-stage reusable launch vehicle.

The design of hypersonic vehicles is so challenging that MDO is the only approach to date that has enjoyed any real success. Most systems in a hypersonic vehicle have to perform multiple functions, and nearly every system interacts with most of the others. For example, the forebody of a hypersonic lifting body not only has to provide aerodynamic lift, but it also has to serve as the primary compressor for the jet engine. Because of intense aeroheating due to atmospheric friction, the fuel system not only pumps prodigious amounts of fuel into the combustor, but it must be plumbed in such a way that circulating fuel can be used to keep the engine cool. There are many other examples. The engineering codes representing all these factors, along with their grid and mesh generation capabilities, must be integrated in such a way that they will run without interactive assistance.

We recognized early on that embedding a parametric geometry generation system in an optimization loop would impose very stringent requirements on the software, requirements that are not met by any of the geometric modeling systems available on the market. Enterprise-scale commercial CAD systems do provide parametric modeling, but the parameterizations and software implementations don't support optimization. Oversimplifying: CAD systems will generate a stack of photos, but what's needed is a movie. This requirement can be met only by properly formulating the math which lies at the heart of the modeling system.

For example, the geometry generation system must be able to embed differential equations describing relevant engineering physics directly into the geometrical lofting algorithms. The forebody of a hypersonic lifting body functions as the primary compressor for the scramjet engine. A desirable goal in shaping the forebody is to maintain the "shock-on-lip" condition, when all of the supersonic compression waves converge on the lower aperture point of the engine cowl. Longitudinal sections of such a forebody satisfy a particular ordinary differential equation at a particular speed, the free-stream Mach number. The boundary conditions of the ODE and the free-stream Mach number are available parameters for optimizing the vehicle shape. The parametric geometry generator we developed using Geoduck has the property that any forebody shape it produces satisfies the condition. Conceptually, the condition could instead be imposed as an optimization constraint, and an advantage would be that the designer could then determine the design cost of satisfying the shock-on-lip condition. As a practical matter, however, the propulsion analysis code will fail when applied to any design which doesn't satisfy the condition. A high-performance optimization code might attempt to evaluate the objective and constraints in infeasible regions, resulting in failure of the propulsion code.

The geometry generator must meet a number of other requirements which CAD systems aren't designed to satisfy. It is often necessary to impose shape requirements (e.g. convexity) on free-form curves and surfaces. External surfaces usually require continuous curvature, precluding the use of many standard CAD techniques such as circular fillets. The optimization code is likely to perform much better if geometry varies differentiably as a function of the design parameters, which is not even a consideration for CAD systems. Finally, in developing a geometry generator for a new vehicle class, we have found that we should parameterize the geometry over a hypercube in such a way that every design point inside the hypercube leads to a valid, analyzable vehicle design. Finding such a parameterization can be a major component of the work.

Developing the multidisciplinary analysis for this project, including the hypersonic geometry generator, led the entire team to new ground. In addition to the geometry generator, eight engineering codes were included in the MDA, e.g.

propulsion, aerodynamics, and aerodynamic heating. They were sequenced so that each input to any of the codes could be obtained as a design parameter or as an output of previous codes in the cascade. For each design point, the entire string of codes was executed, starting with the geometry generator. ModelCenter® was used to integrate and manage the system. We worked with engineers who had vast expert knowledge of hypersonics and a clear idea of the performance they wanted to achieve. Our role was to formulate the problem mathematically and develop the algorithms and software, in the process expanding our knowledge of hypersonics. It was a great example of what's fun about industrial applied math.

With Geoduck and Design Explorer in hand, we were ready to tackle the design challenge. The "baseline" X-30 configuration from 1992 could barely fly its intended mission, which was to place a small payload in low earth orbit. It was very heavy relative to its payload, and it promised to be very expensive to build and maintain, so expensive that the vehicle was never built or tested.

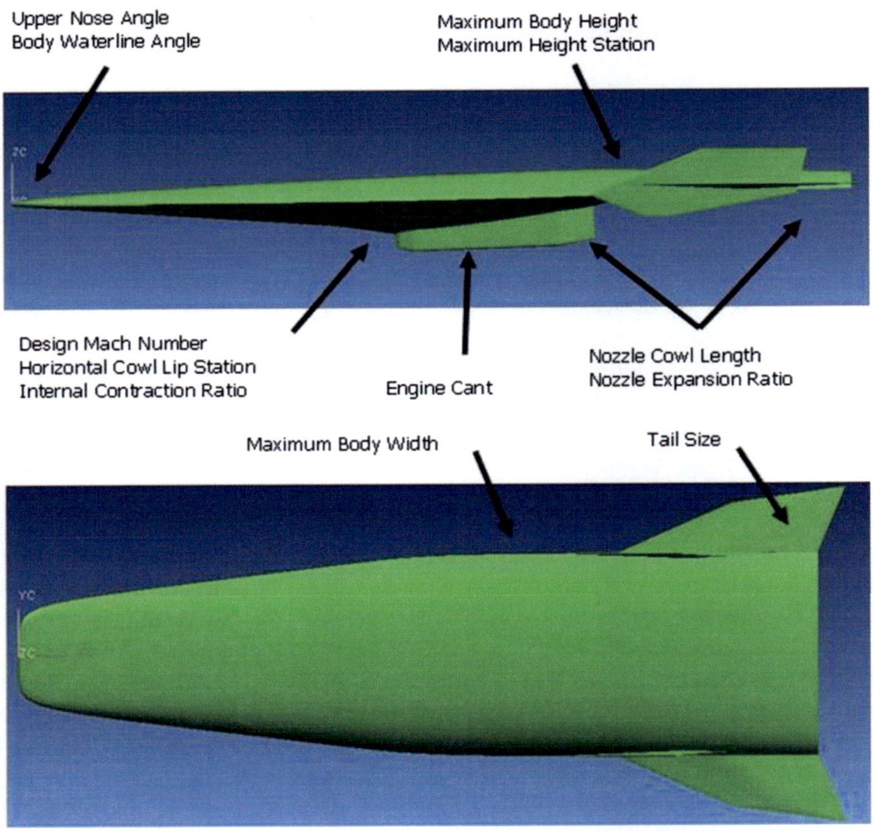

Fig. 1 Original X-30 configuration and design optimization variables

The task was to apply multidisciplinary design optimization to see how much the design could be improved. The first step was to produce a Geoduck model of the configuration, shown in Fig. 1. It depended on some 50 geometric parameters, of which the 12 indicated in the figure were chosen as design variables for the optimization problem. These parameters were selected by the chief scientist, on the basis of long experience in hypersonics, as those most likely to affect performance.

The complete multidisciplinary analysis determined the takeoff gross vehicle weight (TOGW) as well as an estimate of "mass fraction," the fraction of total weight which must be fuel in order to complete the mission. In this exploratory design, the structure was not modeled; instead, a fraction of the mass was allocated for structure. The design optimization problem was then to minimize TOGW, an indicator of mission cost, subject to the constraint that mass fraction not exceed that of the baseline vehicle. The "excess propellant fraction" is the margin by which the mass fraction of the baseline exceeds the mass fraction of the vehicle. Many vehicles produced in the study violated the constraint but were used in building the surrogate models. Some members of the team struggled to assimilate the idea of spending computing time on models which were known to fail. In fact at that stage we were not designing a vehicle, we were designing a system for designing a vehicle.

The initial survey of design space contained 98 vehicles, nine of which are shown in Fig. 2. It's easy to see that there is a lot of variation in size and shape. Even with only 12 parameters, design space is big! Each of these 98 shapes was analyzed in three different sizes, for a total of 294 analysis runs. At a computational cost of approximately two CPU hours each, this represented a substantial investment in computing. Once the survey was complete, surrogate models of TOGW and required mass fraction were constructed by Design Explorer using the 294 points.

We would like to be able to say that at this step in the process we turned Design Explore loose to find the optimum design. Unfortunately, while we now have an asynchronous parallel version of the software, at that time Design Explorer lacked the capability to manage multiple processors efficiently, and instead executed the necessary runs of the MDA sequentially. That limitation made it impossible to use the full optimization capability. Instead we found the optimum of the surrogate model, executed a small number of additional runs near that point and used them to refine the surrogate model, then repeated the optimization.

Figure 3 shows the baseline vehicle, the survey points for each of the three sizes, the initial surrogate model optimum (labeled Iteration 1), the additional vehicles in the pattern search, and the optimum of the improved model. TOGW is shown on the vertical axis as a fraction of the TOGW of the baseline vehicle. Only those vehicles with excess propellant fraction greater than zero satisfy the mass fraction constraint.

The optimal vehicle is the same general shape as the baseline but 13 % shorter and 9 % thinner top to bottom. It would put the same payload in orbit, but with a TOGW 39 %less than was required by the baseline vehicle. The 39 %

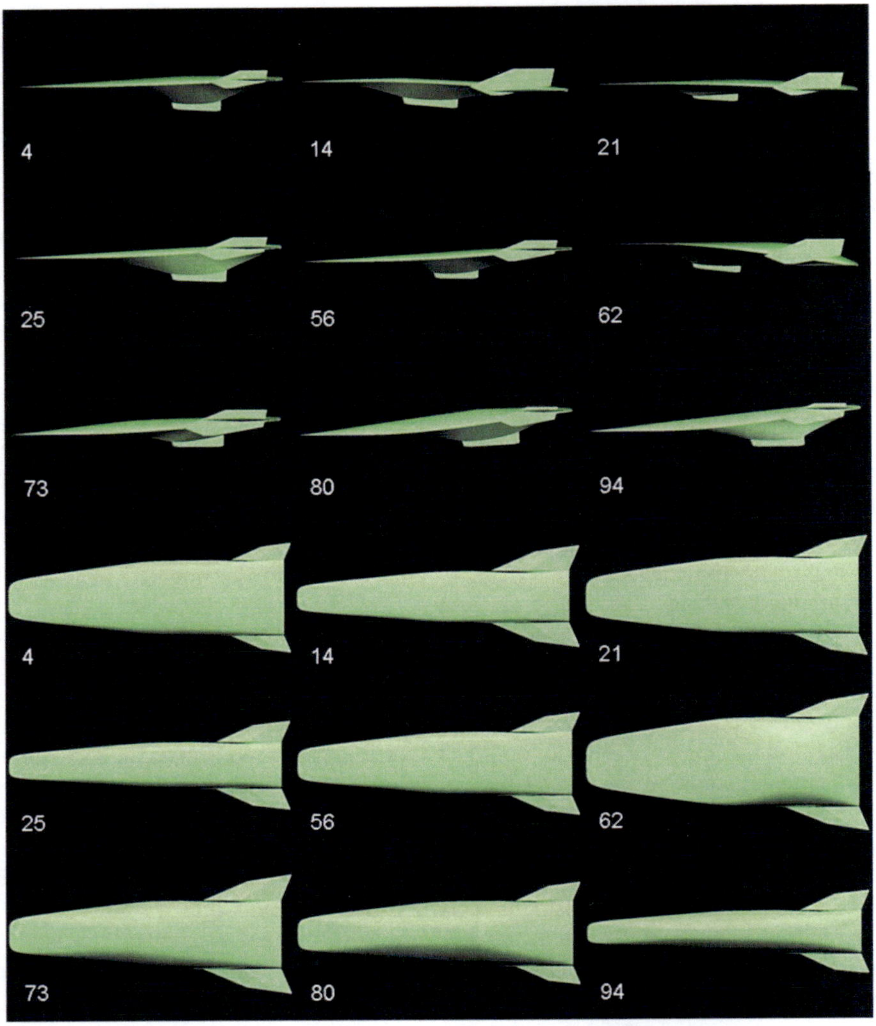

Fig. 2 Examples for the initial survey of design space

reduction in TOGW was accompanied by a 59 % increase in the total aerodynamic drag of the vehicle! For aerodynamicists who spend their working lives trying to reduce drag, it came as a surprise. Analysis of the design tradeoffs made by the optimization software revealed that the propulsion flowpath was reshaped in such a way that the drag penalty was more than offset by a large improvement in propulsion efficiency.

This result is most unlikely to have been obtained in a traditional engineering process, where each discipline works to improve its own performance metrics

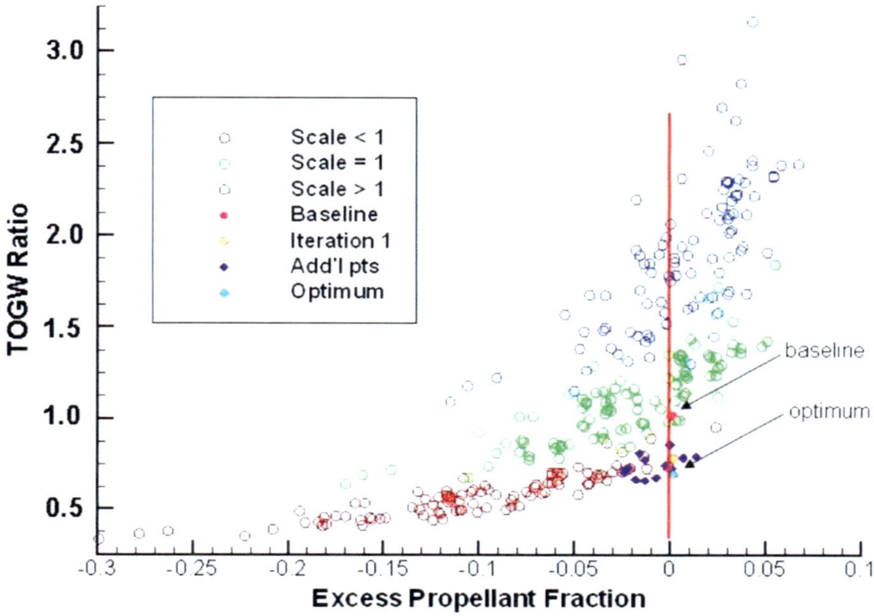

Fig. 3 Results of the design survey and optimization

without regard to total system performance, after which the results are combined through discussion, trial and error. Boeing has been doing multidisciplinary design optimization for 95 years. The mathematics embodied in the engineering codes, geometry system, and optimization software allows us to deal with a much larger design space, explore and understand it more thoroughly, and carry out the process much faster than we could do even 10 years ago.

5 Working with Our Engineering Colleagues, and Helping Them Work with Us

We close with a few observations about the culture we work in. Most often our collaborators are design engineers or manufacturing engineers. While their counterparts in other industries and labs might be physicians or oceanographers, we believe most mathematicians face the same challenges we do in working with their colleagues from other fields. We generalize very broadly, ignoring the fact that at Boeing there are lots of engineers who know a lot of math, and fine mathematicians whose job title is "engineer."

Boeing's customers want to buy the products of engineering design and manufacturing, or in some cases the results of engineering research and development.

With few exceptions they're not in the market for mathematics. For the Applied Math organization to prosper at Boeing, as it has for more than 30 years, we must deliver something that the engineers can't provide for themselves.

What is that something we deliver? Mathematicians are good at abstracting and generalizing, and a concrete result is that an entirely new engineering problem may turn out to be an old math problem. Asked to find an optimal tool path for a numerically-controlled milling machine, we applied the same methods and software we used to compute a missile trajectory with minimum flight time. A contour tape laminating machine lays down a sequence of graphite composite tape courses on a mandrel; we recognized that finding the optimum order of the courses was a traveling salesman problem. Examples like these are commonplace for mathematicians, but may not be part of the engineer's way of thinking. We can often respond quickly to new requirements without substantially increasing the cost of the project, because much of the math has already been done.

Very often what engineers may consider "pure" math turns out to be critically important in applications. An engineering colleague said, "I always liked math in school, but I never understood why we had to do proofs." Mathematicians pay attention to issues of continuity and smoothness. Engineering analysis codes may employ algorithms which assume a degree of smoothness, but in applications no one checks the assumptions. Perhaps no one remembers the assumptions. We're careful in developing our own methods and software, and when we work with engineering codes we're alert for the kinds of failures that can result from violated assumptions. An old code which computed wing surface area used Romberg integration, an algorithm that applies Richardson extrapolation repeatedly to drop successive terms out of the error expansion for the integral. The error estimate after m steps is valid if the integrand has at least $2m + 2$ continuous derivatives. The algorithm was applied to wing surfaces modeled with splines which were only twice continuously differentiable, invalidating the error estimates when even a single Richardson extrapolation step was used. Consequently, the estimate for wing area in a subsequent drag calculation was several orders of magnitude less accurate than assumed.

Another disconnect comes up repeatedly: Engineers and mathematicians define "function" differently. Engineering codes very often produce a list of function values as output, and engineers are used to thinking of those lists as functions, and in particular may overlook the fact that different downstream codes may interpolate differently. As another example of the same problem, for an engineer a surface is an object in space. In cases where it's important to treat the surface instead as a function from 2-space to 3-space—that is, in most cases—some remedial communication may be called for.

A mathematician starting her first industrial job has non-technical skills to learn and a few habits to overcome. A reality of our business model is that we usually can't expect engineers who haven't worked with us to understand how we might help them. There's often a wide gulf between university engineering and math departments—even applied math departments—and a lot of engineers simply don't know there's such a thing as an industrial mathematician. Through their school careers they tackle math problems every day, and that's what they continue to do at work. We accidently encountered a group who were measuring large as-built components in order to compare them with the as-designed components in the CAD system. Their approach, which was a good one, required solving a linear least-squares problem, a term they had never heard. They had been working for several years to develop and improve their solver, not realizing that it's a common problem for which many good software packages are readily available. A corollary is that our staff members must learn to say "yes" to any request for help. Then we follow through on the "yes," even if the help turns out to be brokering a connection with some other organization or expert.

An old joke is, "How do you recognize an extroverted mathematician? When he talks to you, he looks at *your* shoes." We have to get over that habit. Our most valuable contributions happen when we're advocating new ideas to our colleagues, who may well believe they've been getting along fine without us for years. It calls for a degree of cheerful aggressiveness and persistence.

Mathematicians in school study the methods for solving problems. Some of our projects involve improving those methods, but very often the hard work is in understanding and formulating the problem. In school, the professor formulates the problem and you solve it—you hope. In industry, you formulate the problem and the software solves it—you hope.

Mathematicians are trained to get it right. In the engineering setting we often have to deal with the world as it is. It has been roughly estimated that software developed at Boeing represents an investment of twenty billion dollars, and most of it is not getting rewritten any time soon. Beyond the cost, there are codes which are certified by the Federal Aviation Administration and must be recertified if changed. For some of our staff members it's hard to leave things alone when they work but could be improved. We have to resist the drive to streamline, simplify and modernize when the project doesn't need it and can't afford it.

Our collaborators are often our customers, in the sense that they provide the funding for the work, and we pay attention to funding. In spite of that, we have to satisfy ourselves that the project's goals are really valid. We developed a very efficient and practical method of "nesting" part shapes to be cut out of sheet metal. Then we applied it to insulation batts, floor panels, and graphite composite fabric. But when a factory group wanted to use the system to find the best way to hang their tools on the wall, we dissuaded them. Another common example is improvements which aren't on the "critical path." Sometimes the goal is to speed up a production process. We help to shorten one subprocess from 2 months to 2 days. If that success doesn't shorten the overall production time, we haven't really achieved anything.

All these considerations have to do with focusing on the business problem rather than the math problem. That's where the fun, satisfaction and success come from. In the words of John Seeley Brown, then director of the Xerox Palo Alto Research Center, "We have to marinate in the real problems of the corporation."

References

1. Araujo, A., Fernandes, A., Azevedo, A., & Rodrigues, J. F. (2010). *Educational interfaces between mathematics and industry*. Lisbon: Comap.
2. Audet, C., Dennis, J., Moore, D., Booker, A., Frank, P. (2000) A Surrogate-model-based method for constrained optimization. American Institute of Aeronautics and Astronautics AIAA-2000-4891.
3. Bowcutt, K. G., Kuruvila, G., Grandine, T. A., Hogan, T. A., Cramer E. J. (2008) Advancements in multidisciplinary design optimization applied to hypersonic vehicles to achieve closure. American Institute of Aeronautics and Astronautics AIAA-2008-2591.
4. Muhm, J. M., Rock, P. B., McMullin, D. L., Jones, S. P., Lu, I. L., et al. (2007). Effect of aircraft-cabin altitude on passenger discomfort. *New England Journal of Medicine, 357*, 18–27.
5. Phoenix Integration, Inc. ModelCenter. (http://www.phoenix-int.com/software/phx_modelcenter.php), (2011).
6. SIAM Visiting Lecturer Program (http://www.siam.org/visiting/), (2011).
7. SIAM Report on Mathematics in Industry (http://www.siam.org/reports/mii/2012/report.php), (2012).
8. US Bureau of Labor Statistics Occupational Outlook Handbook, (http://www.bls.gov/oco/ocos043.htm), (2011).
9. University of Minnesota Industrial postdoctoral program at the Institute for Mathematics and its Applications, (http://www.ima.umn.edu/people/all-ind-postdocs.html), (2011).

Mathematics in the Workplace: Issues and Challenges

Celia Hoyles, Richard Noss, Phillip Kent and Arthur Bakker

Keywords Technology · Techno-mathematical literacies · Workplace mathematics

In political and educational debate, there is contrary opinion about the mathematical needs of employees. Several influential studies report that, apart from recognizing the need for a small layer of 'symbolic analysts' (Reich 1992), employers do not prioritize mathematical skills (e.g., Department of Labor 1991). In the contrary direction, there are studies in economics and educational policy, which assert that a mathematically literate population is crucial for the economic future of developed countries (Steen 2001; Wolf 2002; Confederation of British Industry 2008). How can this apparent contradiction be understood? We suggest that the contradiction arises from a confusion about what constitutes mathematics and consequently, how mathematical skills are identified (FitzSimons 2002).

Until the middle of the 1980s, research into the mathematical requirements of workplaces assumed that the mathematics used was unproblematically visible to employers and employees alike, so that data on mathematical needs could be obtained by conducting interviews with appropriate personnel and asking for descriptions of workplace activities in mathematical terms. Invariably, such studies demonstrated that little mathematics was used in work and what *was* used was

C. Hoyles (✉) · R. Noss
Institute of Education, University of London, London, UK
e-mail: C.Hoyles@ioe.ac.uk

R. Noss
e-mail: R.Noss@ioe.ac.uk

P. Kent
Former Researcher at London Knowledge Lab, 60 Bramber Road, F46, London N12 9NE, UK
e-mail: phillip.kent@gmail.com

A. Bakker
Utrecht University, Utrecht, The Netherlands
e-mail: arthur@fi.uu.nl

A. Damlamian et al. (eds.), *Educational Interfaces between Mathematics and Industry*, New ICMI Study Series 16, DOI: 10.1007/978-3-319-02270-3_4,

restricted to procedures and calculation, measurement and arithmetic (see, for example, Fitzgerald and Rich 1981; and for a critique of this approach, see Noss 1998). The purpose of this Brief Report is to summarize a more nuanced picture of how workplace mathematics should be understood and developed, as emerging from the most recent research in workplace mathematics (in particular Hoyles et al. 2010).

1 Findings from Ethnographic Studies

One step in this direction has been a series of ethnographic studies of particular work settings. The seminal work by Scribner (1984) on the cognitive (including mathematical) strategies of dairy workers provided important insights into how people regulate and think about their activities by exploiting salient features of their environment. Since then, a range of occupations has been examined, often focusing on disruptions in the routines of work, disagreements between communities as to required action, or problematic communication. It is at these points that mathematical reasoning aligned with workplace expertise has become more visible (Hall 1999; Hoyles et al. 1999; Bakker et al. 2006; Williams and Wake 2007). Many employees have been the focus of such studies, including carpet-layers (Masingila 1994), automotive industry workers (Smith 1999), technicians (Magajna and Monaghan 2003), engineers (Gainsburg 2007; Kent and Noss, 2003), bankers (Noss and Hoyles 1996), nurses (Pozzi et al. 1998; Hoyles et al. 2001; Noss et al. 2002), and employees having 'intermediate-level skills' in sectors such as tourism and food processing (Hoyles et al. 2002).

From this corpus of work, we draw two main conclusions. First, the *visible* mathematics of work tends to be fragmented and associated with routine workplace activities involving measurement and recording, or simple calculations. Although these fragments are meaningful to practitioners in terms of solving particular, well-understood problems, they are finely tuned to specific circumstances, and rarely interpreted as applications of more general mathematical concepts or relationships. Second, the less visible mathematics at stake is routinely tacit, rarely articulated in either written or spoken form. Tools and artifacts regulate activities and scaffold actions, so performance in authentic workplace settings generally outstrips performance on standardized written mathematical tests. For example, a review of 30 studies of practicing nurses (Perlstein et al. 1979) revealed significantly flawed performance on pencil and paper tests of drug calculations, average of 76.6 % (range: 45–95 %). A later study of expert nurses' actual practice on the ward, showed how they eschewed the formal mathematical methods they had learned in training, in favor of error-free strategies tied to individual drugs, their packaging and the organization of clinical work (Hoyles et al. 2001). Information and communication technologies have introduced further layers of invisibility between employees and the mathematical models embedded in the computer systems used as part of routine work practice. The result has been that little trace is evident of the mathematical processes behind results which appear on screen or printout (Kent et al. 2007).

Thus employers might think that any mathematical knowledge required for effective workplace activity can safely be outsourced to the technology. However, such a deskilling assumption takes as read that outputs of computers can be interpreted unproblematically (which is not even the case for expert scientists, see Roth and Bowen 2003). Moreover, interpretation and communication between the different communities in any workplace has become increasingly important, under the pressure of business goals in a global and highly competitive market (Bogni 2010). Technical and analytic information needs to be communicated between shop floor and management, and to customers who are demanding more transparency, more explanations and more flexible responses to their needs (see for example, Victor and Boynton 1998; Zuboff and Maxmin 2004). In other words, the tacit mathematics of the workplace needs to be made explicit in ways that are sensitive to the needs of different audiences.

To name the mathematical knowledge that is required to be effective in an ICT-rich context, we have coined the term 'techno-mathematical literacy,' TmL. It is akin to literacy, in that it involves interpretation as well as the ability to appreciate and communicate with others about mathematical information; and it is mediated by technology—*techno*—in that the information is expressed through symbolic artifacts generated by automated systems.

2 Methods

Our recent work (Hoyles et al. 2010), involved two phases: ethnographic (Phase 1) and developmental (Phase 2). In Phase 1, we developed in-depth ethnographic case studies in a variety of workplaces in the manufacturing and finance sectors, in order to characterize the kinds of techno-mathematical literacies at stake. Methods included work-shadowing, analyses of documentation and semi-structured interviews with managers, team leaders, and a wide range of employees. We progressively focused on the ways in which computer-generated *symbolic* artifacts (charts, graphs, tables of figures, algebraic formulae, and numerical summary measures of performance) served as communicative devices, in the workplace, or between the workplace and customers or suppliers externally. We joined team meetings and listened to conversations around these outputs, to ascertain if problems of communication might be arising and how employees reacted to them; involving 95 person days of ethnographic study in 14 companies, engaging with over 240 personnel, an intensity of work necessary to understand each practice, its distinctive language, and communication requirements.

In Phase 2 computer tools were co-designed with employer partners. These tools modeled elements of the work process, or were reconstructions of the symbolic artifacts identified in Phase 1, as intending to serve a communicative purpose. We called them 'technology enhanced boundary objects' (TEBOs). The aim was that by engagement with TEBOs, a layer of structure could be revealed that we (employers and researchers) deemed to be essential for effective communication, adding, for example, different representations of the output and the functionality to visualize the effects of changing critical variables. In all, we undertook 85 days of development work in 8 companies engaging with over 110 employees.

3 Techno-Mathematical Literacies in Manufacturing and Financial Industry

We found that ambiguity and difficulties of communication tended to stem from the *different meanings* accorded to the symbolic outputs of workplace artifacts: the symbols afforded a visible framework that allowed diverse communities to act and think *as if* they had a common purpose, but this was, in fact, rarely the case. Symbolic outputs were interpreted very differently as, on the one hand representing fragments of mathematical ideas derived from an underlying mathematical system, and on the other, as part of workplace or management culture with little or no inherent logic. This latter interpretation we described as *pseudo-mathematical*, rather like numbers on buses.

In Phase 1, a range of TmL was identified in the manufacturing sector including understanding systematic measurement, data collection and display; appreciation of the complex effects of changing variables on the production system as a whole; being able to identify key variables and relationships in the work flow; reading and interpreting time series data, graphs and charts (Bakker et al. 2006; Noss et al. 2007). Overall, we identified an ubiquitous requirement to understand and reduce variation, as documented in quality control charts and summarized in two process capability indices, Cp, a measure of spread, and Cpk, a measure combining spread and central tendency in relation to specification limits (Hoyles et al. 2007; Bakker et al. 2009). Cp summarizes the spread of a distribution in relation to the required specification for the process:

$$\mathrm{Cp} = \frac{\mathrm{USL} - \mathrm{LSL}}{6\sigma}$$

where
USL upper specification limit
LSL lower specification limit
σ standard deviation

We found that employees across many levels of the workplace interpreted both charts and indices pseudo-mathematically, with weak connections to data or underlying mathematical relationships (Hoyles et al. 2010). In the case of charts, employees, including trainers and managers, failed to understand control limits as artifacts of the distribution, and in the case of process capability indices did not link higher values with reduced variation. There was a tendency to conceive both as arbitrary targets imposed by management.

The TmL at stake in the financial sector included: appreciating the existence of a mathematical model underlying computer output and which variables were critical in determining output; understanding growth (compound interest) and present value of money; interpreting graphs and charts, and making estimates and predictions of the costs of loans based on customer requirements and personal details. Here too, the prevalence of pseudo-mathematical interpretations became evident. A remarkable example was that employees who daily engage with loans or life insurance, rarely appreciated that there was *any* relationship between monthly and annual rates of interest, let alone what it might be.

4 Developing Techno-Mathematical Literacies

The ubiquity of pseudo-mathematical interpretations and the invisibility of mathematical structures suggested a way forward in addressing the problem of developing techno-mathematical literacies in Phase 2, by rendering work process models manipulable through engagement with interactive software tools. These we called, technologically-enhanced boundary objects.

Figure 1 illustrates one TEBO aiming to reveal the fundamental nature of Cp without the need for engagement with the algebraic definition or manual calculation, which we had noted in Phase 1 had diverted employees' attention from underlying mathematical relationships. The TEBO provided employees with an

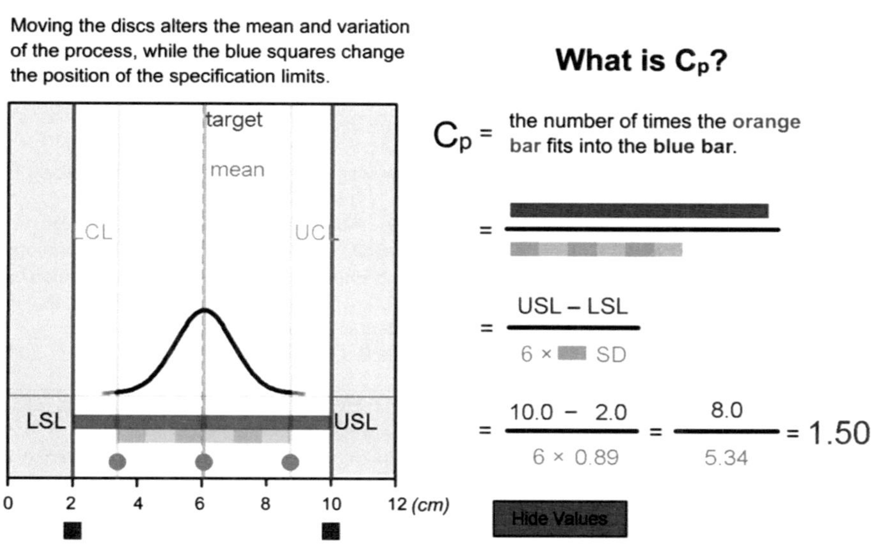

Fig. 1 Screen capture image of the Cp tool

interactive system for manipulating the key variables, the mean and the spread, in relation to the specification limits. A similar TEBO was built for Cpk.

Although our samples were small, we identified in every sector improvement in the relevant techno-mathematical literacies after engagement with the TEBOs, as evidenced in interviews with senior management, and with employees themselves. In the case of the Cp and Cpk TEBOs, there has been remarkable and sustained take-up: In the original factories where the tools were developed, not only with the shop-floor workers but also with supervisors and engineers; and beyond the factories, spreading to SPC courses worldwide (Bakker et al. 2009).

5 Conclusions

These findings have far-reaching implications. First, automated systems create new knowledge requirements—techno-mathematical literacies—an awareness of how models underpin systems and an understanding of how the values of variables define a system's behavior. Second, this knowledge is largely invisible and rarely picked up on the job. Third, numbers, tables and graphs are widely interpreted pseudo-mathematically, as labels or pictures with little if any appreciation of any underlying mathematical machinery. Pseudo-mathematical interpretations of symbolic information clearly impede communication of technical information between communities, and the interpretation and explanation of computer output needs to be informed by sound and appropriate mathematical judgment. However, the design-based approach suggests that the techno-mathematical literacies required can be developed with the help of suitable computer tools.

References

Bakker, A., Hoyles, C., Kent, P., & Noss, R. (2006). Improving work processes by making the invisible visible. *Journal of Education and Work, 19*(4), 343–361.

Bakker, A., Kent, P., Noss R., & Hoyles, C. (2009). Alternative representations of statistical measures in computer tools to promote communication between employees in automotive manufacturing. Technology Innovations in Statistics Education, *3*(2). http://www.escholarship.org/uc/item/53b9122r. Accessed October 15 2010.

Bogni, R. (2010). Data, information and illusion. *Wilmott Magazine*, 45.

Confederation of British Industry. (2008). Taking stock: CBI education and skills survey 2008. London: Confederation of British Industry.

Department of Labor. (1991). *What work requires of schools* (a scans report for America 2000). Washington, DC: United States Department of Labor.

Fitzgerald, A., & Rich, K. M. (1981). *Mathematics in employment* (16–18). Bath: University of Bath.

FitzSimons, G. E. (2002). *What counts as mathematics? Technologies of power in adult and vocational education*. New York: Springer.

Gainsburg, J. (2007). The mathematical disposition of structural engineers. *Journal for Research in Mathematics Education, 38*(5), 477–506.

Hall, R. (1999). Rethinking mathematical practices in design-oriented work. In C. Hoyles, C. Morgan, & G. Woodhouse (Eds.), *Rethinking the mathematics curriculum* (pp. 29–47). London: Falmer Press.

Hoyles, C., Noss, R., & Pozzi, S. (1999). Mathematising in practice. In C. Hoyles, C. Morgan, & G. Woodhouse (Eds.), *Rethinking the mathematics curriculum* (pp. 48–62). London: Falmer Press.

Hoyles, C., Noss, R., & Pozzi, S. (2001). Proportional reasoning in nursing practice. *Journal for Research in Mathematics Education, 32*(1), 4–27.

Hoyles, C., Alison, W., Susan M. –H., & Phillip K. (2002). *Mathematical skills in the workplace*. London: The Science Technology and Mathematics Council. http://www.lkl.ac.uk/research/technomaths/skills2002. Accessed on October 15 2010.

Hoyles, C., Bakker, A., Kent, P., & Noss, R. (2007). Attributing meanings to representations of data: The case of statistical process control. *Mathematical Thinking and Learning, 9*(4), 331–360.

Hoyles, C., Noss, R., Kent, P., & Bakker, A. (2010). *Improving mathematics at work: The need for techno-mathematical literacies*. Abingdon: Routledge.

Kent, P., & Noss, R. (2003). *Mathematics in the University Education of Engineers: A Report to the Ove Arup Foundation*. London: The Ove Arup Foundation. http://www.lkl.ac.uk/research/REMIT. Accessed on October 15 2010.

Kent, P., Noss, R., Guile, D., Hoyles, C., & Bakker, A. (2007). Characterizing the use of mathematical knowledge in boundary-crossing situations at work. *Mind, Culture, and Activity, 14*(1&2), 64–82.

Magajna, Z., & Monaghan, J. (2003). Advanced mathematical thinking in a technological workplace. *Educational Studies in Mathematics, 52*, 101–122.

Masingila, J. O. (1994). Mathematics practice in carpet laying. *Anthropology & Education Quarterly, 25*(4), 430–462.

Noss, R. (1998). New numeracies for a technological culture. *For the Learning of Mathematics, 18*(2), 2–12.

Noss, R., & Hoyles, C. (1996). The visibility of meanings: Modelling the mathematics of banking. *International Journal of Computers for Mathematical Learning, 1*(1), 3–31.

Noss, R., Hoyles, C., & Pozzi, S. (2002). Abstraction in expertise: A study of nurses' conceptions of concentration. *Journal for Research in Mathematics Education, 33*(3), 204–229.

Noss, R., Bakker, A., Hoyles, C., & Kent, P. (2007). Situating graphs as workplace knowledge. *Educational Studies in Mathematics, 65*(3), 367–384.

Perlstein, P. H., Callison, C., White, M., Barnes, B., & Edwards, N. K. (1979). Errors in drug computations during newborn intensive care. *American Journal of Diseases of Children, 133*(4), 376–379.

Pozzi, S., Noss, R., & Hoyles, C. (1998). Tools in practice, mathematics in use. *Educational Studies in Mathematics, 36*(2), 105–122.

Reich, R. B. (1992). *The work of nations: preparing ourselves for 21st century capitalism*. New York: Vintage Books.

Roth, W.-M., & Bowen, G. M. (2003). When are graphs worth ten thousand words? An expert–expert study. *Cognition and Instruction, 21*(4), 429–473.

Scribner, S. (1984). Studying working intelligence. In B. Rogoff & J. Lave (Eds.), *Everyday cognition: Its development in social context* (pp. 9–40). Cambridge, MA: Harvard University Press.

Smith, J. P. (1999). Tracking the mathematics of automobile production: Are schools failing to prepare students for work? *American Education Research Journal, 36*(4), 835–878.

Steen, L. A. (Ed.). (2001). *Mathematics and democracy: The case for quantitative literacy*. Princeton: The National Council on Education and the Disciplines.

Victor, B., & Boynton, A. C. (1998). *Invented here: Maximizing your organization's internal growth and profitability*. Cambridge: Harvard Business School Press.

Williams, J., & Wake, G. (2007). Black boxes in workplace mathematics. *Educational Studies in Mathematics, 64*(3), 317–344.

Wolf, A. (2002). *Does education matter? Myths about education and economic growth*. London: Penguin.

Zuboff, S., & Maxmin, J. (2004). *The support economy: Why corporations are failing individuals and the next episode of capitalism*. New York: Penguin.

Mathematical Modeling Education is the Most Important Educational Interface Between Mathematics and Industry

Tatsien Li

1 Mathematical Modeling is the Most Important Interface Between Mathematics and Industry

First, mathematical modeling is an important bridge between mathematics and industry, and is the inevitable path mathematics has to take toward application.

To solve a practical problem by a mathematical approach, no matter it comes from the field of engineering, economics, finance, or sociology, we need to establish a link between the problem and mathematics. To be exact, we have to first transform the practical problem into a relevant mathematical problem, then do analysis and calculation of the mathematical problem, and finally apply the solution to practice to check whether it can effectively solve the former practical problem. This whole process, especially the first step, is called mathematical modeling, that is, to set up a mathematical model for the practical problem explored. Naturally, for those relatively complicate problems, the process is not likely to be successful on just one trial. If there exhibits great disparity between the final result obtained and the actual situation in the qualitative or quantitative aspect, then we will have to turn back and revise the established mathematical model until we get a somewhat satisfactory result. A truly successful model is one that has been tested by practice and proved effective at length. Therefore, mathematical modeling has to take into account the whole process, that is, to be concerned with not only the 'beginning,' but also the 'ending.' Now it is clear that mathematical modeling is the inevitable path by which mathematics can move toward application, and it plays an indispensable and extremely important role in connecting mathematics with industry.

Second, mathematical modeling is the key to related subjects and application.

T. Li (✉)
School of Mathematical Sciences, Fudan University, Shanghai, 200433, People's Republic of China
e-mail: dqli@fudan.edu.cn

A. Damlamian et al. (eds.), *Educational Interfaces between Mathematics and Industry*, New ICMI Study Series 16, DOI: 10.1007/978-3-319-02270-3_5,

In the third century B.C., Euclid founded Euclidean Geometry by summing up the previous findings of forerunners, which actually proposed a mathematical model for the spatial pattern of the real world. This model turned out to be effective and has been used till today despite various major breakthroughs afterward. Based on a huge amount of data from Tycho's astronomical observation, Kepler summarized three laws of planetary motion which were strictly proved later by Newton using his mechanical principles and the law of universal gravitation that is inversely proportional to the square of distance. This is another brilliant and winning example that brings into full play the function of mathematical modeling. Some important differential equations in mechanics and physics, such as the Maxwell equation in electrodynamics, the Navier–Stokes equation and Euler equation in hydrodynamics, the Schrödinger equation in quantum mechanics, etc., are all mathematical models that reflect the essence of these subjects and create the fundamental theoretical frameworks of related disciplines. As they contain all the important results and potentials for application, these models undoubtedly play a pivotal role and occupy key positions in the fields.

Third, mathematical modeling is the critical breakthrough point and the kernel issue in developing modern applied mathematics.

Nowadays, applied mathematics has entered a phase of advancing rapidly from traditional applied mathematics to modern applied mathematics, while the areas that mathematics can be applied to have expanded to an unprecedented broadness, from traditional mechanics and physics to chemistry, biology, economics, finance, information science, material science, environmental science, energy science, and so on and almost all the high-tech disciplines, even the social field. Since the laws of many new fields are still being explored, the construction of relevant mathematical models is not easy to accomplish, or still faces substantial difficulties which pose great challenge to us today. Hence, the importance of mathematical modeling is ever more eminent than before, and it has become a vital part of modern applied mathematics, which can and will provide more opportunities and broader prospects for applied mathematics as well as for the whole discipline of mathematical science.

Based on the above, we may conclude without the least hesitation that mathematical modeling is the most important interface between mathematics and industry. To manage mathematical modeling is to get hold of the key bond between mathematics and industry which serves as a bridge linking up the two domains, and as a reliable basis and guarantee for effective combination of mathematics and industry, it offers infinite possibilities and paves the way for further development in future. Thus, to form an effectual educational interface between mathematics and industry requires us to concentrate on and strengthen education of mathematical modeling.

2 Mathematical Modeling Education is the Most Important Educational Interface Between Mathematics and Industry

Mathematics is a science that explores quantitative relations and spatial patterns of the real world in a quite broad sense, and as the essential basis of many other sciences, it functions as a sort of think-tank for natural science, engineering science, social science, etc. Meanwhile, it is such a useful tool in economic construction and technological innovation that it is of vital importance to accelerating the modernization process and enhancing the overall national strength. Furthermore, mathematics is also a central component and a solid pillar of human civilization, and mathematical education is of decisive significance to improving the entire people' capacities and cultivating various talents that are badly needed in the construction of modernization.

Considering the features and functions of mathematics, mathematical education should not be separated from other sciences and the external world at large, limiting itself within and circling around only those mathematical concepts, methods, and theories. Otherwise, it will impede in students' understanding of the beginning and subsequent development of mathematical concepts, methods and theories, in inspiring students to voluntarily and actively use mathematical tools to solve all kinds of real-life problems, or in improving students' mathematical accomplishments. However, for quite some time, mathematical courses have formed a system of their own and remain in a state of self-confinement, and we have not been able to find an effective means to associate mathematical learning with the colorful and vivid life in reality, so that students cannot or fail to apply the allegedly extremely important and especially useful mathematical knowledge to real cases, and some may even feel mathematics boring and useless. This kind of situation has undergone noticeable changes only after we emphasized the importance of mathematical modeling, opened relevant courses and organized various teaching activities in this regard which makes a breach in the teaching process to facilitate a link between mathematics and the external world, and helps to develop students' mathematical capability to a large extent.

In addition, the traditional teaching process of mathematics starts from some basic concepts or definitions and then deduces demanded conclusions in a simple and logical way, which of course can allow students to learn as much knowledge as possible in a relatively short period of time step by step, and can enable them to have a taste of the meticulous artistry and flawless beauty of mathematics. Yet, excessive emphasis on this aspect may create a misconception that mathematics is inherently faultless and rightly unexceptionable, which in turn leads to ossification of students' thoughts and their helplessness in the face of lively reality. To foster students' innovative spirit and enhance their mathematical capacities, we certainly need to instill knowledge into their minds, but more importantly, we are to help them know about the creative process of mathematics. This can be fulfilled on the one hand by introducing the modes of thinking and development history of mathematics that can be readily integrated into the teaching contents of mathematical courses, and on the other by creating an atmosphere that would allow students to be personally involved into the discovery and creating process of mathematics. During the teaching process of mathematics, we should take the initiative in encouraging and prompting students to solve some theoretical or practical problems that have no ready answers or fixed solutions, that are not mentioned in any assigned reference book, that cannot be solved by any provided mathematical tool, or that are not even typical mathematical problems. Students would rely mainly on independent thinking, persistent study and exchange of views to generate corresponding mathematical problems, and then proceed to analyze the features of the problems, search for solutions, obtain relevant results, and finally judge the accuracy and falsehood, strength and flaws of the conclusions. In a word, the purpose is to let students have a taste of the pears by themselves and experience in person the creative process of mathematics, so as to gain valuable experience and tangible feelings that can never be acquired from classes or textbooks. This is indeed the distinction and particular advantage of offering mathematical modeling courses and carrying out some related teaching activities.

To sum up, mathematical education should not only focus on introducing mathematical concepts, methods, and conclusions in a spoon-feeding manner, but also lay more stress on letting students grasp the spiritual essence and thinking pattern of mathematics, master the core of the discipline, and willingly receive edification of the mathematical culture so as to make mathematics a versatile weapon that can be used all their lives. Mathematical modeling and related educational activities will change the previous situation of mathematical courses that have established their own system and remain in self-confinement, and will offer a way or an effective teaching approach to link mathematics to the outside world. By taking part in mathematical modeling practices, students can try personally to apply mathematics to the reality, involve themselves into the process of discovery and creation. It is definitely conducive to illuminating their mathematical intelligence and encouraging them to better apply, appreciate, understand and love mathematics. As a result, students can grow quickly in terms of knowledge, capacity and disposition, and stay always in an active and positive state bearing the idea of integrating mathematics with industry.

Since mathematical modeling education is the most important educational interface between mathematics and industry, in order to emphasize the link between mathematics and industry and to improve students' mathematical accomplishments, we should attach great importance to the thinking mode of mathematical modeling in mathematical education, infuse the concept of mathematical modeling into all the mathematical courses, especially the core curriculum of mathematics, and make suitable alterations to the systematic arrangement of mathematical courses. This will be a huge project, a revolutionary change in course offering, teaching contents and teaching methods, and a significant reform of mathematical teaching. However, the former system and basic contents of main mathematical courses are the fruitful and time-tested results of years' accumulation, and we shall not change them totally and in haste without sound reasons. Besides, they are also what most teachers have been acquainted with and got used to, so a radical change is unlikely to be accepted and supported by the vast majority of teachers. Therefore, we would better gradually instill the ideas of mathematical modeling, instead of reaching the goal in one step, and we can strive to combine the ideas with the present teaching contents to form an organic whole, which may consummately exhibit the leading and pioneering role of mathematical modeling as a concept.

To achieve our aim in effect and to avoid occupying too many class hours and increasing students' burdens, the contents of mathematical modeling to be infused into the core curriculum should be strictly and carefully selected, upholding the principles that we should concentrate on the central concepts and essential contents of the main courses, rather than touch upon meticulous details; all the related practical background information should be explained clearly and concisely, rather than be elaborated in a verbose and complicated fashion; we should not seek after establishing a system of our own or attaining self-perfection, but rather be concerned with organically integrating the ideas with the given contents; language should be brief, fluent, plain, and easy to approach, rather than be filled with a large professional vocabulary of abstract concepts.

We do believe that with earnest and unremitting efforts, we will be able to change the present situation of mathematical courses progressively from the aspects of course offering, teaching contents and teaching methods, and the changes will truly help to make mathematical modeling education the most important interface between mathematics and industry.

3 China's Efforts in Strengthening Mathematical Modeling Education and Research

Ever since the 1980s, under the urge of building up a bridge to closely link up mathematics and industry, Chinese scholars in the field of applied mathematics have made persistent efforts to promote education and research on mathematical modeling, and achieved gratifying results of the interfaces between mathematics and industry in the past 30 years (cf. Xie 2010). Here I will simply mention some general facts on the macro-level. In the main, Chinese scholars of applied mathematics have dedicated themselves to the following several tasks these years:

1. Ever since Fudan University first established the course of Mathematical Modeling in 1982 and was followed not long after by some other universities in succession, the two courses, namely "Mathematical Modeling" and "Mathematical Experiments," have been successively set up in many colleges and universities nationwide. The former course usually contains some classic examples of diverse models established for the real world and some necessary preliminaries of mathematics, while the latter focuses on practical training of

students' skills in using computer, software, and other modern mathematical technologies to carry out mathematical modeling. These courses are welcomed by and popular among students, because they on the one hand inspired students' interest in study and facilitated their quest of independent study, and on the other enhanced their capability of and confidence in application and innovation. Up to now, the two courses or other similar courses have been officially listed into the teaching syllabus of more than 1,000 colleges and universities throughout the nation (i.e., over half of the total number of colleges and universities in China) and are regularly open to students, in addition to which around more than 100 textbooks and related teaching materials have been published. Besides, as a supplement to the courses, supporting mathematical experiments laboratories have been built in over 200 colleges and universities. Now, it seems that this positive and healthy trend of development is not only irresistible, but also irreversible.

2. In 1990, the first Undergraduate Mathematical Contest in Modeling was held in Shanghai by Shanghai Society for Industrial and Applied Mathematics (Shanghai SIAM), and ever since 1992, China Society for Industrial and Applied Mathematics (CSIAM) has organized once every year the China Undergraduate Mathematical Contest in Modeling which has been cosponsored by CSIAM and the Higher Education Department, Ministry of Education of China since 1994. Unlike ordinary Olympic Mathematics Competitions, the contest bears the following features: each team consists of three students and a supervising teacher; the contest questions come from areas of engineering, management and so on, which have been suitably simplified into practical modeling problems; within 3 days, that is 72 h, each team will have to choose one question from the alternatives, study into it, set up models, do analysis, and finally submit a conclusive paper for appraisal. During the course, the teams may search into any resources including libraries and the internet, and may carry out thorough discussions within the team of three members, but are not allowed to have any communication in any form with outsiders including the supervisor. This contest aims at cultivating students' creative mind and their ability of handling practical problems, and since it provides a good opportunity and platform for students to develop esprit de corps (team spirit), it is now widely welcomed and supported by students at large. Many participants have summed up their experiences in a word 'to take part in the contest once may benefit one lifelong'. During the 20 years from 1992 to 2011, the number of schools and teams that enrolled in the contest has increased by 16 and 24 %, respectively, each year on average, and in 2011, the number of participant schools reached 1,251 which was more than half the total number of colleges and universities in China, while the participant teams counted up to 19,490 with over 58,000 students involved. What is more? The contest is open to students from all majors, and in fact more than 80 % of the participants major in a subject other than mathematics. It is also worth mention that since 2010, this large-scale and highly successful contest has begun to go international as some foreign teams have participated, and we sincerely hope and welcome more students from more countries and regions to join us in this contest.

3. Since 2002, with the help and financial support of the Ministry of Education of China, we initiated a project of educational reform entitled 'To Infuse the Ideas and Methods of Mathematical Modeling into the Core Curriculum of Mathematics', and our purpose is not to completely change the set courses, such as Calculus, Linear Algebra, Introduction to Probability and Mathematical Statistics, etc., but to design and compile some fitting and applicable courseware of mathematical modeling that can be integrated into and effectively used in the teaching process of the main courses. The courseware will not interfere with the regular and routine practice of teaching and learning, but can on the contrary enhance students' interest and enthusiasm in taking the main mathematical courses and deepen their understanding of the course contents. It has not been long since we executed the project which for the moment is still at the initial stage despite some heartening progress, so there is much more work to be done and furthered in future.
4. Referring to the experience of Oxford Study Group with Industry, the first Study Group with Industry was held in Shanghai in 2000, and ever since then the activity has taken place in China every year, including in Hong Kong. It is an occasion jointly organized and sponsored by the academia and industrial circles, and during the 5 days, people from colleges and universities, research institutions and industrial units will gather together to bring up some imperative issues to be handled in application, and discuss relevant mathematical models and solutions before finally making a brief conclusion of each problem. Some questions proposed can be further studied later on after the activity, and students by participating in it can know about the urgent demands of industries, deepen their understanding of the real world, arouse their interest in and enhance their ability of solving practical problems. Above all, the activity prompts close union of mathematics and industry.
5. Since 2005, the National Natural Science Foundation of China has vigorously advocated 'problem-driven research of applied mathematics' and encouraged outstanding mathematicians and young talents to face the actual demands of national economy, high and new technologies, as well as other scientific fields, and willingly devote themselves to related research projects of application. This type of research mode is driven by the urge to solve problems rather than inspired by papers, and the key is still to establish and study applicable mathematical models. We do believe this will contribute to closely linking up mathematics and industry, and may break a new path to promoting original innovations of mathematical concepts, methods and theories. Furthermore, besides building up a bridge between mathematics and industry, it is also conducive to cultivating interdisciplinary and all-around talents. The National Natural Science Foundation of China has set up a special research fund for its sake, and arranged a corresponding major research project in 2011. At present, the proposal of 'conducting problem-driven research of applied mathematics' has won support among people and has already changed into active actions. It can be predicted that if we persevere in trying toward this direction, this type of research will surely become common practice and produce fruitful results.

To conclude, in the recent 30 years, China's efforts in these five aspects, namely, forming and setting up courses of "Mathematical Modeling" and "Mathematical Experiments," holding Undergraduate Mathematical Contest in Modeling, infusing ideas and methods of mathematical modeling into main undergraduate courses of mathematics, organizing Study Group with Industry, and advocating 'problem-driven research of applied mathematics,' in essence all center around the subject of striving to improve the interface between mathematics and industry. This is the longest, widest, and most successful educational reform of mathematical teaching in China these years, and has won general recognition, warm welcome and strong support from students and all social circles. It has been growing and developing progressively, and we firmly believe that through our concerted efforts, we will witness a broad prospect and a promising future of the interface between mathematics and industry.

Reference

Xie, J. (2010). Mathematical modeling courses and related activities in China Universities. In A. Araujo, A. Fernamdes, A. Azevedo, J.-F. Rodrigues (Eds.), *Proceedings of Educational Interfaces between Mathematics and Industry Conference EIMI 2010*, Lisbon,, Portugal (pp. 597—606), April 19–23 2010.

Models for Industrial Problems: How to Find and How to Use them—in Industry and in Education

Helmut Neunzert

Models are images of objects from the real world—images, which underline features of the object which are considered to be important and which omit features considered to be irrelevant. This is maybe as well true for Giacometti's "models" as it is for mathematical models. These are, however, not made from metals, but from mathematics (Fig. 1).

How do we use them? Mainly for prediction, quite often also for optimization. In practice, for example in industry, we have to solve design problems: We have to design a production process, for example a spinning process for textiles, we have to design a product, for example filter media for many different kinds of filters. In organization we have to design plans, work plans, assignments, etc., for example the assignment of certain tasks to workers.

In order to find a good process, product or plan we make a mathematical model, try to estimate the free parameters in the model; we find or invent an algorithm to evaluate the model, we implement it on appropriate computer platforms in order to get simulations of the behavior of the process or the product or of the realizations of the plan. And we compare it with reality—if we can. But then, we want to change the process, the product, the plan in order to be optimal, optimal with respect to one or even several criteria. And for that, we need many simulations—even, if we are modest enough to ask for something better, not necessarily something optimal. This modesty is called meliorization instead of optimization.

So, we hope to end up with an optimal or, at least, a better design and often we would like to have visualizations since we are not able to cope with so many data. And, finally, we should make clear that our model is not the full truth, it is only justified, if certain conditions are fulfilled, if our parameters are reliable; if we forget that, we may become responsible for errors in the simulation of finance markets, of ash clouds, of evacuation plans (Fig. 2).

H. Neunzert (✉)
Felix Klein Center for Mathematics, Kaiserslautern, Germany
e-mail: neunzert@itwm.uni-kl.de

A. Damlamian et al. (eds.), *Educational Interfaces between Mathematics and Industry*, New ICMI Study Series 16, DOI: 10.1007/978-3-319-02270-3_6,

Fig. 1 Man pointing, MoMA, New York, Photo: S. Mährlein

All these steps—modeling, computing, implementing, optimizing, visualizing, analyzing—are not independent. There is never only one model, but many, and one has to choose one, which is as simple as possible, as complex as necessary in order to allow reliable predictions in a limited time. There are different algorithms that have to balance precision and time consumption; there are good and bad implementations, optimal optimization methods and there are—sometimes forgotten—careful and honest descriptions of the constraints of the model and the simulation. Experience teaches us that it is often quite complicated to define the criteria of

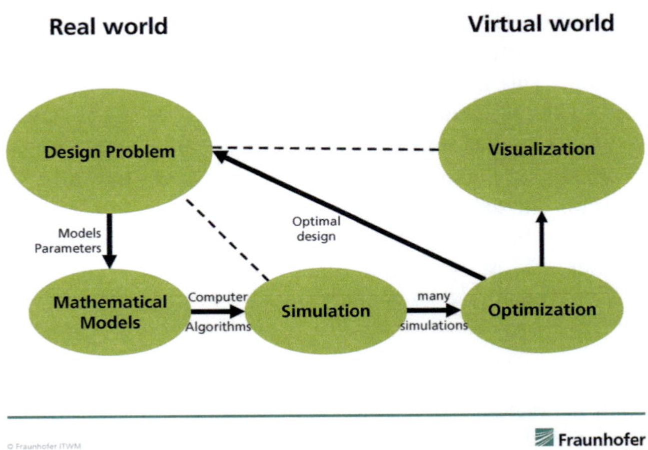

Fig. 2 Real world versus virtual world

optimality applied in practice, therefore, optimization needs many contacts between problem poser and problem solver, and it requests quick simulations.

For this lecture, we want to underline two main messages, which, in my opinion, should be kept in mind when modeling in schools and universities (Fig. 3).

I want to illustrate these two messages with two examples; both are taken from projects, which the Fraunhofer Institute for Industrial Mathematics ITWM has dealt with during the last 5 years. The first one, modeling a production process of textiles, dealt with a very complicated process and needed a series of very high-level publications to be brought to a solution. It is certainly not appropriate for school modeling, but it illustrates very clearly our first message (see Klar et al. 2009).

Figure 4 show (anticlockwise from 22 h) the machine, a scheme of the production of the fibers where polymer is melted, pressed through nozzles to form first liquid, later solidified fibers, which are drawn by air blown in the direction of the fibers and finally a lay-down process on a conveyor belt, a magnified image of the "non-woven" texture produced and a photo of the process itself, where thousands of fibers are curled and pulled down to the belt to form the texture. 20 years ago, I was, first in my life, standing in front of such a machine with an outstanding fluid dynamist from Berlin who told me: "We both will, during our lifetime, not see a reliable simulation of the process." Well, I have seen, mainly due to a fantastic Ph.D. thesis of a young woman, who is still working in the field.
It is quite often a good idea to describe the problem as an input–output system in order to understand what is changeable and what is the goal (Fig. 5).

Let me make a small deviation, discussing the problem of black, white, and gray models, which is often behind hierarchies of models. The general system theoretical setting describes input–output systems by a class of mappings, out of which one mapping is selected either by theory or by observations (Fig. 6).

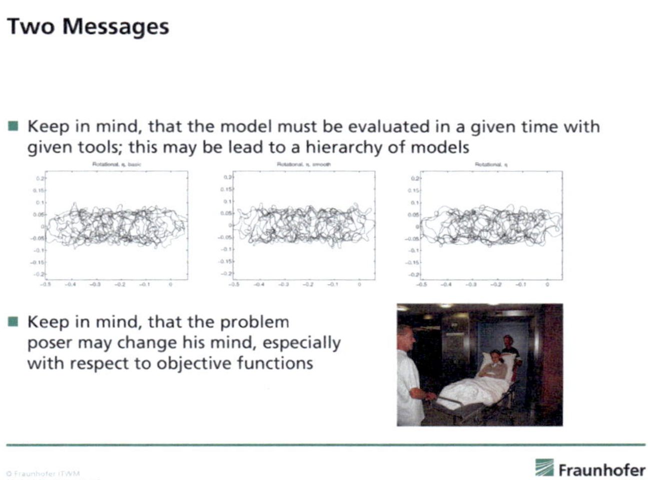

Fig. 3 Two messages

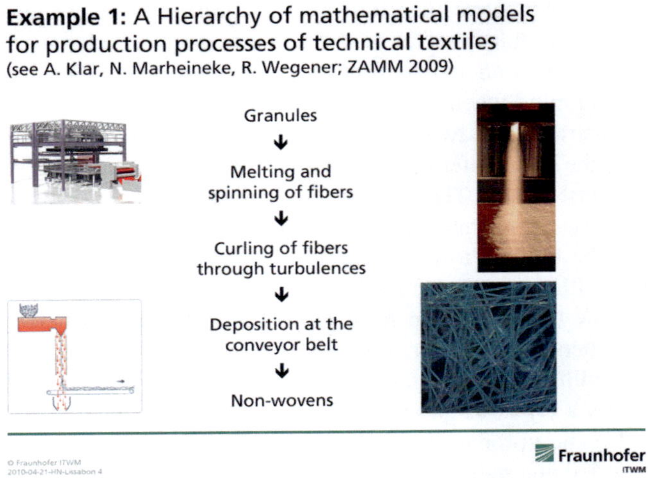

Fig. 4 Example 1: A hierarchy of mathematical models for production processes of technical textiles (see Klar et al. 2009)

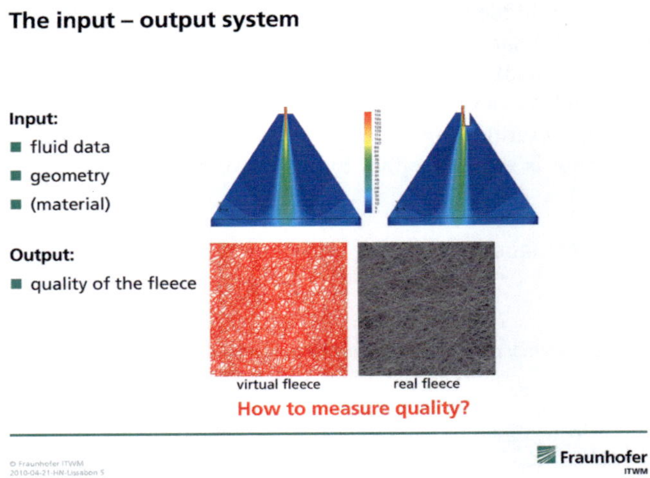

Fig. 5 The input–output system

If we have little theory—as for example in economy or in medicine—we do not know about the structure of the mappings, we need many parameters and, therefore, many observations to identify the parameters. The class consists typically of linear control systems or neuronal networks, and we call these models black box models. To give you an example from our school experiences: We describe

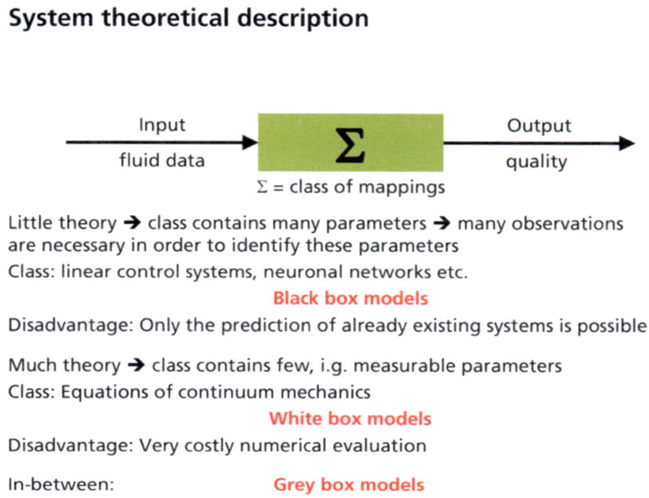

Fig. 6 System theoretical description

windmills as input–output systems with wind data as input, the power generation as output and neuronal networks as mappings. It seems—we are just halfway—that such a 2-year project with 3 h a week may work for 15-year old students. Why do we use black box models in this case? Since we have not enough theory, at least not at this school level. But we have many observations. Black box models, however, have one disadvantage: We can identify the parameters just of one system and we cannot change the design of the system in order to get a different behavior. The parameters of our neural network have no evident physical meaning, and, therefore, cannot be interpreted for a new physical design.

The situation is totally different, when you have theory, for example very special input–output mappings with less parameters that have a physical interpretation, for example some data which describe the geometry, others which express physical quantities, etc.

In our case, the theory may be given by continuum mechanics. We call these models "white box models." They are very useful for finding or optimizing the design, but, quite often, they are very time consuming. It is sometimes said or at least hoped that with improving hardware and improving software, there will be no time constraints anymore. It is the same illusion as it is to say that a better income will lead to the fact that you have no financial constraints anymore. This may be true for Perlman or for Bill Gates—it is most often not true for you and me.

And therefore, we sometimes need something in between: As much "simple" theory as possible, as fast algorithms as possible: Gray box models.

Here we go—and we make it rather short since the purpose is to show how we get the hierarchy of models—for mathematical details please look at the paper cited in Fig. 4.

Of course, a fiber out of liquid or solid, is a 3D-object, which changes its shape and moves in space. This would lead to a "very white" model, but everybody will understand that this model is far too complex to be used to determine the behavior of several hundred fibers. Therefore, we simplify, using a 1d rod theory for the fiber, which is now a curve $r\ (s,\ t)$, parameterized by arclength and depending on time t, and we get (Fig. 7).

The inner tension T is not known, but determined through the constraint of incompressibility. The equation is a wave equation and, together with the constraint, it gives an algebraic-differential system. This equation includes the force term f, describing the forces, which the wind exerts on the fibers.

To get the air forces, we have to calculate the airflow and its influence on the fibers.

This is classical theory, for example Navier–Stokes Equations provide us with the fluid velocity and the difference of it to the velocity of the fiber will determine the force (Fig. 8).

But we have Reynolds numbers above 2000, and we are not able to compute the solution in a "direct simulation" in the time given. Engineers will tell us how to simplify, propose turbulence models as for example k-ε models. Now, it is doable, but we need some hours per fiber—too long!

Therefore, we have to start what British applied mathematics calls "modeling": We have to search for a small parameter in the system, which is "almost zero," so that the asymptotic limit, letting this parameter tend to zero, gives a simpler, hopefully still reliable model. What is "small" or "almost zero" is not a priori clear. Here, some caution is necessary; one is tempted to use the asymptotic limit, if it is scientifically interesting. Whether it is reliable, near to reality—and this is the only criteria, if you want to help practice—must be tested, for example by

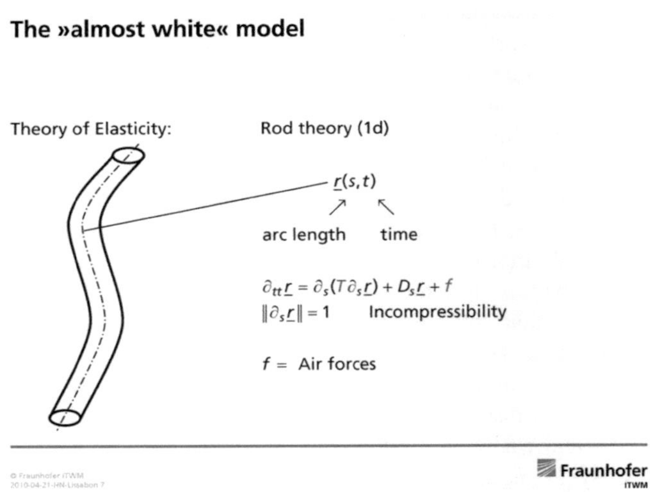

Fig. 7 The almost white model

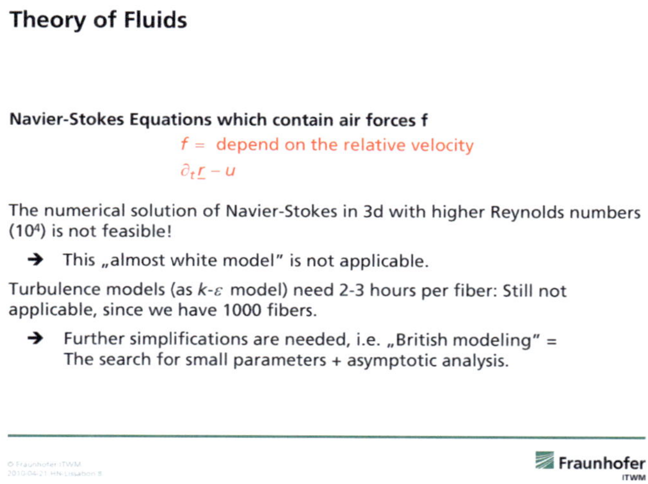

Fig. 8 Theory of fluids

benchmarks. Nevertheless, this kind of modeling is an art, is pleasure and is beneficial. But where is the small parameter here? Not the thickness of the fiber—it is already zero. What else? Look from a bridge into the flowing water of a river: There are, behind the bridge piers, vortices of different sizes, big ones and smaller ones with high rotation at speed. Which air vortices does our fiber feel? Some are too weak or pass over too quickly to bend the fiber. Some have high energy and bend it, so it curls. This curling is important for the quality of the fleece (Fig. 9).

There is a classical theory by Kolmogorov (the probabilist, who did much more than that), which tells us what is shown on (Fig. 10).

The x-axis designs the inverse of a typical size (length, diameter) of a vortex; the y-axis shows the energy in a vortex of a certain size. Large vortices with k small have little energy and very small ones too. In between, there are the vortices with highest energy; for them the interior forces in the fiber and the turbulent (vortex) forces are in balance. They curl the fibers—and they have a typical size l_T (Fig. 11).

Now, l_T turns out to be small compared to the length of the fiber: It is the small parameter and we let it tend to zero. In this limit we get, as effect of turbulence, so-called “white noise,” with parameters, which may be identified from the k-ε model.

Clearly a simple , still reliable model, even more time consuming for our purpose. With this, we can simulate, if we take our time—but for an industrial simulation it is too slow.

Now we simplify “with brute force” and try to model the deposition process directly: We consider the point on the conveyor belt, where the fiber hits it first—we call it the “impact point.” The coordinate vector of this point is ξ, depending on a curve parameter s. Take a (flexible) water pipe and press it down to a plane—

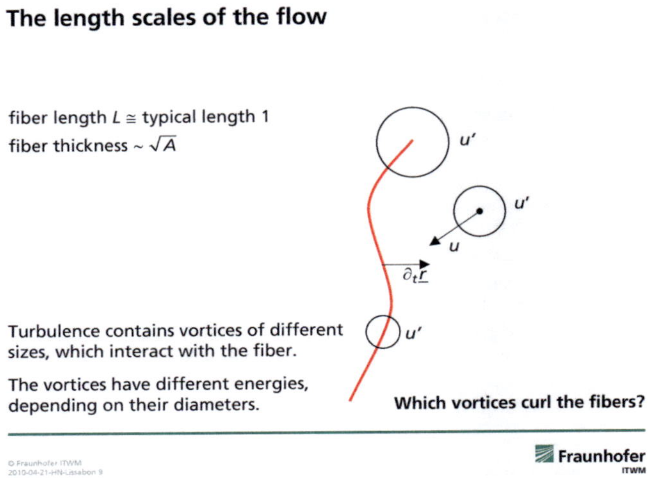

Fig. 9 The length scales of the flow

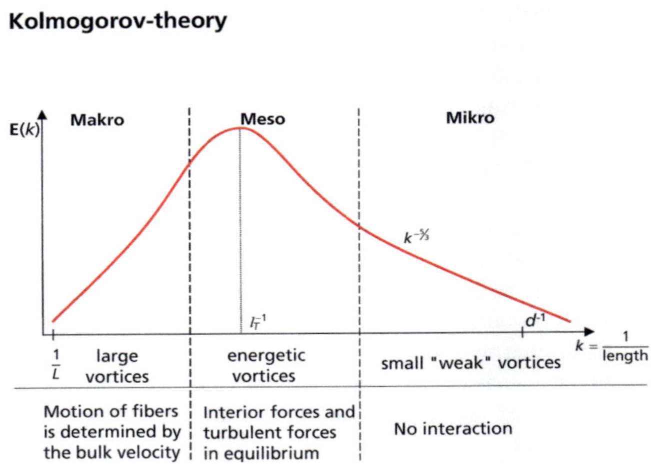

Fig. 10 Kolmogorov-theory

for a moment we assume a non-moving conveyor belt. The pipe always turns towards the direction of ξ—how strong, may depend on the length of ξ. You see the details on Fig. 12, and you see that we added a stochastic term, which models the turbulence forces.

You see the influence of the turbulence, whose strength is given by A (Fig. 13). We assume that $b = 1$ and, still, that the conveyor belt is not moving.

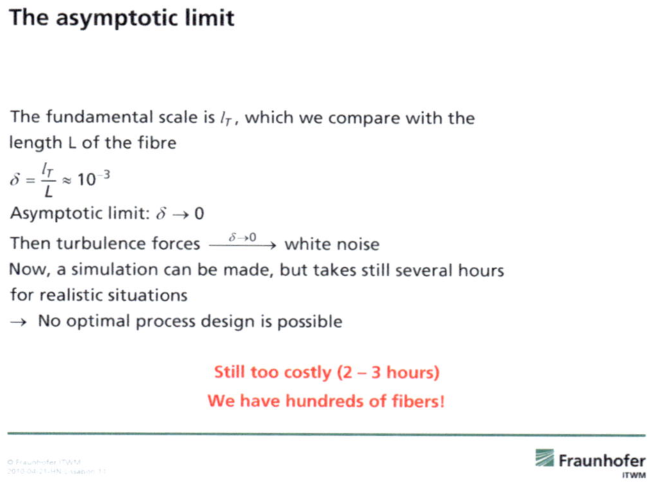

Fig. 11 The asymptotic limit

The more turbulence we have, the more irregular the path is. We can vary a and b and try to identify them in such a way that the result agrees with the previous model, derived on Fig. 11. The agreement is surprisingly good, the time is very short (even for moving belts)—we can simulate almost in real time with good precision.

Everything fine? In industrial mathematics, there is never a "fixed solution." Whenever you believe it is done, your partner will tell you his next wishes—some you can see on Fig. 14.

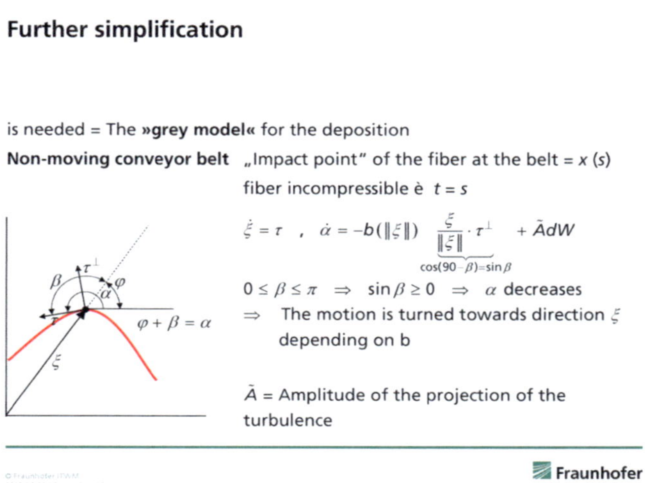

Fig. 12 Further simplification

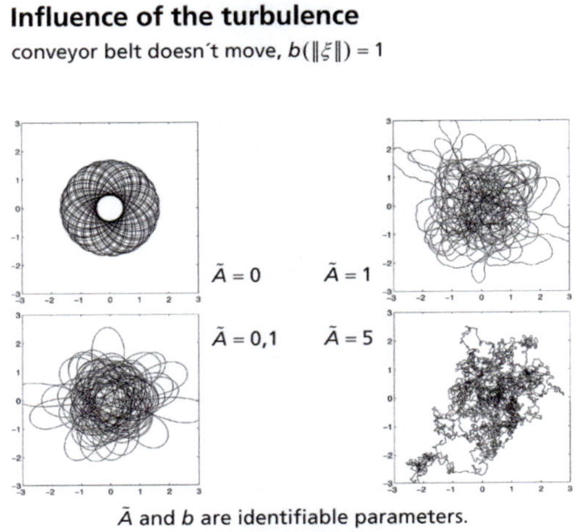

Fig. 13 Influence of the turbulence

What does this first story tell us for modeling in schools? Quite simple:

Do not ask for the model—there are always many different ones. Do not forget that you have to use your model, to evaluate it with the means you have at hand. Therefore, make your model as simple as possible, but as complex as necessary in order to remain reliable. Do not forget to tell the problem poser that your model has limitations, it may give wrong simulations, when you leave the area of its admissibility. Teach your students that models are not robust with respect to wrong applications—look at simulations of financial markets, of evacuation plans or of behavior of ash clouds. Sometimes small errors in parameter values create simulation differences, especially for long-time behavior.

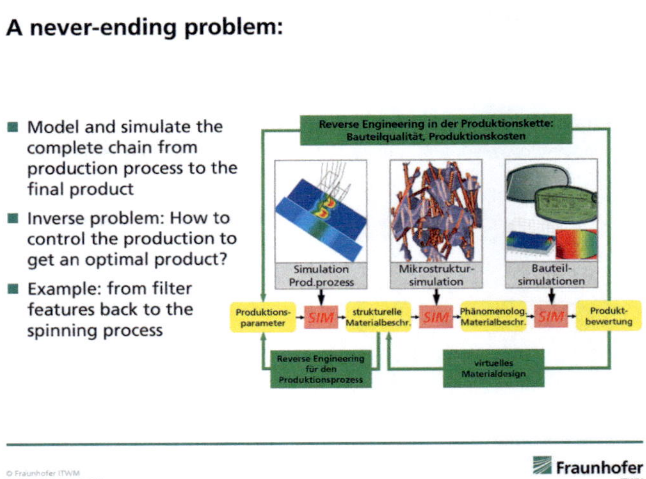

Fig. 14 A never-ending problem

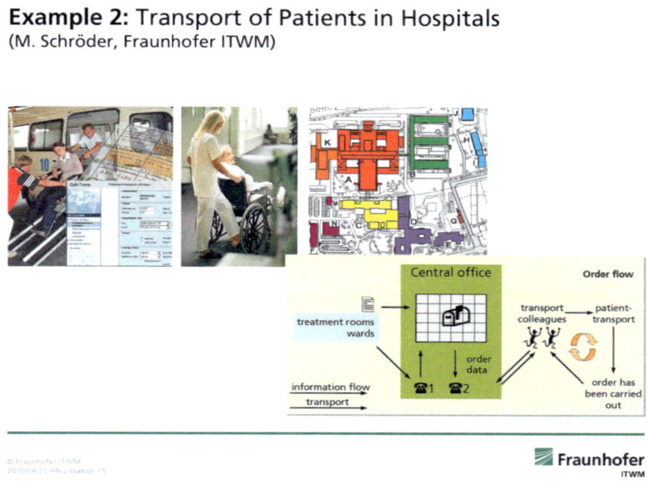

Fig. 15 Example 2: transports of patients in hospitals (M. Schröder, Fraunhofer ITWM)

The second example tells us a quite different story. The mathematics we need here is rather simple, so that it could be done in schools. But the task to understand the problem well enough has psychological components. What is "optimal" is sometimes hard to define.

I take an example from our optimization department treated by Michael Schröder, and I am grateful to him for providing the material (See Beauty et al. 2010). It is a problem most of us know: Patients in hospitals must be moved to the

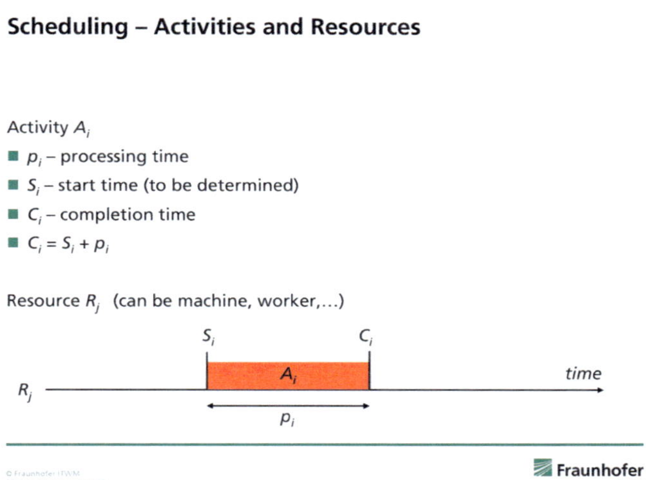

Fig. 16 Scheduling—activities and tardiness

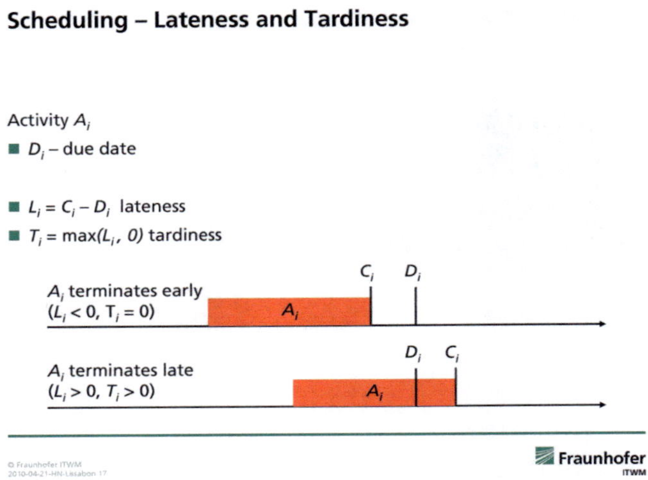

Fig. 17 Scheduling—lateness and resources

X-ray department or to another doctor. There is staff that transports them. The orders for transport come in a stream, an order flow, and must be assigned to the personal (Fig. 15).

Therefore, it is an assignment problem—you have to assign resources to activities, so that the total performance is optimal. What "optimal" means is the modeling task. Of course, the patients should not be too late; to be too early, is not as bad, but not useful either (Fig. 16).

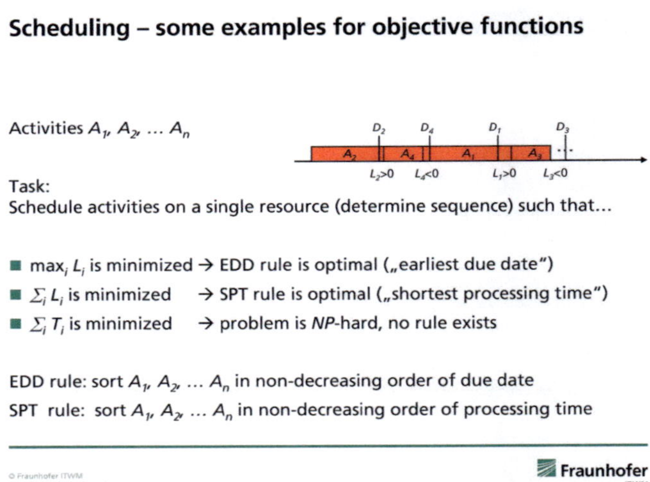

Fig. 18 Scheudling—some examples for objective functions

So first we consider a scheduling task: How do we order the activities in time? Look for the moment on one resource, for example one person, who is responsible of carrying out the activities. We need some definitions, which you understand from.

Now we define concepts like being too late or being retarded (Fig. 17).

Next, we consider several patients—or activities—and ask for different objective functions. We still have a single resource and ask to schedule the activities with their different processing times and their due dates. We can ask the schedule to care for the largest lateness to be minimized; or for the sum of all latenesses being minimized—but we are aware that lateness may be negative, which would reduce the sum of all latenesses. Or—what seems most natural—one minimizes the sum of all tardinesses (Fig. 18).

It is not hard to see—even in school—that maximal lateness is minimized, when you schedule according to the order of the due dates: What should be finished first is executed first. The sum of all latenesses is minimized, when you schedule according to the length of the processing time: What takes the least time is executed first. Finally, for the most appropriate objective function, the sum of all tardinesses, we do not find a simple rule. There is, as the optimizers say, no polynomial algorithm, it is NP-hard. But of course, there are algorithms, which approach the minimum—you will find them in the internet.

But, finally, you have not one "transporter," you have several. And now, on top of the scheduling problem, we have the assignment problem.

Normally, scheduling and assigning is done "by experience," for example by people with telephone, paper and pencil (Fig. 19).

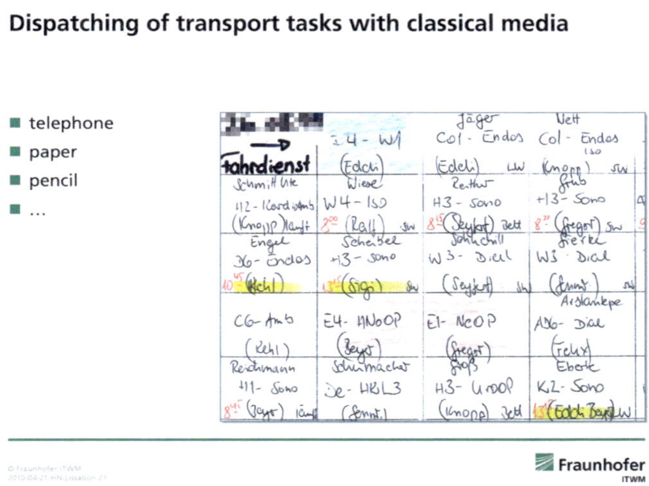

Fig. 19 Dispatching of transport task with classical media

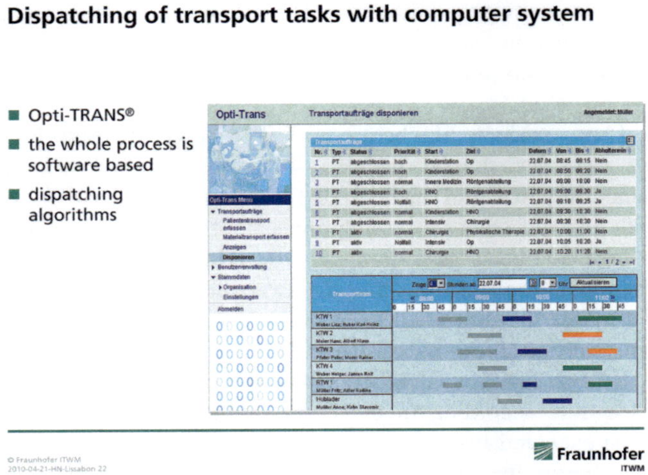

Fig. 20 Dispatching of transport task with computer system

If we model the situation, solve the problem approximately with a single computer and put it in a software (which we may call Opti-Trans from Optimal Transport), we may do it better (Fig. 20).

But, sometimes, we realize quite late that we have forgotten some important aspects. It may be that not we have forgotten them, but the people in the hospital, with whom we are talking, had not thought about them. Here, it was the workload for the staff, which should be almost the same for everybody. It is creating irritation, if x is running all the time, while y sits with a cup of coffee in the cafeteria (Fig. 21).

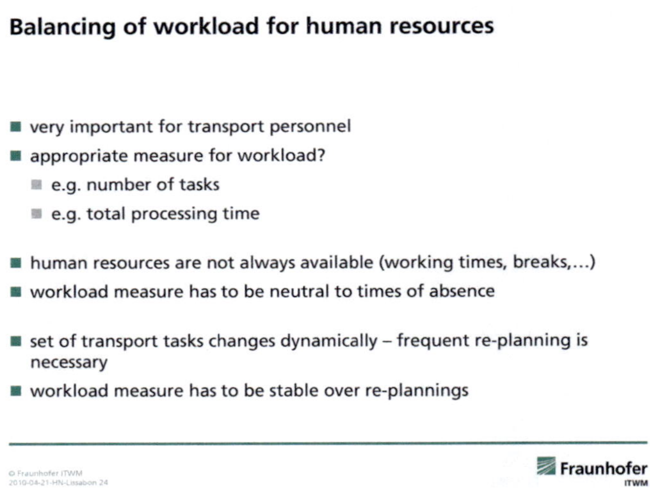

Fig. 21 Balancing of workload for human resources

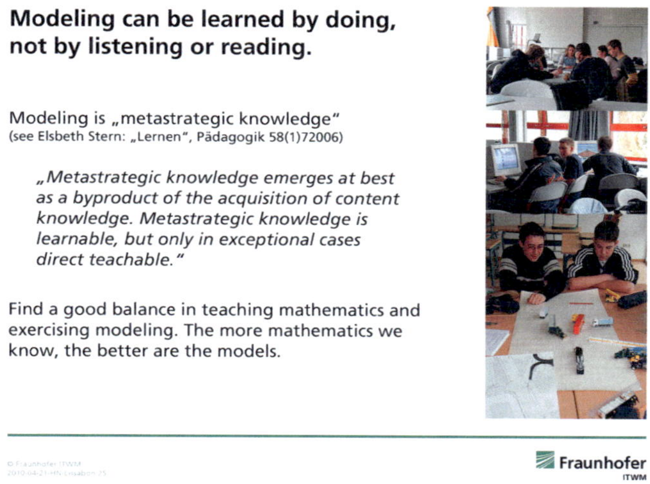

Fig. 22 Modeling can be learned by doing not by listening of reading

It is not very difficult, but should be included. As I said in the beginning: In problems, where psychology is playing a role, modeling is sometimes more complicated—not from a mathematical point of view, but from a "modeling point of view."

So, is modeling a mathematical activity or is it something else? It is certainly more than mathematics, it means understanding certain aspects of the world, to give a structure to this understanding and to translate it into mathematics. From a pedagogic point of view, it is a metastrategic knowledge, as E. Stern wrote in 2006.

And the most important thing is—I repeat: One cannot learn it in watching what others have done. You learn it only by doing. Therefore, books on modeling are useless in my eyes, reading them, you may learn about transports of patients in hospital—but you do not learn what you should: To transform the problem into mathematics. What you can do is describing interesting real world problems. And maybe one can create platforms in internet, which give a chance for interactive modeling. I do not yet know such a platform, but I believe it could be possible (Fig. 22).

Our two examples in this lecture are only carriers of messages and I repeat these messages in the Figs. 23, 24.

Let me finish with a few more personal remarks.

Together with some young people I started modeling in the early 70s. Oxford and Linz began a bit earlier, focussed on university education, but it encouraged us in Kaiserslautern to go in the same direction. We soon realized that modeling should also be done in schools, and 10 years later, around 1985, we organized our first modeling weeks for high school students. The young people and their successors (!) continue this activity until today—they have the experience of circa

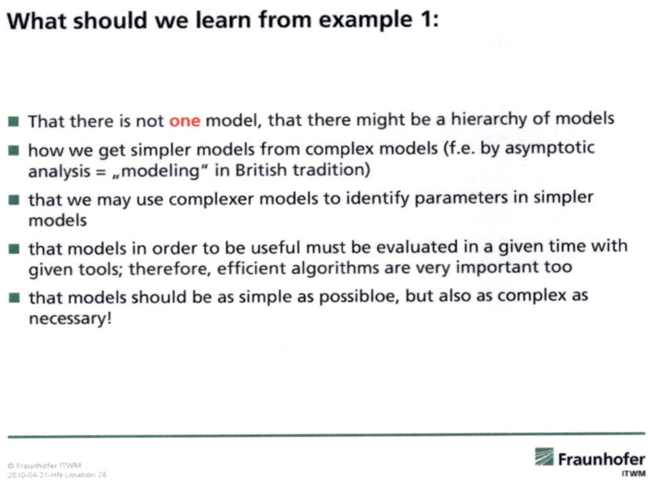

Fig. 23 What should we learn from example 1

100 weeks in Germany, in Italy, etc. The activity was exported to many European, but also to Asian and Latin American countries. We have changed the format of the weeks, included teachers as participants. We got many problems from the activity of a Fraunhofer Institute for Industrial Mathematics; I myself have visited more than 300 companies all over the world, discussing the question, whether mathematics is able to solve their "nonstandard" problems. This gave us the deep conviction that real mathematical problems are everywhere, on each level of

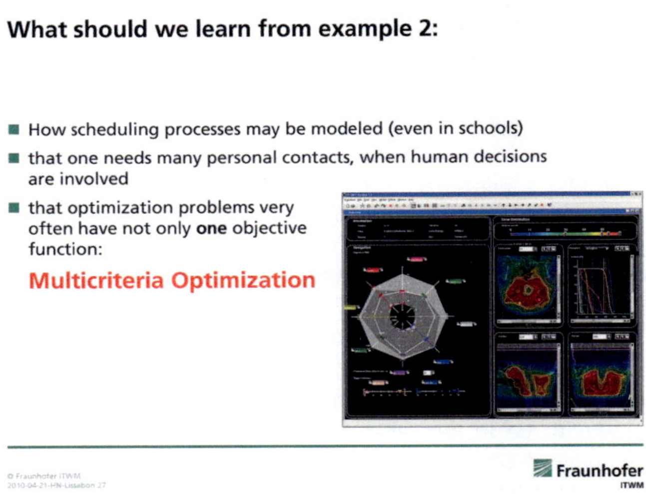

Fig. 24 Multicriteria optimization

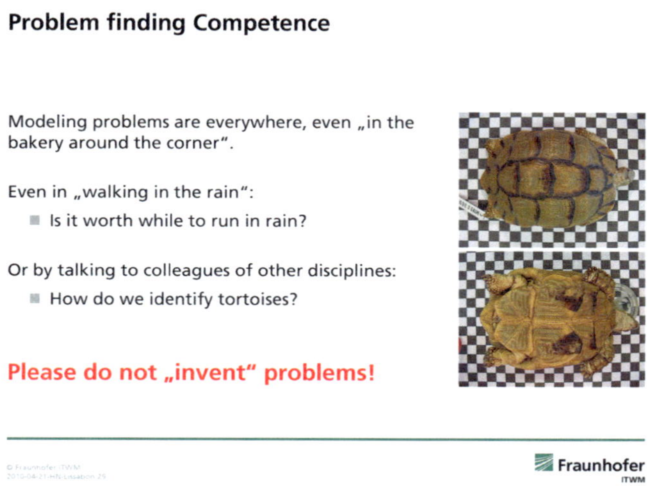

Fig. 25 Problem finding competence

mathematical competence. You just have to see them and then to dare! Even when you walk in the rain and ask yourself, whether it is better to move faster. Or when a colleague from biology comes asking you for a pattern in the shield of a tortoise, which is invariant with respect to age but characterizes an individual. There are problems even "in the bakery around the corner"—so, please, move around the corner and do not invent toy problems at your desk (Fig. 25).

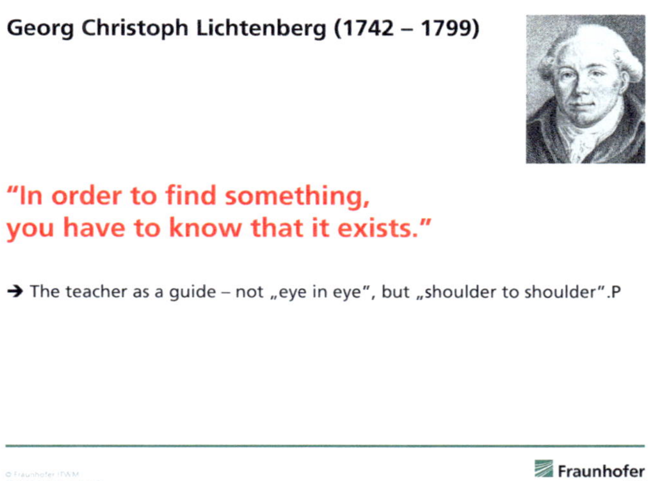

Fig. 26 Georg Christoph Lichtenberg (1742–1799)

After most of the modeling weeks we have collected questionnaires from the participants. And all of them show without any doubts: Modeling gives a “meaning” to mathematics, it creates enthusiasm, the students show creativity, they were surprised how efficient mathematics can be, how innovative they themselves can be, they want to do it more often.

And the teachers? Of course, they lost their lead, their superior competence. They were partners, in best cases guides, shoulder to shoulder, but not eye in eye. Their deeper knowledge of mathematics, their larger experience may help them to realize a sentence of the famous eighteenth century philosopher and physician Georg Christoph Lichtenberg (Fig. 26).

References

Beauty, A., Laporte, G., Melo,T., & Nickel, S. (2010). *Dynamic Transportation of Patients in Hospitals, OR Spectrum*, 70–107.

Klar, A., Marheineke, N., & Wegener, R. (2009). Hierarchy of Mathematical Models for Production Processes of Technical Textiles. *ZAMM - Journal of Applied Mathematics and Mechanics*, *89*, 941–961.

Interfacing Education and Research with Mathematics for Industry: The Endeavor in Japan

Masato Wakayama

1 A Brief Historical Observation for Mathematics in Japan

1.1 Historical Comments

Mathematics could be likened to an inextinguishable light that illuminates dark or unclear regions in our extremely sophisticated modern societies. Galileo Galilei once wrote that "nature is written in the language of mathematics," and needless to say, Descartes and Newton are also followed this tradition. Even now, mathematics continues to be the common language of science. Furthermore, even more than in the past, mathematics today plays the role of a compass by indicating the directions that research in science and technology should follow. Without mathematics, progress in all fields could only be made by blind exploration. In fact, we find mathematics at the heart of nearly all the advanced technologies that drive modern society, including information security, networks, medical technology such as CT scans and MRIs, control in chemical plants, blast furnaces, and nuclear reactors, the development of airplanes and automobiles, design in robotics, scheduling in transportation, logistics in industries, finance and insurance, risk management, searching for resources, weather and earthquake prediction, and entertainment, etc. No matter how different these fields are in appearance, many of them have identical structures in terms of their mathematics. This is frequently acknowledged as "the universality and the general applicability of mathematics." They are major characteristics of mathematics. For example, while it is a fact that the designs of CT scanners, MRI, and the control of blast furnaces for manufacturing iron each entail difficulties unique to their respective fields, mathematically, each involves the same activity—the solution of an inverse problem.

After the age of the revolutionary development of physics based on quantum mechanics and the theory of relativity in the first half of the twentieth century,

M. Wakayama (✉)
Institute of Mathematics for Industry, Kyushu University, Fukuoka, Japan
e-mail: wakayama@imi.kyushu-u.ac.jp

A. Damlamian et al. (eds.), *Educational Interfaces between Mathematics and Industry*, New ICMI Study Series 16, DOI: 10.1007/978-3-319-02270-3_7,

mathematics has also gradually taken on a key role in the dramatic developments in biology, especially in molecular biology, in the latter half of the twentieth century to the present day. While existing mathematics was used in the initial creation of quantum mechanics, quantum mechanics in turn promoted the subsequent development of new mathematics. Even today, new research in mathematics that began because of a stimulus from research originating in physics, such as superstring theory and Witten's theories, is flourishing. In addition, in the biological sciences, mathematical methods have become important. Fundamental concepts are formalized as numerical models and biological phenomena are investigated through the analysis and simulation with such models. What is of note is that through its involvement in these areas of research, mathematics itself undergoes new development. In this way, scientific, technological and industrial applications have provided sustenance to mathematics. Mathematics is a marvelous, mysterious or somewhat unusual science without analogies in other fields of science. Without doubt, if anywhere in the universe there is a planet on which a sophisticated civilization has developed, mathematics will be playing a key role.

The traditional/historical relationship between mathematics and technology principally involved the physical and chemical sciences, and mechanics. After the Industrial Revolution in Europe, this relationship fostered both an enhanced understanding of nature and an increase in mathematical analysis research of scientific and technological problems. Among other things, this involved analysis of differential equations of Newton's mechanics. However, from the middle of the 1990s, when high capacity computing became available, modeling and simulation of the actual processes being studies became the focus. In this way, modern society gradually, and in recent years, dramatically, increased its dependence on mathematics. For example, mathematics, including statistics in addition to the mathematical analysis mentioned above, has become necessary in a greater variety of sciences and technologies. In addition to that the role of providing new techniques (design, construction, synthesis) for incorporation into a variety of problems, including, for instance, subject matter that does not necessarily have governing equations, has become stronger. The contributions of mathematical analysis, modeling and simulation to a wide range of sciences and technologies will become increasingly important in the future. In terms of its exploratory role, it is thought that mathematical analysis will become even more essential. In fact, the formulation of the theories of particle physics and the understanding of the cosmos, the role of mathematics in biological sciences and ecology, and in particular, in financial engineering, encryption and information security, computer graphics and the like, has already broadened the impact of mathematical analysis, modeling and simulation. For example, while the virtual reality in the movies "Jurassic Park", "Harry Potter" and the recent "Avatar" are heavily dependent on advanced mathematical algorithms related to data compression and image processing, they have abundant use of the synthesis aspects of mathematics. The functions of both the analysis and synthesis in a variety of studies are inseparable, and, in any cases, we can observe that the applications of mathematics have dramatically expanded into new unexpected fields. In July of 2008, in the Global Science Forum of the

OECD (Organization for Economic Co-operation and Development), a report entitled "Mathematics in Industry" was published that clarified how necessary mathematics is in the everyday activities of modern society (OECD). In spite of this, there is a major shortage of mathematicians and mathematical scientists capable of responding to this need. Indeed, this represents a challenge for a society heavily dependent on mathematics.

Since the ancient Greeks, mathematics has become one of the intellectual pinnacles of civilization. Motivated by curiosity and the history of their training, many mathematicians see their own research as a pure logic independent of the real world. However, reflecting on the history of mathematics, we understand how useful the discipline of mathematics has been for human beings even when separated from immediate applications. Examples include the use of non-Euclidian geometry, which was essential for Einstein's construction of the theory of relativity; the role of the factorization by prime numbers and the study of elliptical curves in Number theory, which plays a central role in information security; and Dr. Kiyoshi Ito (1915–2008) formulation of stochastic differential equations, which were the origin of the foundation of financial engineering. When Koch's snowflake curve was discovered, it was thought to describe only to "pathological" phenomena. However, this idea was refined into fractals, and now, as is widely known, fractals have significantly contributed to the development of computer graphics. Thus, even if they appear to be pure theory, all fields of mathematics always have important applications hidden therein. In addition, compared to studies that originally aim at application, once pure mathematics has been discovered to be useful the impact of the results of pure mathematics is something extraordinary (Related to this historical remarks, see also Wakeyama 2013).

1.2 Mathematics in Japan

In this connection, in Japan, there are several misunderstandings and mistaken assumptions concerning mathematical studies. There are occasions in which mathematicians appear to be as if inhabitants from another planet. In the background, one can imagine that, in the early Meiji era in Japan, there were difficulties—limitations arising from geographical conditions—in transferring and developing involvement with the spectrum of Western mathematics, including the various aspects of applied mathematics. As a result, the European traditions of applied mathematics, directed toward solving practical problems in society, did not come to Japan. In this situation, only limited fields of application were developed. The reason for this was that the centers for applied research in Europe were mainly located at technical universities. Because applied research was not relevant to Japan's situation, a role for applied mathematics was not envisaged. In other words, we can infer that systematically importing applied mathematics was much more difficult than importing pure mathematics, which did not require the infrastructure associated with the mathematical analysis of practical problems. Ultimately, the pure mathematics of Göttingen in Germany, which was said to have been at the highest level in the world, was first chosen for dissemination during the opening up of the country at the end of the Meiji Era. But having said this, fortunately, the import of this pure mathematics suited the mental tenor of Japan, and was welcome. Actually, even leaving aside the use of mathematical puns by the oldest Japanese poets (the Manyo-shuu anthology) who enjoyed word play, Japan had a social worldview in which mathematics was valued (including a game element), as can be seen with the spread of calculation using the abacus (e.g. accounting), the mathematical votive tablets featuring mathematical puzzles mainly in Euclidean geometry offered to the Shinto Shrines and Buddhist Temples (Sangaku) and the like. In addition, as represented by Takakazu Seki (1642–1708), the rise of Japanese-style mathematics (Wasan) in the late 17th century during the Edo Era, which had reached a world-class level, is even now well known internationally. Naturally, in Japan, it can be said that in terms of world history, during the peaceful Edo Era, the need for mathematical studies related, for example, to the calculation of the trajectories for cannons, was very limited. In addition, the ancient national character, that enthusiastically imported leading-edge sciences and technologies and that furthermore spared no effort in perfecting them, also played a role. Under the prevailing stable political conditions, a very unique culture developed. Consequently, after the opening up of Japan during the Meiji era, its mathematical capabilities enabled a rapid absorption of the imported Western pure mathematics traditions. In addition, this was the source of the energy that allowed globally advanced centers to be established at a surprising speed in pure mathematical studies that focused on algebra, and in particular, number theory, as represented by the completion of class field theory by Dr. Teiji Takagi (1875–1960). In addition, many remarkable results were attained in various fields of pure mathematics that brought recognition by the international research community.

1.3 Applied and Industrial Mathematics in Japan

At the same time, the high level of mathematical applications in leading fields of engineering, such as mechanical engineering, electrical engineering, and precision machinery engineering, which supported and produced Japan's period of advanced growth, should be single out for special mention. The difficulty with importing applied mathematics along with pure mathematics in the Meiji Era has already been described above. Accompanying the transfer of leading-edge technology at the time, the sprouts of applied mathematics were also being transferred (perhaps surreptitiously) to newly established university engineering schools. There it was subsequently cultivated and, helped by Japan's sophisticated pure mathematical tradition, applied mathematics took root in engineering. In fact, one can immediately find the many high level technologies in the old Japanese history, e.g., in civil engineering (Japan is indeed a country of the mountains and rivers), architecture such as Horyuji Temple, the oldest standing wooden architectures in the world which was built in the beginning of the seventh century, etc. Behind these, it is no doubt about the existence of developed practical/non-abstract arithmetic computation and measurement techniques.

Furthermore, in the latter half of the Showa Era, when the performance of computers was still not very high, in order to use computers and obtain meaningful results, first a problem had to be formulated mathematically and numerically. Researchers in mechanical and electrical engineering at the time were probably not consciously aiming to be applied mathematicians, but essentially they drew on high level mathematics, and when necessary, developed new mathematical tools/ techniques by themselves, and then carried out research on the resulting topics. Indeed, similar to theoretical physicists, it is thought that because there were many mathematically talented engineers, they could also respond to the great needs of industry, which had provided explosive economic growth. The situation described above is one reason that Japan is unusual in thinking that "mathematics" is pure mathematics. However, ironically, entering the era of fast and inexpensive high performance computing, for engineers in particular, the applied part of mathematics has been placed, through the extensive use of general-purpose software, in a black box. As a result, although there are still superior "applied mathematicians" involved, the emphasis on the mathematics in engineering (the principal object of which is making things) has rather declined. It is worrisome that most general-purpose software is imported from abroad. This general-purpose software is used, for example, for motion capture machines, which are indispensible for making, among other things, movies, animation, and games. Fortunately, in computer science and informatics (which, in Japan, is frequently situated in engineering department due to its relationship with the development of hardware as a subject in electronics), mathematics is even more widely and fundamentally considered to be an important subject of research.

Even in mathematics, the pronounced tendency to create a division between pure and applied mathematics appears in both the East and the West. Because of their different roles, the distinction between the two may be convenient if used appropriately. We can, however, broadly perceive that this distinction, which places a wall between mathematicians is diminishing. Consider Nineteenth Century mathematicians. Great mathematicians such as Gauss, Fourier, Cauchy, Riemann, who will continue to influence present and future mathematicians, did not stop at research in pure mathematics. For example, the "Riemann hypothesis" was conceived 151 years ago and is the greatest unsolved problem in pure mathematics, still defying the efforts of mathematicians. But its results are also a focus of attention from the point of view of information security and research on cryptography. This is because this hypothesis makes clear the ultimate form of the distribution of prime numbers. Thus, in cryptography research, there is an interest in mathematics that transcends boundaries between pure and applied mathematics. In fact, we should remember that the outcome and attraction of applied mathematics is that it frequently stimulates the desire in pure mathematicians for pure mathematical studies which do not aim at application.

2 Education and Research Hub for Mathematics for Industry

2.1 Mathematics for Industry

The phrase "Mathematics-for-Industry (MI)" denotes a new research field in mathematics that will serve as a foundation for creating future technologies. This phrase/notion was born from the integration and reorganization of pure and applied mathematics into a fluid and versatile form capable of responding to the needs of industrial technologies.

2.2 Graduate School of Mathematics, Kyushu University

Going beyond the wall between pure and applied mathematics, and fully expressing the freedom and variety that mathematics possesses for fostering young researchers, the Graduate School of Mathematics, Kyushu University has initiated a twenty-first Century Center of Excellence Program called "Development of Dynamic Mathematics with High Functionality" (FY2003-2007) which was promoted by the Ministry of Education, Culture, Sports, Science, and Technology (MEXT). The same graduate school and Faculty of Mathematics have made constant pioneering efforts, leading the whole of Japan, in developing an environment that promotes integrative research between mathematics and other fields, and fosters cooperative research between mathematics researchers and industry. These efforts dedicated to organizational reformulation have gained the support of the MEXT Program for Enhancing Systematic Education in Graduate Schools, "Raising Doctorates and New Masters in Mathematics Required by Industrial Technology" (FY2007-2009), the Global Center of Excellence (Global COE) Program "Education & Research Hub for Mathematics-for-Industry" (FY2008-2012 GCOE), and a Special Fund by MEXT "Project on Graduate Education in Mathematics for Globalization" (FY2010-2015). Through the activities associated with these programs, we find that the acceptance of long-term internships for students in the doctoral program is spreading to a variety of industries in Japan and abroad. A new career path is being opened up for the mathematics doctorates, and in addition, and cooperative research prompted by this has also commenced. The significance of such activities is spreading to other universities, and the attitude of the faculty members is changing along with the graduate students majoring in mathematics.

2.3 Background About the Programs: From the Report by Nistep

Advances in mathematics are essential for future technologies, and many companies feel a lack of personnel with advanced mathematical skills and expertise. In fact, according to the report by the National Institute of Science and Technology Policy (NISTEP) of MEXT in 2006, the percentage of people in research and development in the private sector in Japan, who possess mathematical background, is planned to be around 65 %, which is the same as that of advanced countries of Europe and North America. The current percentage in Japan is only 26 %. As Japan aims for sustainable development in science and technology, there is an urgent demand to train people to fill this almost 40 % gap (see Hosotsubo 2009 and references therein).

2.4 Various Activities

a) Long-term Internship for PhD Students in Mathematics: The Graduate School of Mathematics, Kyushu University, introduced in 2006 a program "Long-term internship for PhD students in industries" as part of the curriculum for the new doctoral course Functional Mathematics. This long-term internship provides a joint research opportunity for the students for 3 months or more at industrial R&D laboratories. About 45 students have already completed the program. The long-term internships have turned out to be far more successful than expected. They have not only attracted attention from domestic universities but also from people working in industry. Some students obtained patents, papers, and even initiated new collaborative researches opportunities between enterprises and faculty members. After getting PhD degrees in Mathematics, almost 15 % have obtained positions in academia, 60 % in industry, and the remaining have post-doc positions in several universities and public research institutes.

A: To perform a necessary improvement continuously for this program, we have been trying to observe and understand the real situation before and after the students spent months at the industry. As a result of the observations, the points to assist the students are:

Change their current views

1. to recognize the usefulness of mathematics in industry and social life
2. to recognize the integrated power of the company to tackle a problem
3. to have responsibility and to experience the joy of achievement
4. to recognize the usefulness of logical thinking intrinsic to mathematicians
5. to recognize the power of math graduates
6. to experience a regular daily life

Develop the required skills or recognitions

1. the importance of skill of computer programs
2. the necessity of communication ability
3. the shortage of knowledge of physics, etc.

B: Companies where we have sent PhD students are:

Hitachi Ltd., NTT (Nippon Telegraph and Telephone Corp.), IBM Japan, Toshiba Corp., Ube Industries Ltd., DIC Corp., Mitsui Engineering & Shipbuilding Co. Ltd., Fujitsu Ltd., Zetta Tech. Inc., Panasonic Corp., Nishinn Fire & Marine Insurance Co. Ltd., ING Insurance (the Netherlands), Nippon Steel Corp., Mazda Motor Corp., OLM Digital Inc., Nessan Heat Corp. (also, the National Institute of Communication Technology.)

C: Students' fields of specialization include:

Statistics, Fluid dynamics, Algebraic geometry, Number theory, Representation theory, Optimization theory, Nonlinear analysis (PDE), Theoretical computations and Computer algebra, Operator theory, Differential geometry, Topology, Combinatorics, Numerical Analysis, Game theory, Cryptography, Mathematical physics, Probability theory.

D: Research themes include:

Face recognition, Analysis of electromagnetic fields, Information processing of visually movements, Cryptography, Fluid analysis, Information processing, Fluid analysis in the chemical device, Tree structure data analysis, Liquid crystal simulation, Optimization, Electronics system modeling, Risk management, Date analysis, Time series analysis (Fig. 1), Processor analysis, Data compression, Video encoding, Image compression, Data processing, Codes research by using Chaos, Machine learning, 3-dim heating control.

b) Master of Mathematics Administration (MMA): The MMA program is a mathematics version of the MBA program and designed to cultivate (Master course) graduate students to be experts in coordinating research and development activities, that is, building bridges between R&D and planning sections in industry, based on mathematics from broad and long-term perspectives. It is also expected that graduates under this program will make a significant contribution as MI research personnel along with traditional Master's in mathematics.

c) Study Group Workshops: From 2010, a short-term camp-style "Study Group" that tackles unsolved problems that industries are faced with has been initiated on a nationwide scale. This is the first full-scale Study Group Workshop in Japan organized jointly with the Global COE program "The Research and Training Center for New Development in Mathematics" in Graduate School of Mathematical Science, the University Tokyo. More than 120 people joined this one-week workshop in 2010. The list of companies and their proposed themes in 2010 were:

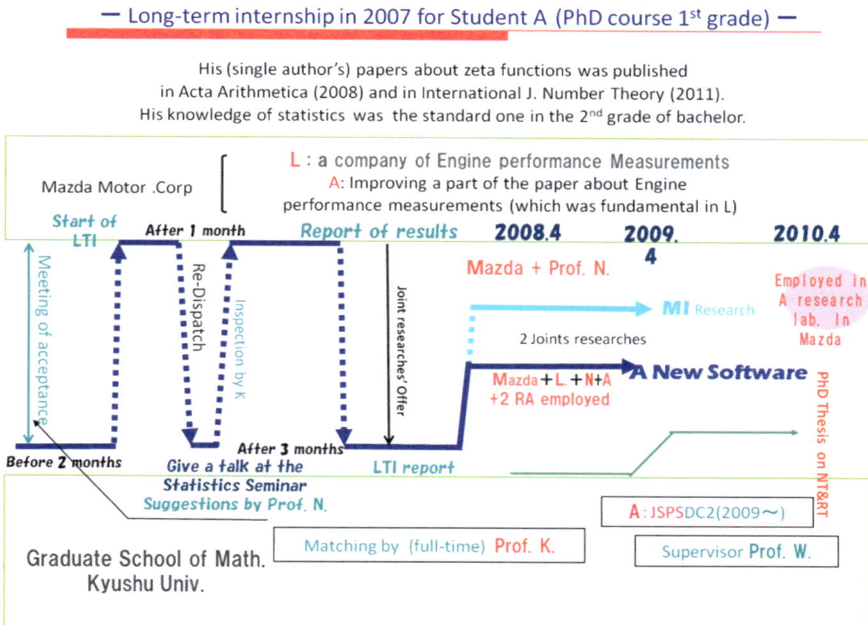

Fig. 1 Internship in Mazda

- Hitachi Ltd.: Validity of lattice reduction algorithms for CVP
- Toshiba Corp.: The section finding problem and algebraic surface cryptosystems
- OLM Digital Inc.: Interactive editing of light and shade for 3DCG
- WETA Digital: Open problems in visual effects
- Nippon Steel Corp.: Inverse problem from multi-scale viewpoint utilizing a combination of stochastic, analytic, and geometric modeling
- Kao Corp.: The problem of spread of communication on the net space
- Takeda Pharmaceutical Company, Ltd.: (1) Development of prediction method for solubility of crystalline compound via 2D-structure, (2) Development of structural calculation system associated with chemist' sense, (3) Development of practical algorithm to generate theoretically possible structures under restriction conditions

Also, two mathematicians from Indian Institute of Technology and CSIRO in Australia have proposed, respectively, the following problems.

- Challenges in provable security of cryptosystems
- Analysis and utilization of spectroscopic data

This year in August, at the Study Group Workshop organized jointly also by Kyushu University and the University of Tokyo, we will have the theme-presentations from NTT, NEC Corp., Mitsubishi Chemical Corp., OLM Digital Inc., Fujitsu Ltd., Fujitsu Lab. Ltd., Nippon Steel Corp., etc.

d) Forum "Math-for-Industry": Every year we hold this forum. The last two were as follows:

2009: Casimir Force, Casimir Operators and the Riemann Hypothesis—Mathematics for Innovation in Industry and Science—November 2009, Nishijin Plaza, Kyushu University

The year 2009 marks the 100th birthday of Casimir and the 150th birthday of the Riemann hypothesis. Actually, the chapter in which Riemann proposed the hypothesis was published in November 1859. It was also the year when he was appointed as full-professor at Göttingen. Casimir, known for the Casimir force in physics and Casimir operators in mathematics, was director of Royal Philips research for a long time in his career. Casimir got his PhD in November 1931 at Leiden University and the named operators appeared in his thesis. He is remembered for his outspoken opinion on the importance of fundamental science for industry. There is a nice connection between Casimir and Riemann. The Casimir force was first proven to exist theoretically, using the analytic continuation of the Riemann zeta function. Recently, the force was measured experimentally.

2010: Information security, visualization, and inverse problems, on the basis of optimization techniques, October 2010, Hilton Fukuoka, Sea Hawk

At the forum, we will mutually relate and discuss the three topics: the information security, visualization, inverse problems, on the basis of optimization techniques, and at the succeeding 5-days study group, after proposals of concrete real problems from companies and public institutes, we will work toward solutions by divided groups of participants.

The Forum "Math-for-Industry" 2011 will be held in the East–West Center, University of Hawaii at Manoa (24–28 October) under the title:

2011: "TSUNAMI—Mathematical Modelling" Using Mathematics for Natural Disaster Prediction, Recovery and Provision for the Future

Our objective is to consider what help can mathematics provide against natural disasters; to understand their fundamental mechanisms, to increase the accuracy of predictions, to minimize damages and risks, to establish more robust and efficient social systems for recovery, and so on. Thus, the lecture theme includes the following research topics: tsunami, earthquake, environmental pollutions, logistics, optimization, control theory, statistics and data analysis, information security, CG visualization, PDE, inverse problems, numerical analysis, fluid dynamics, integrable systems (solitons), geometry and topology, etc.

e) Journal of Math-for-Industry: We launched an electronic journal (but have published a hard copy once a year), the Journal of Math-for-Industry (JMI), which is published at the institutional repositories. This journal JMI is dedicated to the broadening of the horizons of Mathematics for Industry and swiftly and internationally publicizes achievements in education and research on MI. JMI presents original research papers and survey papers with original view-points in all scientific disciplines in which the mathematics or that in industry plays a basic role. Articles by scientists in a variety of interdisciplinary areas are published.

Research areas include significant applications of mathematics to industry, including feedback from industry to mathematics, new developments in Mathematics for Industry, and new developments in mathematics.

3 Mext's 2009 Project "Investigation and Estimation of Promotion of Cooperation of Mathematics and Mathematical Science with Other Fields"

This was the Investigation project commissioned by MEXT in 2009 and promoted by the group consisting of Kyushu University (the representative institute), the University of Tokyo, Mathematical Society of Japan and Nippon Steel Company. In this project, we investigated and estimated the activities of mathematics and mathematical science, and those of their cooperation with the other fields that had been implemented in Japan, and thereby gained ideas for making a proposal to the Japanese government for promoting mathematics and for strengthening cooperation with various fields surrounding mathematics, which will achieve creation of new values and promote innovations. We made the final proposal in the end of March 2010 as follows:

After the investigation, we have submitted the policy consisting of the following five proposals:

1. Support an establishment of hub (governmental funds),
2. Establish a section for the promotion of mathematics and mathematical science in the government (so far, no such section has existed in the government),
3. Renew or continue to the JST (Japan Science and Technology Agency) program "Alliance for Breakthrough between Mathematics and Sciences", which is currently being promoted, for mathematics with other discipline and technology,
4. Put the clear description for promoting mathematics/mathematical science at the governmental important document such as "the fourth governmental science and technology master plan",
5. Establish an official committee and organization for promoting mathematics and the applications in a flame in the government.

It was conceptually and in principle accepted and the government would make a policy to promote researches in the field of mathematics and mathematical science in the following few years. In fact, the results we have now are; 2. the establishment of the "Mathematics Innovation Unit" (MIU) in MEXT, 4. in "the fourth governmental science and technology master plan" of the Prime Minister's Council of Science and Technology Policy (CSTP), the "mathematical sciences" are specifically mentioned for the first time, and 5. a new committee "Committee of Mathematics Innovation" has been established under the Council for Science and Technology in the government, while 1. has not been accomplished although MIU has been trying but looks pretty difficult especially in the current situation in

Japan and 3. is under discussing. Actually, this year the government would provide small funds for holding such workshops. There will be hence held 23 workshops jointly organized by MEXT and several groups of mathematicians or institutes of mathematics in the academic year.

4 Foundation of the Research and Education Hub "Institute of Mathematics for Industry" in Kyushu University

The origin of the idea for establishing this Institute can be traced back to the establishment of the Mathematics Research Center for Industrial Technology (MRIT, April, 2007–March, 2011) and subsequently, a concrete concept was delineated when Kyushu University applied to the Global COE Program. During the University's participation in the Global COE Program project, in order to organize PhD education and research in MI in a more systematic manner, the Faculty of Mathematics was reorganized, with strengthening of the function of the MRIT, to found the IMI. About one-third of the members of the old Faculty of Mathematics are participating in the Institute.

The Institute of Mathematics for Industry (IMI) has been established in April 2011 as the university's fifth research institute. IMI is a deliberately planned offshoot of the Global COE Program "Education and Research Hub for Mathematics-for-Industry." From its birth in 1939, the Department of Mathematics in the Faculty of Science has enjoyed a long tradition of respect for striking a harmonious balance between pure and applied mathematics. This balance serves as the framework of designing IMI.

The basic activity of IMI is to promote industrial mathematics on and for the basis of various mathematics researches and to grow human resources in this field. By Mathematics for Industry we mean a new field of research for creating industrial mathematics of the future. The history of the development of mathematics has led us to the belief that an essential idea will someday find its utility. The history shows, on the other hand, examples of the inability of research to respond to the serious requests from the society even in the cases of actual demand. For that reason, with a focus on mathematics, IMI aims to cooperate more closely with industry through joint research and to advance, over the pure–applied barrier, to a wider range of mathematics in amalgamated form. IMI focuses on basic research while respecting its contact with modern society. IMI also needs to focus on the facts like, for example, the great pioneers of the nineteenth century who combined theoretical research and numerical calculations. While focusing on both theoretical and computational/experimental research, IMI will incorporate deterministic considerations, and statistical and probabilistic methods to continually promote research from a broad outlook. In addition, IMI will step-by-step establish itself as a center of interactive cooperation between mathematics and various industry and scientific fields. Internationally, IMI will forge industry–mathematics collaboration in the

Pacific and East Asia regions as a global partner in activities associated with Mathematics for Industry.

The activities of IMI include (1) promoting collaboration with industry (2) organizing workshops and study groups (3) holding seminars for industry-academia partnerships and mathematical tutorials, and (4) raising PhDs and future human resources in Real World with mathematical background. The institute has 25 regular faculty members and consists of three main divisions; Advanced Mathematics Technology, Applied Mathematics, and Fundamental Mathematics. Beside, in order to efficiently promote the above projects/activities, the Visiting Scholars Division, and Partnership Promotion and Technical Consultation Room have been established.

Kyushu University has instituted Graduate School/Graduate Faculty System. According to the rule of this system, university faculty members, in principle, belong to a Graduate Faculty or a Research Institute. A Graduate School is an educational organization (education body) of a graduate school studies. IMI and the Faculty of Mathematics are the partner bodies responsible for the Graduate School of Mathematics.

Today, the importance of mathematics is better recognized than ever even in Japan, with accumulation of evidences that mathematics contributes directly and crucially to deepen things and to solve problems in the society. As a consequence, the roles of experts dedicated to mathematics are growing for the development of the society. The day is not far off when mathematicians—the describers of scientific foundations—will be the navigators for the real world and industrial technology. While aiming to train researchers for positions at universities, research institutes, and in industry, IMI will make every effort to develop excellent professionals who can meet the needs of society. IMI's mission is to become a center on the cutting edge of mathematics in the real world and world of science.

In the field of mathematics, IMI is the third research institute established in Japan after the Institute of Statistical Mathematics and the Research Institute of Mathematical Science (RIMS), Kyoto University, but has a distinctive feature. As such, the birth of IMI marks the formation of a unique international institute. Actually, this Institute aims to form a world-wide preeminent research center for industrial mathematics based on the idea of MI. Furthermore, by maintaining and developing cooperative relations with the Faculty of Mathematics in Kyushu University, we shall achieve a global educational research center for mathematics that will represent Japan (Fig. 2).

Mathematical Research

- There are no limits,
- A gradual broadening will be seen in the "universe of the real world",
- The ordered sequence, say the pure and then followed by the applied, in the progress and deepening of mathematics is not necessarily fixed, and
- Fundamental research in mathematics having the present progressive form turns out to be the future industrial mathematics.

The mathematical research area viewed from such a perspective is the Mathematics for Industry.

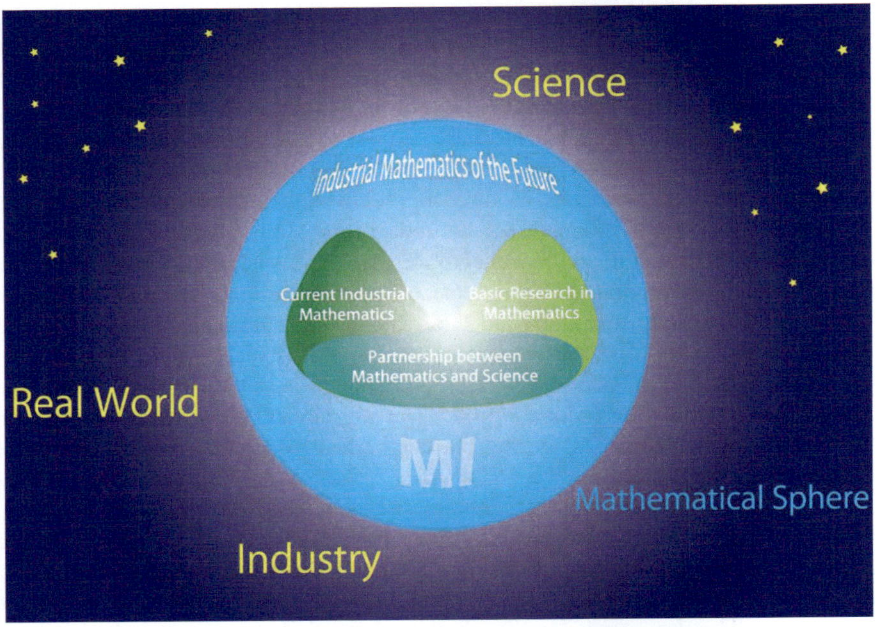

Fig. 2 The results of basic research or pure mathematics, though not in a form that can be imagined or predicted in advance, will contribute in the future to the development of many sciences and will be useful for resolving the problems of the real world. In addition, new mathematics will be born and a deepening of mathematics will be generated by the applied research in mathematics that is developed to solve real-world problems. This figure shows

5 Asia-Pacific Consortium of Mathematics for Industry

We will inaugurate the Asia-Pacific Consortium of Mathematics for Industry (APCMI) in the East Asian and Pacific areas in October 2011, which has been planned from the fall of 2009. The APCMI aims to create an international stronger platform for the worldwide developed activities of the mathematics for industry.

Acknowledgments The author thanks Dr. Bob Anderssen in CSIRO for his helpful suggestion on the first draft of this article. Dr. Anderssen has been actively involved with the Australian Mathematics in Industry Study Group (MISG) meetings since their inception more than a quarter century ago. The author has enjoyed very much a fruitful discussion with him about what we call "Mathematics-for-Industry." The author also thanks the organizers of the EIMI 2010 conference for the invitation leading to this article.

References

M. Hosotsubo (2009). Survey analysis of Japanese mathematics research. *Journal of Mathematics-for-Industry, 1*(A-10), 73–80.

M. Wakeyama (2013). Importance and unpredictable effectiveness of mathematics in the real world and for industry. *What mathematics can do for you*, Giga, Yoshikazu, kobayashi & Toshiyuki (Eds.), Springer, 101–121.

http://gcoe-mi.jp/english/.

http://www.oecd.org/document/8/0,3746,en_2649_34319_42626653_1_1_1_1,00.html.

Part III
WG Education/Training with Industry Participation

Education/Training with Industry Participation

Gail FitzSimons and Tom Mitsui

1 Introduction

Because of the nature of the broad theme of our working group—*Education/training with industry participation*—we believe that it is necessary to provide the context for our presentations and discussions. In this Working Group we were privileged to learn of and to discuss many interesting and innovative practices addressing the education/industry interface.

Working Group 6 encompassed a diverse group of people with many different experiences at different levels and sectors of education across the spectrum of life-long learning. Among the participants were people who are working ith students and teachers at primary and secondary schools, in vocational and adult education, with university students of mathematics (undergraduate & postgraduate), as well as with teacher education students. Some are in teaching roles, some are researchers, and some are both. Some Group members are working directly with industry as professional mathematicians, finding themselves in situations of mutual coaching with engineers, or as facilitators of graduate mathematics or post-doctoral students, as well as secondary teachers participating in industry-based internships.

Co-Chairs: Gail FitzSimons (Australia) and Tom Mitsui (Japan).Working Group Members: Cinzia Bonotto (Italy), Martin Bracke (Germany), Matthias Brandl (Germany), Lars Gustafsson (Sweden), Corinne Hahn (France), Gert Hana (Norway), Ragnhild Hansen (Norway), Inger Elin Lilland (Norway), Richard Millman (USA), Lars Mouwitz (Sweden), Junichi Nakagawa (Japan), Sarah Petersen (Canada), Kathrin Winter (Germany). In absentia: Diana Coben (UK), Guenter Toerner (Germany).

G. FitzSimons (✉)
University of Melbourne, Melbourne, Australia
e-mail: gfi@unimelb.edu.au

T. Mitsui
Department of Mathematical Sciences, Doshisha University, Kyoto, Japan
e-mail: tom.mitsui@nagoya-u.jp

A. Damlamian et al. (eds.), *Educational Interfaces between Mathematics and Industry*, New ICMI Study Series 16, DOI: 10.1007/978-3-319-02270-3_8,

Our Group's overarching goal is to enable people working in the education sector and industry (where industry is understood in the wider sense as any non-educational enterprise) to develop an understanding of each other's cultures, locally and globally. Three of our explicit goals are to help learners at all levels experience how mathematics is, or can be, useful in industry, and to help industry recognize the value of mathematics to help meet their needs, not only through working with graduates and post-graduates in mathematics but also through interaction with people associated with the education sector at all levels. The third goal is to foster partnerships between industry and education which enable teachers to bring mathematical activities from industry to the classroom at all levels.

From an industry-based mathematical perspective, one objective is to understand problems, formulate them in mathematical terms, and solve them in a manner that is useful to the particular industry or workplace concerned. Underpinning this objective is the understanding that the best mathematical solution does not necessarily mean the optimal or most practical solution in an industrial context. In fact, the complex and ever evolving nature of many industrial problems contributes to the development of new fields of mathematics and must accommodate what is known as *Mode 2* knowledge production (Gibbons et al. 1994) due to the interconnected and transdisciplinary qualities of many activities, rather than the traditional *Mode 1* disciplinary approach. Whereas Mode 1 mathematical knowledge is produced solely within academic institutions and subjected to scientific peer review processes before being accepted and possibly applied elsewhere, Mode 2 mathematical knowledge is produced in the context of its application by people from mathematics to other disciplines (e.g., engineering, physics) working together in a dynamic problem solving process, bringing both theoretical and tacit knowledges to bear as they communicate with one another to understand each other's perspectives and collaborate to find a workable solution within the economic, environmental, physical, and even political constraints of the task at hand.

From an educational perspective, one important objective is to enable people young and old, preparing for the world of work or already employed (whether paid or unpaid), to participate meaningfully and with dignity as citizens in their local and global communities. Activities discussed within the Group included school learners completing projects or solving problems offered by or found useful to industry, especially at the local level. Discussions also included people qualified in mathematics at the higher education level working as professionals collaborating in multidisciplinary teams and in carrying out new research, opening up the field of Applied Mathematics to others.

Among the many reasons given for participating in these activities are, from a mathematical perspective, to create new knowledge, to apply existing knowledge, and to introduce interested parties (from education and from industry) to knowledge new to them personally. In fact, knowledge can be generated which contributes simultaneously to the needs of educators and of industry, albeit in different ways. From an educational perspective, reasons are to motivate students to continue their studies of mathematics, to experience and understand how mathematics works in real and authentic contexts, and ultimately to change the image of mathematics held by students and by the general public. From an industrial perspective, the most important consideration is to get the job done as efficiently as possible. However, other goals include increasing industry awareness of the most recent developments in mathematics and of the potential of academic participants (i.e., teachers and learners at all levels) to offer them new ideas and techniques. Industrial-based or -oriented projects and research, involving learners and graduates intending to move into the labour market, can offer an opportunity to gain experience and to learn how to fit into a new community of practice—one which is vastly different from that of the typical education institution. This is a point that is often overlooked by many who assume that for a suitably qualified person to move from school or university to the workplace is relatively simple and uncomplicated. In fact, as research demonstrates (e.g., Wood 2010) this is far from the case, especially where mathematics is concerned, whether the new worker is employed as a mathematics/statistics specialist or elsewhere in the workplace, from entry-level through to management.

According to the Study Discussion Document, intended beneficiaries include, among others: students enrolled in formal education systems across all sectors; pre-service teachers and practicing teachers involved in continuing education, professional development programs, or industrial internships; teacher educators for the above categories; learners undertaking workplace education—from low-skilled workers through to management—and their workplace teachers/trainers; industry decision makers; mathematicians working in industry; and policy makers. This Working Group also identified education administrators, scientists working in industry, researchers from the disciplines of mathematics and education, and the general community as beneficiaries.

The Group considered what each partner could contribute to the process. From an industrial perspective, contributions could include authentic material and intellectual resources (e.g., artefacts, equipment, manuals, mentors), the commissioning of research, allowing a field of experimentation, and even the possibility of opening up a new research field altogether. From an educational perspective, governments and education institutions could contribute through commissioning funded research or supporting collaborative work with industry in terms of policy. This means offering encouragement to new initiatives, giving permission to staff and students to move outside the normal physical and regulatory boundaries, and through providing the necessary human and material resources when needed. The sites of collaborative work could be located in industry, at school or university, virtually, or some combination of these.

Finally, the Group identified the following list of intended outcomes:

1. The stimulation of mutual and respectful dialogue and collaboration between education and industry by establishing professional learning communities at the interface of mathematics education and industry.
2. In formal education, the development of curricular content (including new mathematical methods, teaching & assessment practices) as well as improved personal and communication skills in learners—critical skills which are often overlooked in mathematics education.

3. The general acceptance of the principle that if mathematics is used as a selection mechanism in the workplace it should be appropriate to the specific work roles and functions.

2 Theoretical Framework

The main underpinning research theme of our group was the notion of *boundary crossing* as industrial and educational groups each moved beyond their "comfort zones" in an effort to communicate and collaborate with each other. However, even the concept of boundary crossing itself was contested in our discussions.

2.1 Informal Learning

According to Eraut (2004) informal learning draws attention to the learning that takes place in the spaces where there is an over-riding purpose, such as production of goods or services, and goes beyond learning from experience alone. He recognized that workplace situations are generally in a state of flux, affected by external conditions as well as by the actions of people as they become engaged in the problem at hand. They need to be aware of the context and to correctly read the initial and developing situation and respond to changing conditions. They need to think, act, and communicate with others, drawing upon past experience and new insights of their own and of others. The outcome could then include a new product or process, decisions or transactions, and learning by self and others. In the following examples of mathematics education with industry participation it can be argued that the learning by all will encompass much more than the traditional academic mathematics curricula found in schools, vocational colleges, or universities.

Eraut (2004) acknowledges the complexity of transfer of knowledge from school (or university) to workplace, and identifies five interrelated stages:

1. the extraction of potentially relevant knowledge from the context(s) of its acquisition and previous use;
2. understanding the new situation—a process that often depends on informal social learning;
3. recognizing what knowledge and skills are relevant;
4. transforming them to fit the new situation;
5. integrating them with other knowledge and skills in order to think/act/communicate in the new situation (p. 256).

He points out, 2, 4, and 5 tend to be ignored in formal institutional education and workplace training processes. However, their importance is underlined in the papers contributed to this WG.

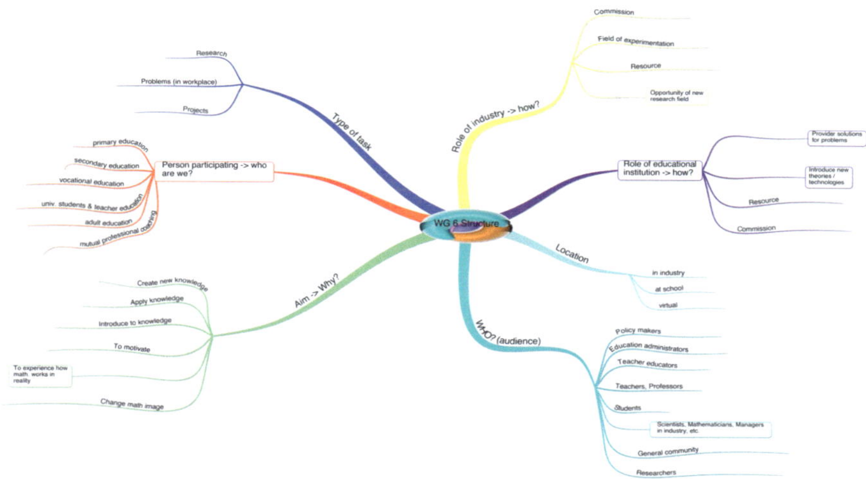

Fig. 1 Mind map created to capture WG 6 discussions

Figure 1 was prepared by Matthias Brandl as part of his contribution to our Group.

Eraut (2004) reminds us that, although codified academic knowledge is prominent in higher education, in workplaces there is much codified non-academic material in the form of textual resources such as manuals, plans, and historical records. He claims that there is also a wealth of non-codified *cultural knowledge* acquired gradually and implicitly learned through social participation in the work process. Further, he defines *personal knowledge* as "as what individuals bring to situations that enables them to think, interact and perform" (p. 263). It includes "everyday knowledge of people and situations, know-how in the form of skills and practices, memories of episodes and events, self-knowledge, attitudes and emotions" (p. 264).

2.2 Boundary Crossing

In designing work-responsive programs or curricula, Garraway (2010) identified the need for participants from the academic and the industrial worlds to go beyond their generally recognized boundaries of expertise to collaborate in productive ways. Different communities of practice have different knowledge bases which may prevent the effective communication between them, so that boundaries need to be crossed. Boundary objects that are meaningful to both industry and the academy could include, for example, a curriculum document which specifies for industry the intended capabilities of prospective employees while meeting the regulatory requirements of the academy. Statistical control charts are used in business by management to make decisions, based on data collected on production by workers and/or their machinery. They are also used in formal education to provide practical examples of the use of statistical theory and reasoning. Mathematical models can also be shared between engineers and mathematicians (see below). Boundary crossing between the academy and the workplace offers the possibility of innovative developments. Garraway sums up: “Boundary recognition involves making difference which may have previously been implicit, explicit” (p. 220) and, depending on the quality of interactions, explicit differences can become a resource.

3 Practical Examples of Educational Interfaces Between Mathematics and Industry

Our Group originally planned to have six presentations but, due to circumstances beyond their control, two presenters were unable to take part in the Discussions. Since these were regarded as valuable to our work, and were included in the original proceedings (Araújo et al. 2010), they will be included in this report. This chapter draws upon papers contributed to the original EIMI 2010 conference proceedings. Each illustrates a different facet of the interfaces between mathematics and industry, as do the two papers considered prior to the meeting: Coben & Hutton, and Toerner & Grotensohn. Clearly, there is a strong need for the long-term evaluation of projects discussed below.

Cultivating an Interface Through Collaborative Research between Engineers in Nippon Steel and Mathematicians in University (See Chap. 44) by Junichi Nakagawa and Masahiro Yamamoto, focuses on the Japanese steel making process which requires control of a diverse range of phenomena involving mathematical applications for problem solving and modelling. Their paper describes ongoing collaboration between industrial engineers and academic mathematicians resulting in better understandings developed by each group, leading to mathematical innovations supporting diverse industrial applications and initiating new developments in the field of mathematics.

The six phases of collaboration identified by Nakagawa and Yamamoto bear a striking similarity to Eraut’s (2004) five stages of transfer, noted above. All agree upon the critical importance of high quality communication between participants.

1. Intuition and expertise from industry. Insight is based on observation of phenomena occurring in the manufacturing process and should be enhanced by mathematical reasoning.
2. Communication. Engineers understand real problems on site, express them in the language of physics, and offer possible model equations to mathematicians. Mathematicians explore the underlying mathematics to the model equations. This forum is extremely important in order that engineers and mathematicians may reach common understanding of the nature of the problem and the mathematical components.

3. A logical path. This corresponds to the extraction of mathematical principles from phenomena assisted by good communication.
4. The analysis of data. The reasonable and quantitative interpretation of observations carried out on site enables extraction of the essence of phenomena from a mathematics perspective
5. Manufacturing theory. This results from the integration of logical paths from viewpoints of operation and economic rationality on site.
6. Activation to mathematics. The motivation of mathematicians has launched new mathematical research fields.

The starting point is clearly to understand the context of the problem in all its complexity, including relevant historic and current information. Engineers and mathematicians then express their developing understandings of the new problem until they reach a common understanding. This requires not only mathematical and scientific knowledge but also social skills and learning. The mathematicians then recognize what knowledge and skills are relevant, in a logical manner, and then analyze the data to interpret their previous observations. Transforming their mathematical knowledges and skills to fit the new situation requires integration with the operational and economic realities of the site. This analysis and transformation of mathematical knowledges and skills in the resolution of the original problem offers a starting point for innovation in the field of mathematical research. The mathematicians have brought to bear their cultural knowledge of the workplace and its processes as well as their personal skills; they have needed to develop a collaborative and holistic approach from understanding the nature of the problem in the beginning through to communicating the outcomes at the end. Each group, engineers and academics, is required to cross boundaries in order to understand the developing problem and find a solution, in the process creating new areas of mathematical research.

As boundary objects, the mathematical models derived allow mathematicians and engineers to communicate effectively and to conceive possible innovations. However, Nakagawa and Yamamoto stress that a communicative framework is necessary from the outset to discuss the phenomena under consideration, to define suitable targets and markers of progress; they note that mutual confirmation of progress is essential. They are also looking for continuous improvement in the skills of combining appropriate mathematical methodologies for the phenomenon under consideration, and this requires people who are suitably qualified, with the personal skills and cultural knowledge, identified by Eraut (2004), appropriate to the conditions and the role. In Nakagawa's opinion, mutual and respectful dialogue is the most important thing and the most challenging one. While the culture of education has traditionally implied teaching to be uni-directional, and this has been of great value to industry, he stresses the importance of teachers and mathematicians in academia learning from industry personnel. For him, mutual coaching means fostering human resources and solving problems simultaneously, in order to bring out hidden talents as well as to plan and implement mutual efforts towards a shared goal. In his presentation, Nakagawa also highlighted the *multiplier effect* of interdisciplinary collaboration between mathematics, industry, and engineering.

Linking Professional Experiences With Academic Knowledge. The Construction of Statistical Concepts by Sales Manager Apprentices by Corinne Hahn, offers an illustration of a partnership in mathematics education with industry participation through the French *alternance* system for 16–26-year-olds. Both the firm and the vocational school (a business school in this case) have the common goal of preparing young adults for future work, and Hahn describes the *alternance* system as a boundary space between work and school that supports sense-making by learners. However, she notes that linking disciplinary knowledge with work practices is not straightforward as work situations are always multidisciplinary: The workplace and the academy operate under two different forms of logic.

Already employed as junior staff in the workforce, these young people bring to the learning situation their current experience of workplace practice and, in one sense, this is a strength because they have a meaningful context for their studies. However, this can also be a disadvantage in that they tend to approach problems which require a sophisticated level of conceptualization, statistically or otherwise, more simplistically—in line with their relative inexperience as managers. This presents a challenge for educators such as Hahn to raise learners' statistical awareness above their current levels of experience and expertise so that it will be available to them in the future. The paper contains an important theoretical insight, later supported by the data, that decision-making is shaped not only by scientific rationality (i.e., explicit logical reasoning), but also by the values and beliefs associated with other social forms of rationality, often tacitly acquired through participation in different communities of practice, and which can have an impact on what the problem means to each person and how they go about solving it. Hahn also raises the issue of the learner's identity: They need to be able to move from the local to the global, particularly in terms of their statistical expertise.

Points raised in subsequent discussion included the following:

1. Scientific tools, such as standard deviation, are disconnected from the learner's reality.
2. Each different kind of rationality (pragmatic, technical, and scientific) exists within the classroom and these need to be connected through school/work activities.
3. Learners have two identities—as students and as workers; they are actually bicultural
4. There are language boundaries: conceptions of reality; categorizations, ethical, aesthetic, and practical approaches and considerations which differ between the workplace and the school. This includes the difference between mathematics as an object and mathematics as tool.

Learning Conversation in Mathematics Practice—School-Industry Partnerships as an Arena for Teacher Education by Gert Monstad Hana, Ragnhild Hansen, Marit Johnsen-Høines, Inger Elin Lilland, and Toril Eskeland Rangnes of Norway.. The authors describe a school-industry partnership where school pupils cross boundaries to act as industry consultants, supported by student teachers who are themselves crossing boundaries between the school, the university, and the workplace. Their first example concerns lower-secondary school pupils acting as statistical consultants for a local company producing valves. It illustrates how an assignment made possible by an authentic industrial context, altering the regular conditions of learning and teaching, influences the intentionality, functionality and empowerment of pupils and student teachers. Their second example is concerned with mathematical modelling of regression equations relevant to an industrial context, especially with regard to the development of critical democratic competence in the pupils. The authors believe that these connections with the world outside of school mathematics may support engagement by pupils and teacher education students in developing mathematical literacy. Of course this project is also dependent upon key industry personnel, such as the manager in example 1, crossing into the educational sphere and participating respectfully as they would with adult consultants in giving professional critical feedback from the company to pupils. The pupils themselves were able to cross into the industrial sphere to ask critical questions of the company. Thus communicative competence appears to be an essential aspect of successful interfaces between mathematics and industry throughout this report. This paper offers another strong example of the concept of boundary crossing. The student teachers having their practice connected to school-industry partnership enabled them to learn and teach in (and between) different contexts. Changing the teaching and learning conditions for pupils, students, and informally for industry participants, leads to a different type and quality of pupil conversation and different interactions in the learning loops in which the pupils and students move.

How is it Possible to Make Real-World Mathematics More Visible: Some Results From Two Italian Projects by Cinzia Bonotto. This paper presented results from two ongoing projects. The first focused on fostering a mindful approach towards realistic mathematical modelling and application work in primary schools. The second, discussed at greater length, concerned high school students in an equitable partnership between school, industry, and university. It was intended to change school students' images of mathematics and to encourage them to continue their studies of mathematics at higher levels. Several examples were given of school children working on industrial problems. In the Veneto region, problems are suggested by local industry, public administration, and other non-scholastic institutions. These problems are clearly of importance to industry, with the results from the school students actually being of value. Workplace managers visited the schools and offered data, themes, statistics, linear programming, Operations Research, and modelling; also cryptography tasks.

Apart from students having their work taken seriously, one positive outcome was an actual increase in mathematics enrolments at university: In Padua, university mathematics enrolments doubled! Another unexpected result was the a mutual appreciation between industry and the university and high school teachers of each other's competence and dedication, which in some cases produced scientific collaboration and resulted in reciprocal knowledge development. In concluding her presentation, Bonotto noted the importance of interdisciplinarity.

MITACS ACCELERATE: A Case Study of a Successful Industrial Research Internship Program (See Chap. 42) by Rebeccah Marsh and Sarah Petersen, is an example of collaboration between Canadian industry and universities. It links graduate students and postdoctoral fellows in the mathematical sciences, among others, through applied research internships, matching researchers with relevant projects in 4-month programs where they interact with people in industry. In this further example of boundary crossing between the academy and industry, Marsh and Petersen identify four major challenges for the program: (a) convincing industry of the benefits of research expertise, (b) encouraging the academic community to consider the needs of industry in their research, (c) persuading the different levels of government to provide financial support, and (d) recruiting and placing significant numbers of trainees in industrially-relevant and academically-sound research projects. A critical role in this boundary crossing is played by the business development team in building connections between the academy and the industrial sector, identifying opportunities, and matching academic expertise to public and private sector needs. Also important is the rigorous peer review process for the research projects which often form part of the interns' graduate theses. Apart from the interns and the partner organizations, the faculties concerned also have the opportunity to build long-term collaborations with industry. For the interns themselves, communication is clearly important; they develop the 'soft skills' at university before, during, and after the placement. This participation in the workplace offers interns possible employment opportunities while industry can assess their suitability over time in their own workplace setting. While the research projects are commissioned by industry, the university acts as a resource to industry in supporting the creation of new knowledge and transfer of existing knowledge.

The Project "Mathe-Meister"—A Mathematical Self Assessment Centre With Diagnostic Feedback For Vocational Trainees by Kathrin Winter. One of the main problems in advanced vocational education in Germany is the abolition of formal mathematics lessons in favour of situated learning: Mathematical competences are now embedded in the context rather than taught as an explicit and isolated topic. As a consequence, students are responsible for their own mathematical competences. However, many vocational trainers lack special mathematical didactical education and although they notice gaps in the mathematical competences of the students they cannot diagnose the mistakes and problems in any detail. For trainers, the problem is to know what competences are needed and which are actually held.

The aim of the project 'Mathe-Meister' is the development of a web-based self-assessment centre for vocational education. It is intended to make people aware of the importance of mathematics in vocational training, to show hidden mathematics in the workplace, and to give mathematical or diagnostic information to vocational teachers. It will help students to appreciate their own mathematical competences in view of their chosen vocation, and detect and remediate their individual mathematical deficits. Students receive (a) help towards getting the correct answer to each item; (b) advice on helpful textbooks, educational computer programs, and tutoring; and (c) an individual diagnostic report. However, realizing a web based test including a real-time analysis report with individual diagnostic information represents an actual and yet not solved problem in an interrelated field of computer science, didactics of information technologies, didactics of mathematics and vocational training.

4 Papers not Presented but Relevant to Our Group

Mathematics in a Safety–Critical Work Context: The Case of Numeracy for Nursing by Diana Coben and Meriel Hutton from the UK, offers a different perspective on mathematics education with industry participation. In the UK, there is no single recognized benchmark for the numeracy requirements of professional nurses at entry level, or even for those updating their skills. Coben's team has worked with the nursing industry to determine the nature of, and justification for, an appropriate benchmark as well as an authentic and comprehensive simulation-based teaching–learning-assessment tool that is coherent with current professional practice. This project has innovatively tried to harness the potential offered by developing technologies in education for the benefit of all concerned in the 'production' of better educated nurses, ultimately leading to a safer health system.

The Threefold Dilemma of Missing Coherence—Bridging the Artificial Reef Between the Mainland and Some Isolated Islands by Guenter Toerner and Volker Grotensohn, discusses issues surrounding the employment of school leavers as apprentices in the German steel industry. The educational interface is centred on the university and the industry-based instructors responsible for the mathematics education of new workers, taking into account recent changes to vocational preparation in schools, as discussed by Winter. At the time of writing, the authors were attempting to cross the boundary between the university and a steel company with the hope of making some positive interventions. They attempted to build an atmosphere of trust with the instructors and to avoid implying that they were deficient in some way. They were also hopeful of engendering interest in mathematics and self-confidence among the apprentices. This paper raises important issues of mathematics curriculum in vocational preparation, the transfer of situated mathematical knowledge to new contexts, vocational instructors' philosophies of mathematics and of mathematics education, and the effect on students leaving school with limited experiences of success in mathematics in relation to the affective domain (i.e., beliefs, attitudes, and emotions) especially in relation to further mathematics education in the workplace. These are significant and ongoing problems worldwide and are ideally suited to boundary crossing work such as begun here.

Several other Group members who contributed to the discussions presented their papers in the General section of the meeting.

The Project "Ways to More MINT-graduates" of the Bavarian Business Association (vbw) with Focus on the M (=Mathematics) at the University of Augsburg, Germany by Matthias Brandl, follows an incremental-evolutionary approach to change mathematics students' and teachers' attitudes and beliefs. Several actions that interact in a synergetic way in the form of an integrated approach are taken to attract more students to study mathematics at school and to reduce the dropout-rate at university. These actions consist of information concerning studies and the vocational field of mathematics for high school students, "early studies" at university for highly gifted high school students, and a one-week "Preparatory Course for Mathematics" as a bridge between high school and university. There is also information concerning the vocational field of mathematics for students at university, and the development of learning environments addressing mathematical issues in the transition between high school and university.

A Meta-Analysis of Mathematics Teachers of the GIFT Program Using Success Case Methodology (See Chap. 43) by Richard Millman, Meltem Alemdar, and Bonnie Harris from the U.S.A., illustrates the creation of a successful collaboration between mathematics education and industry. In this cross-sectional study, which featured the views of teachers, the analysis of the experience from an industrial internship for teachers (GIFT) shows that internships can increase mathematical understanding and skills on the part of mathematics teachers, and research can be integrated into the classroom to the benefit of teachers and students. Three goals of the GIFT program were achieved during the three-year study: (1) The GIFT facilitators and industrial researchers assisted teachers with creating an "Action Plan" for implementing their summer experiences into their classrooms; (2) the development of an extended professional community of learners was fostered; and (3) extended partnerships for communication and collaboration between teachers and industry mentors resulted.

4.1 General Discussion

Boundary crossing: What is it that we want to cross? Speech? Thought? How? Who has the power to define the boundary? Mathematicians define it for academic mathematics. What about the mathematics developed in the workplace, at any level? Boundary crossing can be regarded as a contradiction: How this is resolved is important. Tensions give rise to the possibility of creating development: for example, *alternance* where people change and develop.

In relation to issues of diversity within the Group, there were many commonalities and differences. However, differences were welcomed, not feared or seen as problematic. Adults returning to study bring significant practical experience, compared with younger people. The Group noted that it is important that mathematics be seen as more than merely useful: as aesthetic. On one hand, mathematical reasoning is strong and beautiful in the human sense; also powerful.

On the other, there are aesthetics in fields of design and architecture, chemical, physical, biological sciences and the environment; and in a job well done in any field.

It is important to have the global involvement of industry in education in order to have the expertise and experience to help teachers; and to enable industry to work with teachers. It is also important to overcome school students' negative views of mathematics, and to improve the quality of teaching internationally. There are multiple forms of collaboration between school and industry aiming to encourage students to continue to study mathematics in higher education and increase their awareness of career possibilities; also to address young peoples' perceptions of mathematics—as a solution to the problem of its general invisibility. There is a need to recognize that mathematics is not only about written tests and the application of existing models; that there is a tacit rationality that is part of each of our beings, mentally and physically, in our cultural practices and social interactions. In summary, at the heart of each interface is communication, mutual respect, and an underlying need for interdisciplinarity.

Acknowledgements We would like to thank Prof. Tine Wedege for her insightful comments on an earlier draft of this chapter.

References

Araújo, A., Fernandes, A., Azevedo, A., & Rodrigues, J. F. (eds.) (2010). *EIMI 2010 conference: Educational interfaces between mathematics and industry: proceedings*. Lisbon: Centro International de Matemática.

Eraut, M. (2004). Informal learning in the workplace. *Studies in Continuing Education, 26*(2), 247–273.

Garraway, J. (2010). Knowledge boundaries and boundary-crossing in the design of work-responsive curricula. *Teaching in Higher Education, 15*(2), 211–222.

Gibbons, M., Limoges, C., Nowotny, H., Schwartzman, S., Scott, P., & Trow, M. (1994). *The new production of knowledge: The dynamics of science and research in contemporary societies*. London: Sage Publications.

Wood, L. N. (2010). Graduate capabilities: Putting mathematics into context. *International Journal of Mathematical Education in Science and Technology, 41*(2), 189–198.

How it is Possible to Make Real-World Mathematics More Visible: Some Results from Two Italian Projects

Cinzia Bonotto

1 Mathematical Modelling and Problem Posing

The first project described in this contribution regards the primary school level and is articulated in some teaching experiments aimed at showing how an extensive use of suitable artifacts could prove to be useful instrument in creating a new tension between school mathematics and real world with its incorporated mathematics. The teaching/learning environment designed in these teaching experiments is characterized by an attempt to establish a new classroom culture also through new socio-mathematical norms, for example norms about what counts as a good or acceptable response, or as a good or acceptable solution procedure, are debated (Bonotto 2005). The focus is on fostering a mindful approach toward realistic mathematical modelling, mathematics applications and also a problem posing attitude, even at the primary school level.

The term mathematical modelling is not only used to refer to a process whereby a situation has to be problematized and understood, translated into mathematics, worked out mathematically, translated back into the original (real-world) situation, evaluated and communicated. Besides this type of modelling, which requires that the student has already at his disposal at least some mathematical models and tools to mathematize, there is another kind of modelling, wherein model-eliciting activities are used as a vehicle for the development (rather than the application) of mathematical concepts (Greer et al. 2007). This second type of modeling is called 'emergent modeling' in Gravemeijer (2007), and its focus is on long-term learning processes, in which a model develops from an informal, situated model ("a model of"), into a generalizable mathematical structure ("a model for").

Although it is very difficult, if not impossible, to make a sharp distinction between the two aspects of mathematical modelling, it is clear that they are

C. Bonotto (✉)
Department of Mathematics, University of Padua, Padua, Italy
e-mail: bonotto@math.unipd.it

A. Damlamian et al. (eds.), *Educational Interfaces between Mathematics and Industry*, New ICMI Study Series 16, DOI: 10.1007/978-3-319-02270-3_9,

associated with different phases in the teaching/learning process and with different kinds of instructional activities (Greer et al. 2007).

We deem that an early introduction in schools of fundamental ideas about modelling is not only possible but also indeed desirable even at the primary school level. We argue for modelling as a means of recognizing the potential of mathematics as a critical tool to interpret and understand reality, the communities children live in, and society in general. An important aim for compulsory education should be to teach students to interpret critically the reality they live in and understand its codes and messages so as not to be excluded or misled (Bonotto 2007).

As regard the problem posing, this process is of central importance in the discipline of mathematics and in the nature of mathematical thinking and it is an important companion to problem solving. Recently many mathematics educators realized that developing the ability to pose mathematics problems is at least as important, educationally, as developing the ability to solve them and have underlined the need to incorporate problem posing activities into mathematics classrooms (e.g. English 2003; Christou et al. 2005).

Problem posing has been defined by researchers from different perspectives (see Silver and Cai 1996). In this contribution we consider mathematical problem posing as the process by which, on the basis of mathematical experience, students construct personal interpretations of concrete situations and formulate them as meaningful mathematical problems. It, therefore, becomes an opportunity for interpretation and critical analysis of reality since: (1) they have to discern significant data from immaterial data; (2) they must discover the relations between the data; (3) they must decide whether the information in their possession is sufficient to solve the problem; and (4) to investigate if numerical data involved is numerically and/or contextually coherent. These activities, quite absent from today's Italian school context, are typical of the modelling process and are similar to situations to be mathematized that students have encountered or will encounter outside school, for examples in the work situations.

In our approach in and out of school mathematics, even with their specific differences, in terms both of practices and learning processes, are not seen as two disjoint and independent entities. Furthermore we think that the conditions that often make out-of-school learning more effective can and must be re-created, at least partially, within classroom activities. Indeed, though there may be some inherent differences between in and out of school mathematics, these can be reduced by creating classroom situations that promote learning processes closer to those arising from out-of-school mathematics practices (Bonotto 2005).

The classroom activities present in the teaching experiments we have conducted fall under the type of "rich context", in the Freudenthal sense, i.e. context which not only serves as the application area but also as source for learning mathematics, and in particular for the emergence of a mathematical modelling disposition (Freudenthal 1991).

2 About Artifacts

The artifacts introduced in our teaching experiments (for example receipts, advertising leaflets containing discount coupons for supermarkets and stores, the weather forecast from a newspaper, a weekly TV guide, an informational booklet issued by "Poste Italiane", and so on) are materials, real or reproduced, which children typically meet in real-life situations and then are relevant and meaningful. In this way we offer the opportunity of making connections between the mathematics incorporated in real-life situations and school mathematics, which although closely related, are governed by different laws and principles. In this way we present mathematics as a means of interpreting and understanding reality and increase the opportunities of observing mathematics outside the school context. The usefulness and pervasive character of mathematics are merely two of its many facets and cannot by themselves capture its very special character, relevance, and cultural value. Nonetheless, these two elements can be usefully exploited from a teaching point of view because they can change students' common behaviour and attitude (Bonotto 2005).

Furthermore the use of suitable artifacts allows the teacher to propose many questions, remarks, and culturally and scientifically interesting inquiries. The activities and connections that can be made depend, of course, on the students' scholastic level. These artifacts contain different codes, percentages, numerical expressions, and different quantities with their related units of measure, and hence are connected with other mathematical concepts and also other disciplines (chemistry, biology, geography, astronomy, etc.).

Furthermore by asking children (1) to select other artifacts from their everyday life, (2) to identify the embedded mathematical facts, (3) to look for analogies and differences (e.g. different number representations), (4) to generate problems (e.g. discover relationships between quantities) the children should be encouraged to recognize a great variety of situations as mathematical situations, or more precisely mathematizable situations. A "re-mathematization" process is thereby favoured, wherein students are invited to unpack from artifacts the mathematics that has been "hidden" in them, in contrast with the demathematization process in which the need to understand mathematics that becomes embodies in artifacts disappears (see Gellert and Jablonka 2007). In this way we can multiply the occasions when students encounter mathematics outside of the school context, "everydaying" the mathematics (Bonotto 2007).

3 Some Results

The results of this project (which involves about 20 primary teachers and about 400 children) show that, contrary to the practice of traditional word problem solving, children do not ignore the relevant and plausible aspects of reality, nor did they exclude real-world knowledge from their observation and reasoning. They confronted with this kind of activity also show flexibility in their reasoning processes by exploring, comparing, and selecting among different strategies (see e.g. Bonotto 2003b and 2005). These strategies are often sensitive to the context and number quantities involved, and closer to the procedures emerging from out-of-school mathematics practice; so mathematical reasoning needed in extra-scholastic contexts, for example in work places, is favoured (see e.g. Bonotto and Basso 2001; Bonotto 2003a and 2005). Finally also creativity and a problem critiquing process is favoured; the children attempted to criticize and make suggestions or correct the problems created by their classmates or the results obtained (see e.g. Bonotto 2006 and 2009).

Regarding the mathematical content we also laid the basis for overcoming some conceptual obstacles, for example the misconception that multiplication always produces a larger result than the factors (see Bonotto 2005). This confirms the hypothesis that this kind of classroom activities can give support for accessing more formal mathematical knowledge and promote a process of "abstraction-as-construction", in according to the "emergent modelling" perspective (Gravemeijer 2007).

The positive results obtained in this project can be attributed to a combination of closely linked factors: (a) an extensive use of suitable artifacts that, with their incorporated mathematics, played a fundamental role in bringing students' out-of-school meaningful reasoning and experiences into play, and allowed a good control of inferences and results; (b) the application of a variety of complementary, integrated, and interactive instructional techniques (involving children's own written descriptions of the methods they use, work in pairs or small groups, and whole-class discussions); (c) the introduction of particular sociomathematical norms that played an important role in giving meaning to new mathematical knowledge, in reinforcing previous knowledge and in paying systematic attention to the nature of the problems and the classroom culture; (d) an adequate balance between problem-posing and problem-solving activities, in order to promote also a mathematical modelling disposition.

But is there a reverse of the coin, if the word "reverse" can be used? On the basis of the experience of our studies, we entirely agree with Freudenthal (1991), that the main problem regarding rich context is implementation requiring a fundamental change in teaching attitudes. The effective establishment of a learning environment like the one described here makes very high demands on the teacher, and therefore requires revision and change in teacher training, both initially and through in-service programs.

We do not suggest that the classroom activities present in the teaching experiments of this project are a prototype for all classroom activities related to mathematics, although we think that the presence of realistic mathematical modelling activity, as well as of problem posing activity, should not emanate from a specific part of the curriculum but should permeate the entire curriculum.

4 The Project Aimed at High Schools: Progetto Lauree Scientifiche

The second project described in this contribution is part of a national project launched in 2005. The Italian Ministry of Education, in collaboration with the Faculties of Science of the Italian Universities and the Confederation of Italian Industries (Confindustria) started a special project (PLS = Progetto Lauree Scientifiche) whose main purpose was to encourage high-school graduates entering Universities to apply for degrees in the "hard" sciences, including Mathematics, Physics, and Chemistry.

This project turned out to be successful and presently represents the most effective initiative of collaboration between high-school students and teachers, university teachers, and Industries, over the last two decades.

The project management was divided into geographic regions, dealing separately with Mathematics, Physics, Chemistry and Science of Materials.

5 Procedure

With regard to Mathematics in the Veneto region (Venezia, Verona, Padova, Vicenza, Treviso, Rovigo, Belluno), over the past 4 years PLS has involved about 30 University teachers, 80 high school teachers and over 1,000 students. The Director was the prof. Benedetto Scimemi of the University of Padova.

To start the project activity, in the Fall of 2005 the Project coordinator selected in the Region 15 public high schools, each enjoying a good reputation as educational institution and having a reasonable amount of math in its curriculum (mainly "Licei scientifici"). The choice was made so that all the provinces were represented.

In each of these institutes a reliable math teacher was asked to join the Project and to enrol two more colleagues of his school. The proper choice of these teachers was crucial for the Project success. In fact, most of them had previous contacts with the Universities, having being either former students or participants of teacher-training activities. These 3–4 people in each school would be in charge of selecting a number from 15 to 30 volunteer students of their school (17–18 years of age), who would be willing to attend supplementary classes in Mathematics, 5–6 times in the afternoons, 2–3 h each.

At the same time, within the Universities of Padova and Verona, the Project Coordinator contacted a number professors and assistants, about 25 people, whose competence and dedication to teaching activities was well known. The coordinators were invited to choose a one of their favorite topics in Mathematics, which would be both understandable by good high school students and suitable for a short term "Math Laboratory". They were invited to prepare a plan for delivering their material within 5–10 teaching hours, followed by an adequate number of practice hours.

These two groups of mathematicians—high school and university teachers—joined in a two-hour meeting in Padova in September 2005. First the university people were asked to briefly describe their proposals; then each high school was invited to choose a subject and its expert. As a result of this meeting and further personal contacts, in each school a "workgroup" was formed, made of 3 local teachers plus one or two from the university staff. In later months each work-group met separately to plan the details and decide its own calendar.

A typical schedule would be: for 5 weeks, each Thursday, from 3 to 5 p.m.

In most of these Labs the role of the university teachers was very important during the first meeting, to explain the main theory involved. Then more time was left to the high-school teachers until, in some cases, the final work—normally an application to a real life problem—was entirely made by the students, who often used their computers.

In parallel, a teacher in-service training program [Corso di perfezionamento in Metodologia e Didattica della Matematica, Director Cinzia Bonotto] was initiated. It aimed at providing in-service secondary mathematics teachers with new training in the background (mathematical or not) needed for the applications discussed in the project.

6 Contents

Various subjects were discussed in the different schools including, when possible, problems that were suggested by local industries, public administrations or other non-scholastic institutions. Various subjects were discussed in different schools, including, when possible, problems suggested by local industries, public administrations or other non-scholastic institutions. In some cases the managers of the firm or institution which had provided the data wanted to take part in the preparatory seminars, either visiting the school themselves or inviting the students to visit the firm offices. At the end, the students' work was reported to the Confederation of the local industries, which also assigned a prize to a school which had best interpreted the spirit of collaboration between school and industry.

Here are some examples of the mathematical themes presented, some of which were later published (see Languasco and Zaccagnini 2006; Centomo et al. 2007; Carminati et al. 2006, 2007 and 2008; Chignola et al. 2006; Burato et al. 2007):

- Statistics, in particular, Cluster analysis: data to be treated were collected from a provincial tourist office, a pharmaceutical industry, and an agency specializing in quality control of industrial production.
- Linear programming and operations research: optimization problems were suggested by a garbage-collecting firm, a telephone call-center, a stock exchange trading agency and others.

- Modelling: a model for the growth of tumour tissues was suggested by a biology researcher; a number of mathematical subjects, such as Fourier analysis, were suggested by the cardiology department of a public hospital.

Other subjects, although not directly suggested by local industries or agencies, were also chosen with special attention to applications (Cryptography, Dynamics of populations, Dynamical systems).

The mathematics themes chosen seem to please the audience much more than the traditional subjects of a standard school program. However, in addition to these choices of topic, an important role was surely played by the selected audience, all students attending the seminars having indicated a potential interest in mathematics.

7 Some Results

In reviewing the activities carried out by the various schools under the Project, the following examples of collaborations with non-scholastic institutions may be considered specially interesting:

(a) at Padova (Liceo Cornaro) a number of statistic elaborations were made by the students, regarding the therapeutical efficiency of some drugs produced by a national pharmaceutical industry (Sigma–Tau). A scientific consultant of this industry visited the school and explained how they test their drugs, then made comparisons with the students' methods.
(b) at Belluno (Liceo Galilei) more statistical studies were made on the wild animal populations in the near region. The lecture of a scholar from the University of Vilnius (Lithuania) permitted the students to learn and use a special free software (called R) to elaborate their data for finding the principal components.
(c) at Thiene (Liceo Corradini) a class visited the local Hospital, where a special computer program, produced by the national industry Exprivia, was currently used for cardiological purposes. The students' work contributed to improving this mathematical program, which was then adopted. This school was the winner of a special prize for the most interesting PSL activity of the year.
(d) at Bassano (Liceo Da Ponte) a local firm in charge of collecting public garbage was looking for criteria to optimize the place and time schedule of their collecting cars and trucks. The students studied Linear programming and the Simplex method, enough to produce a final proposal. The firm sponsored the publication of a book describing the whole content of this Math-Lab (see Carminati et al. 2006).

8 Final Comments

In conclusion, PLS can be considered a very successful project, in terms of both encouraging mathematics teachers and increasing enrolments into scientific faculties: The number of mathematics students at the University of Padova has doubled in the last five years. Beyond this target, unexpected result was a mutual appreciation of the competence and dedication of University and High-school teachers, which in some cases even produced scientific collaboration.

The Veneto branch of the Confederation of Italian Industries welcomed the work done within the project. Indeed, it established a prize for the participating schools that most actively collaborated with local industries in applying mathematical techniques to one or more problems of interest to the industries themselves.

The PLS project has recently been extended for another next two years.

Acknowledgments The author wishes to thank all of the teachers and students who took part in these two projects. Special thanks go to Prof. Benedetto Scimemi for his contribution concerning the Progetto Lauree Scientifiche, and to Prof. Frank Sullivan for his precious help in the translation of this paper into English.

References

Bonotto, C. (2003a). About students' understanding and learning of the concept of surface area. In D. H. Clements, & G. Bright (Eds.), *Learning and teaching measurement, 2003 NCTM Yearbook* (pp. 157– 167), Reston: National Council of Teachers of Mathematics.

Bonotto, C. (2003b). Investigating the mathematics incorporated in the real world as a starting point for mathematics classroom activities. In N. A. Pateman, B. J. Dougherty, & J. Zilliox (Eds.), *Proceedings of the 27th Conference of the International Group for the Psychology of Mathematics Education* (II, pp. 129–136). Honolulu: Hawaii University.

Bonotto, C. (2005). How informal out-of-school mathematics can help students make sense of formal in-school mathematics: The case of multiplying by decimal numbers. *Mathematical Thinking and Learning. An International Journal, 7*(4), 313–344.

Bonotto, C. (2006). Extending students' understanding of decimal numbers via realistic mathematical modeling and problem posing. In J. Novotná, H. Moraová, M. Krátká, & N. Stehlíková (Eds.), *Proceedings of the 30th Conference of the International Group for the Psychology of Mathematics Education* (II, pp. 193–200). Prague: Prague Charles University.

Bonotto, C. (2007). How to replace the word problems with activities of realistic mathematical modeling. In W. Blum, P. Galbraith, M. Niss, & H. W. Henn (Eds.), *Modelling and applications in mathematics education* (pp. 185–192). New ICMI Studies Series no. 10. New York: Springer.

Bonotto, C. (2009). Artifacts: influencing practice and supporting problem posing in the mathematics classrooms. In M. Tzekaki, M. Kaldrimidou & C. Sakonidis (Eds.) *Proceedings of the 33th Conference of the International Group for the Psychology of Mathematics Education* (II, 193–200), Thessaloniki, Greece: PME.

Bonotto, C., & Basso, M. (2001). Is it possible to change the classroom activities in which we delegate the process of connecting mathematics with reality? *International Journal of Mathematics Education in Science and Technology, 32*(3), 385–399.

Burato, A., Chignola, R., Castelli, F., Corso, L., Pezzo, G., & Zuccher, S. (2007). *Matematica e radioterapia dei tumori—Sviluppo e applicazioni di un modello predittivo semplificato*. Verona: I quaderni del Marconi—Quaderni didattici.

Carminati, R., Gheno, G., & Mattarolo, M. (2006). *Matematica in Azienda*. Bassano del Grappa: Editrice Artistica Bassano.

Carminati, R., Gheno, G., & Mattarolo, M. (2007). *Matematica in Azienda: il Dado o Montecarlo*. Bassano del Grappa: Editrice Artistica Bassano.

Carminati, R., Gheno, G., & Mattarolo, M. (2008). *Matematica in Azienda: Curve Mirabili*. Bassano del Grappa: Editrice Artistica Bassano.

Centomo, A., Gregorio, E., & Mantese, F. (2007). *Crittografia per studenti*. Milano: MiMiSol Edizioni.

Chignola, R., Castelli, F., Corso, L., Pezzo, G., & Zuccher, S. (2006). *La Biomatematica in un problema di oncologia sperimentale*. Verona: I quaderni del Marconi—Quaderni didattici.

Christou, C., Mousoulides, N., Pittalis, M., Pitta-Pantazi, D., & Sriraman, B. (2005). An empirical taxonomy of problem posing processes. *Zentralblatt für Didaktik der Mathematik, 37*(3), 149–158.

English, L.D. (2003). Engaging students in problem posing in an inquiry-oriented mathematics classroom. In F. Lester, & R. Charles (Eds.), *Teaching Mathematics through Problem Solving* (pp.187–198). Reston: National Council of Teachers of Mathematics.

Freudenthal, H. (1991). *Revisiting mathematics education. China lectures*. Dordrecht: Kluwer.

Gellert, U., & Jablonka, E. (2007). Mathematization—Demathematization. In U. Gellert & E. Jablonka (Eds.), *Mathematization and demathematization: Social, philosophical and educational ramifications* (pp. 1–18). Rotterdam: Sense Publishers.

Gravemeijer, K. (2007). Emergent modelling as a precursor to mathematical modeling. In W. Blum, P. Galbraith, M. Niss, & H. W. Henn (Eds.), *Modelling and applications in mathematics education* (pp. 137–144). New ICMI Studies Series no. 10. New York: Springer.

Greer, B., Verschaffel, L., & Mukhopadhyay, S. (2007). Modelling for life: mathematics and children's experience. In W. Blum, P. Galbraith, M. Niss, & H. W. Henn (Eds.), *Modelling and applications in mathematics education* (pp. 89–98). New ICMI Studies Series no. 10. New York: Springer.

Languasco, A., & Zaccagnini, A. (2006). *Crittografia*. Padova: C.L.E.U.P.

Silver, E. A., & Cai, J. (1996). An analysis of arithmetic problem posing by middle school students. *Journal for Research in Mathematics Education, 27*(5), 521–539.

The Project “Ways to More MINT-Graduates” of the Bavarian Business Association (vbw) with Focus on the M (=Mathematics) at the University of Augsburg, Germany

Matthias Brandl

1 The Project

The project “Ways to more MINT-graduates” was launched by the Bavarian business association (vbw) in January 2008 to attract more students to the subjects mathematics (M), informatics (I), natural sciences (N), and technology/engineering (T). In the last year, the Bavarian economy lacked of 7,500 jobs for engineers; and while the demographic change makes this situation worse, the college drop-out-rate of more than 30 % in the subjects mentioned has to be reduced.

At the University of Augsburg, the project is concentrated on the attraction and fostering of students in mathematics (project title: “Studying Mathematics!”). Because of its success, in the meantime, the project in Augsburg was honored by a visit of the Bavarian minister for science in May 2009.

2 Empirical Facts to College Dropouts in Germany

In order to reduce the dropout-rate, the first step is to analyze the situation. According to Heublein et al. (2008), the current dropout-rate averaged over all subjects in Germany is 21 %, but in MINT-subjects it is often beyond 30 %. A dropout in this scenario is a former student, who left the system of higher education without a first degree. Students interrupting their studies or changing university are not included. The biggest problem results from the fact that universities neither are allowed to follow the course of the students’ studies, nor to give this information away. So the numbers concerning the dropout-rate are only based on an estimation and can only be limitedly differentiated.

M. Brandl (✉)
Department of Informatics and Mathematics, University of Passau, Passau, Germany
e-mail: matthias.brandl@uni-passau.de

A. Damlamian et al. (eds.), *Educational Interfaces between Mathematics and Industry*, New ICMI Study Series 16, DOI: 10.1007/978-3-319-02270-3_10,

The causes for a college dropout are regularly evaluated by the Higher Education Information System (HIS) in Hanover. The results of the last surveys were published in Heublein et al. (2003) and Heublein et al. (2009). In Fig. 1, taken from Heublein et al. (2009) p. 153 and translated into English, the crucial causes for a dropout in the subject group mathematics and science at universities are listed in comparison for the two evaluated years 2000 and 2008.

Most problems mentioned by the students cannot or hardly be influenced by the universities. As the most promising fields remain problems of performance, problematic study conditions and exam failure. As systematic offers concerning the first two moments, especially knowledge-based supporting methods are necessary to help the student getting along with the amount and standard of the learning matter.

According to Fig. 1, the step to the bachelor/master-system in the context of the Bologna process has increased the dominance of problems concerning the performance requirements severely. Together with exam failure, the value for dropping out because of insufficient performance is two-fifth of all exmatriculated students without an exam. The challenging study and examination requirements already in the first semesters are experienced as a performance concentration that is hardly manageable without adequate support.

Besides these factors, a lacking study motivation may also be a starting point for the university to act. The lack of interest on the study subject can be the result of insufficient information about the content and the requirements, while still attending school. So a broadening of the range of information possibilities seems reasonable.

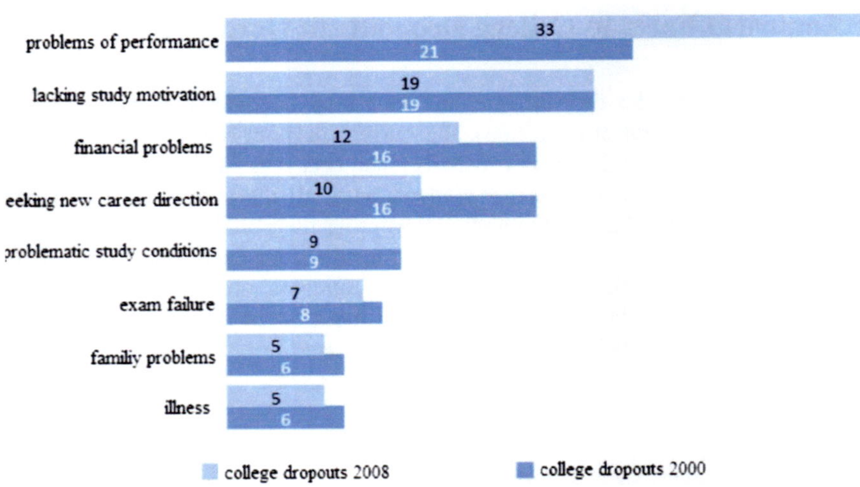

Fig. 1 Crucial causes for a dropout in the subject group mathematics and science at universities (in %)

3 Students' Opinions About a Dropout in Mathematics

To adapt the project "Studying Mathematics!" at the University of Augsburg best possible to the situation, we did an additional survey among 83 students of different math study courses in December 2007. The students were asked to give free answers to the following questions:

When is the most possible time to drop out?
Answer: 82 out of 83 respondents referred to the first three semesters, which corresponds to the official dropout statistic at the institute for mathematics at the University of Augsburg, where the main loss of maths students is observed in just this period.
Why are there dropouts?
Answers: There were different aspects in the students' answers that can be grouped according to "Insufficient vision of the kind what studying mathematics is like", "Large discrepancy between mathematics at school and at university", "Unfamiliar extent of personal responsibility and self-organisation" and "Barely recognizable relation to a subsequent profession".

4 Incremental-Evolutionary Changes on the Meta-Level

In order to achieve the aim of attracting more students to the subject math, the Chair for the Didactics of Mathematics follows an "integrated approach" that is based on insights from the theory of cybernetics.

According to the definitions given in (Malik 1992) or (Vester 1999), mathematics education in Europe and even mathematics education at a concrete school must be seen as "complex" systems. Such are networks of multiple connected components. One cannot change a component without influencing the character of the whole system. The same holds for the variety of industry itself and the interface between the system of mathematical education and the subsequent industrial employers.

With reference to (Malik 1992), two dimensions of steering complex systems can be distinguished. While the first dimension concerns the manner, the second one deals with the level of steering activities [see Fig. 2, taken from Ulm (2009)].

Hierarchical-authoritarian systems, for example, are founded on the method of analytic-constructive steering. This principle needs a controlling authority that defines ways for reaching certain aims. However, complex systems are defined as a network that can potentially be in so many states that nobody can cognitively grasp all possible states of the system and all possible transitions between the states. So this first approach fails by the fact that it would afford information about the system that cannot be gained in reality. So this first approach fails by the fact that it would afford information about the system that cannot be gained in reality.

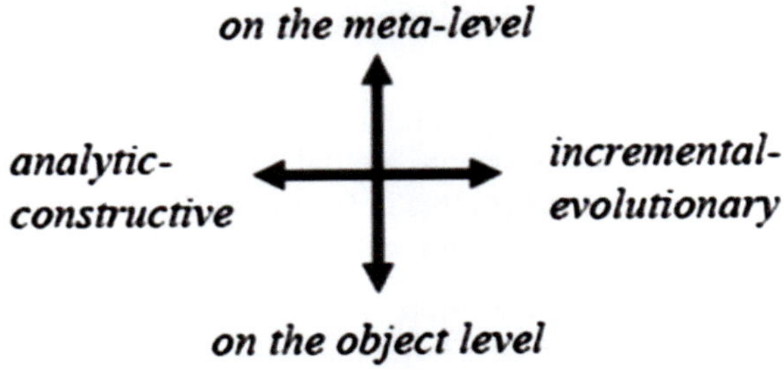

Fig. 2 Steering of complex systems

On the other hand, one can try to focus on the natural growing and developing processes and claim that the changes in complex systems only result from those. This incremental-evolutionary steering tries to influence these systemic processes by accepting the fact that complex systems cannot be steered entirely in all details. Instead, this approach is satisfied with only little steps, i.e., incremental changes, in promising directions. And as to the metaphor of the butterfly's wings that may change the weather far away, every small step may cause unpredictable consequences. So with respect not to endanger the soundness of the whole system, only small changes are essential.

Perpendicular to this dimension, Fig. 2 illustrates the dimension that distinguishes between the object and the meta-level. As the name implies, the object level consists of all concrete objects of the system. In the system of higher mathematical education these would be the teachers, the students, the materials as books, computers, software, the buildings, and so on. In contrast, the meta-level comprehends organizational structures, social relationships, notions of the functions of the system etc. In the educational system, e.g., notions of the subject mathematics and beliefs concerning learning and applying the subject matter are on this level.

For the question, how substantial innovations in the complex system "mathematics education with respect to the interface maths—industry" can be initiated successfully, the theory of cybernetic says: attempts of analytic-constructive steering will fail in the long term, since they ignore the complexity immanent in the system; changes on the object level do not necessarily cause structural changes

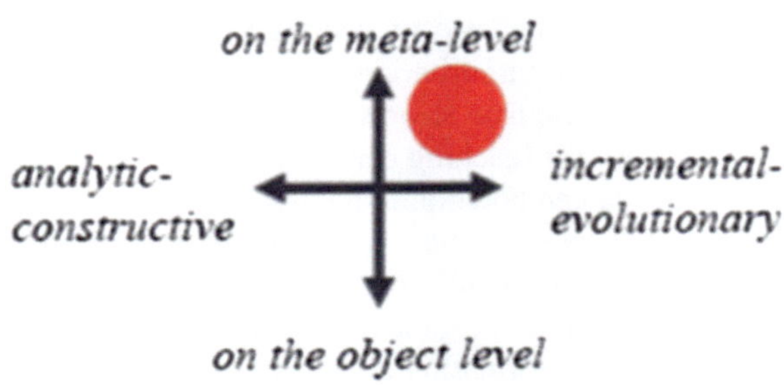

Fig. 3 Innovations in complex systems

of the system; in contrast, it is much more promising to initiate incremental-evolutionary changes on the meta-level of beliefs and attitudes of the learners, see Fig. 3. On the one hand, these are in accord with the complexity of the system and do not endanger its existence, and on the other hand they can cause substantial changes within the system by having effects on the meta-level, especially when they work cumulatively (see Ulm 2009).

5 Integrated Approach

As a consequence of the integrated approach with respect to incremental-evolutionary changes on the meta-level of beliefs and attitudes of the students, the project "Studying Mathematics!" is based on two columns that interact in a synergetic way:

1. Attraction of more students to study a MINT-subject (at school)
2. Reduction of the dropout-rate (at university)

Column 1:

- information concerning studies and vocational field of mathematics for high school students.

  The answers of the survey in December 2007 showed that students do not have appropriate ideas about studying mathematics—neither with regard to contents, nor to organization. Hence, we offer and perform information events concerning both studying mathematics and working as a mathematician in industry and economy for students in upper secondary school. These events, which up to now always had a positive resonance, serve as a motivational help for occupational orientation and science propaedeutics.

- "early studies" at university for highly gifted high school students In order to foster mathematical gifted students of upper secondary school, we offer free "early studies". These students take part in regular lectures and the credits earned can be transferred to regular studies after finishing high school. Hereby, potential later students of MINT-subjects are lead to the university in a very early stage.

Column 2:

- the one-week-lecture "Preparatory course for mathematics" as a bridge between high school and university.

  Within a one-week-lecture prior to the first semester, the characteristics of studying mathematics are presented: less calculation of routine tasks (that is what often happens in maths lessons at school), more developing of and working with theories. Concrete contents are "What is mathematics?," symbolism, mathematical texts, logic, proofs, set theory, mappings and relations, construction of the number system. The course was started in 2008 and regularly has over 100 participants each year that evaluated it very positively.

- information concerning the vocational field of mathematics for students at university Students often have problems seeing the relevance of study contents for jobs aimed at. The consequence is a lack of motivation or a "crisis of meaning" while studying. Therefore, we offer in collaboration with the center for further training and transfer of knowledge (ZWW) at the University of Augsburg regular talks by employed mathematicians as well as excursions to companies that employ mathematicians.

- development of learning environments addressing mathematical issues between high school and university.

  In cooperation with high school teachers, we develop learning environments for students that aim at two things: first, the methodological concept is such that it fosters self-dependent, autonomous, and cooperative working of the students; second, contents are chosen with respect to building bridges between mathematics in high school and university. By developing the aforementioned skills at high school already, students will do much easier at university level later on. Several examples [as Brandl (2008) or Brandl (2009)] were published on the Internet at the online portal "Begabte fördern" (i.e., Program for Gifted) at www.lehrer-online.de.

6 Further Specifics of the Project Framework

In order to provide a potential success of all the small actions taken, to alter the students' opinions in a positive way, there are several supporting structures of the projects, which are

- Academic support services by the Bavarian State Institute for Higher Education Research and Planning (Bayerisches Staatsinstitut für Hochschulforschung und Hoch-schulplanung, IHF)
- Support of the participants by workshops such as construction of questionnaires or project management
- Documentation of the experiences and results by the project members themselves and the IHF in order to collect a pool of good and bad practice examples for other interested universities
- Gaining knowledge by integration; realized by biannual network meetings and internal newsletters

7 Positioning Within the EIMI-Study

So for the aims of the ICMI Study 20 on "Educational Interfaces between Mathematics and Industry" (EIMI) in (Damlamian and Sträßler 2009), i.e., among others

- to attract and retain more students, encouraging them to continue their mathematical studies at all levels of education through meaningful and relevant contextualized examples, and
- to improve mathematics curricula at all levels of education

we strongly refer to the research questions in paragraph 9 Teacher training (ibid.), for example,

- "What are good practices that support this new direction in teacher training?" and
- "How to implement these changes in an efficient way?",
- but also try to answer the questions
- "How can we attract more students to study mathematics?" and
- "How can we efficiently reduce the drop-out-rate in mathematics?"

References

Brandl, M. (2009). Vom Lotto zum Pascalschen Dreieck—eine etwas andere Kurvendiskussion, learning unit at the Lehrer-Online portal Begabte fördern, lo-net GmbH Köln.

Brandl, M. (2008). Von Kegeln zu höheren algebraischen Kurven und wieder zurück, learning unit at the Lehrer-Online portal Begabte fördern, Schulen ans Netz e. V. Bonn.

Damlamian, A. & Sträßler, R. (2009). ICMI Study 20: educational interfaces between mathematics and industry, Discussion Document in ZDM Mathematics Education (2009) 41:525–533.

Gensch, K. & Börensen, C. (2009). MINT—Wege zu mehr MINT-Absolventen. Zwischenbericht 2009, Bayerisches Staatsinstitut für Hochschulforschung und Hochschulplanung.

Heublein et al. (2003). Ursachen des Studienabbruchs. Analyse 2002, HIS Hochschulplanung Band 163, Hannover.

Heublein et al. (2008). Die Entwicklung der Schwund- und Studienabbruchquoten an den deutschen Hochschulen. Statistische Berechnungen auf der Basis des Absolventenjahrgangs 2006, HIS-Projektbericht Mai 2008, Hannover.

Heublein et al. (2009). Ursachen des Studienabbruchs in Bachelor- und in herkömmlichen Studiengängen. Ergebnisse einer bundesweiten Befragung von Exmatrikulierten des Studienjahres 2007/2008, HIS-Projektbericht Dezember 2009, Hannover.

Malik, F. (1992). *Strategie des Managements komplexer Systeme*. Bern: Paul Haupt.

Ulm, V. (2009). Systemic innovations of mathematics education with dynamic worksheets as catalysts, Proceedings of CERME 6, Universite de Lyon

Vester, F. (1999). Die Kunst vernetzt zu denken. Ideen und Werkzeuge für einen neuen Umgang mit Komplexität. Stuttgart: Deutsche Verlags-Anstalt.

Mathematics in a Safety–Critical Work Context: The Case of Numeracy for Nursing

Diana Coben and Meriel Hutton

This chapter draws on our interdisciplinary research on the relationship between mathematics and nursing, a safety–critical work context in which mathematics really matters—for nurses, their patients, their employers and the public at large. There is a growing international literature indicating the problematic nature of this relationship in many countries. Scare stories of mathematical errors by nurses dent public confidence and may impede recruitment into nursing from those who, with the right education and continuing professional support and development, have the potential to become good nurses. Against this background, we are working on the teaching, learning and assessment of numeracy for nursing. In one project, we aim to create an evidence-based benchmark in numeracy for nursing, capable of being operationalised as part of nursing students' preparation for professional practice and used in qualified nurses' continuing professional development. In this chapter, we outline our work towards the creation of such a benchmark, focusing on our study of the assessment of student nurses' skills in medication dosage calculation against our proposed benchmark, undertaken with the aim of ensuring that all concerned may have confidence in nurses' ability to manage the mathematical demands of this key area of nursing safely and effectively.

D. Coben (✉)
National Centre of Literacy of Numeracy for Adults, University of Waikato, Hamilton, New Zealand
e-mail: dccoben@waikato.ac.nz

D. Coben · M. Hutton
King's College London, London, United Kingdom
e-mail: meriel.hutton@btinternet.com

A. Damlamian et al. (eds.), *Educational Interfaces between Mathematics and Industry*, New ICMI Study Series 16, DOI: 10.1007/978-3-319-02270-3_11,

1 Background

Mathematics matters in nursing: to patients, nurses, their employers and nurse educators. Successive studies reveal a lack of proficiency amongst both student and registered nurses (Sabin 2001) and alarming headlines periodically fuel public fears (Hall 2006). The development of appropriate mathematical competence by healthcare staff and students is a key area for concern, but there is no consensus on the nature and scope of what is usually termed numeracy for nursing, nor on ways of improving the situation. The relationship between mathematics and nursing—the scope and nature of numeracy for nursing and how to ensure that nurses are well prepared for and periodically updated on the mathematical demands of their work—is still poorly understood (Coben et al. 2008). The EIMI Discussion Document statement that 'there is a need for a fundamental analysis and reflection on strategies for the education and training of students and maybe the development of new ones' certainly applies to numeracy for nursing. This need is made more urgent by the safety–critical nature of nursing generally (Cooke 2009), particularly with respect to aspects of nursing involving mathematics (e.g. ISMP 2008). For example, nurses need to be able to calculate drug dosage, estimate a patient's fluid balance and nutritional status and interpret and act appropriately on data shown by machines used to monitor a patient's condition or dispense treatment: a mistake in any of these could be life-threatening for the patient and end the nurse's career. Through our work on numeracy for nursing, we are working to reduce this risk.

We decided to focus initially on one area of nursing responsibility that is high risk and closely associated with the use of mathematics: medication dosage. Medication errors have been highlighted recently by the National Patient Safety Agency (NPSA) in England and Wales and targeted for remedial action (NPSA 2006, 2009). The number of injuries and deaths attributable to medication error in the National Health Service (NHS) in the UK is unknown, but 9 % of incidents reported to the NPSA in its 2003 audit involved medicines (NPSA 2003), a figure consistent with historical data. The extent of calculation error encompassed within medication error is not specified in the report, but the NPSA state that:

> Miscalculation, failure to titrate the dose to the patient's needs, miscommunication and failure on the part of all team members to check the dose before dispensing, preparing or administering a dose are the most common factors contributing to dosing errors. (NPSA 2009, p. 19).

The Department of Health report on Improving Medication Safety (Smith 2004) highlighted inadequacies in the education and training of both doctors and nurses as contributory factors in medication errors. In response to growing concern, from September 2008, the body regulating the nursing profession in the UK, the Nursing and Midwifery Council (NMC), has required students to be assessed in numerical competence in the practice setting with 100 % passmark before being allowed to register as nurses (NMC 2007).

However, there are currently no national standards for teaching or assessment of numeracy during preregistration nurse education, and in the absence of a robust evidence-based standard (a benchmark), a relativistic position has emerged with a multiplicity of tests, processes and criteria being developed and deployed locally. We have recently investigated this situation with respect to the assessment of numeracy for nursing in one university in England (Coben et al. 2008).

We argue that without such a benchmark, any measure of numerical competence is:

> … in the eye of the recipient of evidence of that competence, be it higher education institutions, regulators, employers or service users. (Hutton 2004).

Given the link between the required competence and its public consumption, the nature of any benchmark requires not just an assessment method which is reliable and valid in educational terms, but one which directly and authentically represents the purpose and context in which it will be performed. Such a process cements the relationship between the desired expectation of competence in the workplace and the governance of its development and subsequent performance.

Once established, a robust assessment benchmark provides not just some assurance of baseline professional standards but for the first time an opportunity to move away from relativistic interpretations of mathematical competence in relation to nursing and explore the relationship between entry qualifications, in-programme preparation, placement experience, remedial support and subsequent achievement. Thus, it would be possible to work backwards from the benchmark to key stage indicators and the probability of achieving the required standard in the time available. Such data would be extremely powerful in supporting subsequent education. It would also allow us to look forward to subsequent stages in the nurse's career where either higher-level skills are needed or a refreshing of core skills can be facilitated by recourse to the core standard.

2 Developing an Evidence-Based Benchmark in Numeracy for Nursing

Against this background, NHS Education for Scotland (NES) brought together an interdisciplinary group of subject experts to explore the key issues associated with determining the achievement of competence in numeracy for professional practice in nursing[1]. This followed a review of relevant literature (Sabin 2001),

[1] The interdisciplinary team comprised Professor Diana Coben, Professor of Adult Numeracy, King's College London; Dr Carol Hall, Associate Professor, School of Nursing, University of Nottingham; Dr Meriel Hutton, Senior Visiting Research Fellow, King's College London; Dr David Rowe, Reader in the Department of Sport, Culture and the Arts, University of Strathclyde; Dr Keith Weeks, Reader of Health Professional Education, Faculty of Health, Sport and Science, University of Glamorgan; Norman Woolley, Head of Learning and Teaching and Associate Head of the Department of Professional Education and Service Delivery, Faculty of Health, Sport and Science, University of Glamorgan. The project is sponsored by Michael Sabin, Learning

a consultation on healthcare numeracy (NES Numeracy Working Group 2006) and the subsequent strategy developed by NES (Sabin 2006a).

The overall aim of this work is to develop a proposed benchmark assessment for numeracy for nursing in Scotland. In the first instance, we propose the establishment of a benchmark at the point at which students become registered nurses.

In our first-stage project towards this end: 'Benchmark assessment of numeracy for nursing: Medication dosage calculation at point of registration'[2] (Coben et al. 2010), funded by NHS Education for Scotland http://www.nes.scot.nhs.uk/, we aimed to evaluate empirical evidence of the reliability and convergent validity of a computer-based learning and assessment tool of medicine dosage calculations by comparing its outcomes with the outcomes of a practical activity requiring the same calculations and to determine learner-perceived acceptability of the assessment tools in relation to authenticity, relevance, fidelity and value.

Our research questions were as follows:

1. What is the internal consistency reliability of the computer-based assessment and the practice-based assessment?
2. What is the criterion-related validity of the computer-based assessment and the practice-based assessment?
3. How acceptable to nursing students are the assessments in terms of authenticity, relevance, fidelity and value?

The research began with preliminary work undertaken between November 2006 and April 2007 (reported in Coben et al. 2008). This comprised

- The adoption of a definition of numeracy applicable to nursing and capable of being operationalised in our research, as follows:

  > To be numerate means to be competent, confident, and comfortable with one's judgements on whether to use mathematics in a particular situation and if so, *what* mathematics to use, *how* to do it, *what degree* of accuracy is appropriate, and what the answer means in relation to the context. (Coben 2000, p. 35, emphasis in the original);

- The development of evidence-based principles for numeracy for nursing;
- The development of criteria for a benchmark assessment of numeracy for nursing;
- The development of an evidence-based prototype benchmark assessment tool covering practical knowledge in medicine dosage calculations.

(Footnote 1 continued)
Teaching and Assessment, Programme Director, NHS Education for Scotland, on secondment as Nursing Officer Workforce Development, Chief Nursing Officer's Directorate, Scottish Government.

[2] The project report and associated materials are online at http://www.nursingnumeracy.info/index.html. This chapter draws on this report and other publications from the work of the project listed on the Website.

Having reviewed the literature, we developed the following criteria, recognising that an authentic numeracy assessment tool should be

- Realistic: Evidence-based literature in the field of nursing numeracy (Hutton 1997; Weeks 2001) strongly supports a realistic approach to the teaching and learning of calculation skills, which in turn deserve to be tested in an authentic environment. Questions should be derived from authentic settings.
- Appropriate: The assessment tool should determine competence in the key elements of the required competence (OECD 2005; Sabin 2001).
- Differentiated: There should be an element of differentiation between the requirements for each of the branches of nursing (Hutton 1997).
- Consistent with adult numeracy principles: The assessment should be consistent with the principles of adult numeracy learning teaching and assessment, having an enablement focus (Coben 2000).
- Diagnostic: The assessment tool should provide a diagnostic element, identifying which area of competence has been achieved and which requires further intervention (Black and Wiliam 1998).
- Transparent: The assessment should be able to demonstrate a clear relationship between 'test' achievement and performance in the practice context (Weeks et al. 2001).
- Well-structured: The assessment tool should provide a unique set of questions with a consistent level of difficulty and a structured range of complexity (Hodgen and Wiliam 2006).
- Easy to administer: the assessment should provide the opportunity for rapid collation of results, error determination, diagnosis and feedback (Black and Wiliam 1998). (Coben et al. 2008, pp. 96–97).

In response to these requirements, an evidence-based benchmark computer assessment tool was designed by Weeks and Woolley to develop nurses' medication dosage calculation skills. It uses a framework derived from the Authentic World® computer program based on Gulikers et al's framework for authentic assessment (Gulikers et al. 2004). Graphics and participant interaction create close proximity to real-world practice. It provides the full range of complexity of dosage calculation problems which Adult Branch nurses are likely to meet at the point of registration. This includes unit dose, sub- and multiple-unit dose, complex problems and conversion of systeme internationale (SI) units (Weeks et al. 2001). The 28 practical simulation items were a subset of 50 computer simulation items selected on the basis of their high internal consistency reliability in an earlier study (Clochesy 2008; Weeks and Woolley 2007).

The practical (real world) simulation assessment was designed and developed to mirror the computer-based assessment.

A full account of the research design, the development of key concepts, the research process and its outcomes is given in the project report (Coben et al. 2010). For the purposes of this chapter, we report our findings, as follows.

We found high congruence between results from the two methods of assessment in the tablets and capsules section, suggesting that for the determination of calculation competence in management of this type of prescription, an authentic computer assessment is equivalent to an assessment through practice simulation.

In assessing calculation of liquid medicine doses, both the authentic computer environment and the practice environment assessments facilitated the detection of technical measurement errors associated with the selection of inappropriate measurement vehicles and measurement of incorrect liquid doses. The definition of numeracy which we have used includes competence in knowing what the transfer of a calculated answer to a technical measurement vehicle means in practice, but the computer model did not allow for technical measurement errors such as failing to displace air from syringes, an error manifested by several students in the practice environment. This was a measure of numeracy competency which has not been widely considered in the literature and was apparent in all the sections of the practice assessment which involved liquids. Some students made arithmetic or computational errors, but the majority of errors were made in practical technical measurement and were apparent regardless of the accuracy of the original calculation of dose. This element was appropriately identified in the practice simulation assessment, but would not have been detected via the computer assessment. However, this competence could be assessed as a practical skill with any prescription requiring liquid medicine without recourse to repeated measures across the range of complexity and if coupled with the authentic computer assessment would be adequately assessed. The same argument would apply to prescriptions for injections.

We conclude that for calculation of medicine dosage, the major advantage of the authentic computer environment was to provide prescriptions covering the full range of calculations likely to be met in practice. It allowed easy assessment of the mathematical element of these calculations with large numbers of students in a short time, and given that marking and feedback generation were entirely automated, the process was quick, easy and totally objective. In assessing nurses' calculation of medicine doses, an authentic computer model that presents dosage problems within an agreed rubric is invaluable in providing assessment of the full range of calculations likely to be met in practice as a newly qualified nurse.

In the practice context, it would be impossible to ensure that all third-year student nurses encountered and were reliably assessed in the full range of dosage calculation problems they might meet as a qualified nurse. However, the assessment of the numeracy element of a nurse's competence in medicine management needs to include assessment of both the full range of calculations likely to be required and the measurement vehicle manipulation and measurement skills available in most clinical settings and/or able to be simulated in a practical environment.

We propose that if used together, the assessment tools and processes identified within this report provide a robust form of assessment that meets the needs of regulators, educators, employers, practitioners, students and public in reliably identifying conceptual, calculation and technical measurement competence in the context of medicine administration.

Further, we propose that this research provides a benchmark against which other researchers and interested stakeholders can measure the impact of other innovations in learning, teaching and assessment strategies, and of recruitment, development and support/retraining strategies.

3 Summary of Our Findings

The main overall focus of this study was to determine the validity of the computer simulation format of delivering dosage calculation problems. The validity of the computer simulation format was tested against the gold standard practical simulation format. The underlying rationale was that the practical simulation format is not feasible for mass testing, particularly across the full range of question type (tablets, liquid medicines, injections, etc.) and complexity (single unit, multiple unit, etc), whereas computer simulation enables testing across the range of question type and complexity of large numbers of people at the same time, in remote locations, with limited costs. If computer-simulated testing could be shown to operate similarly (provide similar results) to practical simulation testing, this would validate the use of computer-simulated testing for future research.

From the results presented above, the criterion-related validity of the computer simulation format has been supported, both in terms of putting participants in a similar order of competence and in terms of participants obtaining similar absolute results (getting the same number of questions correct on the computer simulation as they would on the practical simulation). These results supplement Weeks' (2001) more detailed item-by-item comparisons, which produced similar results in terms of confidence in the computer-simulated testing format as a substitute method for practical assessment.

Some caveats remain, however,

- Computer simulation does not test certain elements of the real-world dosage calculation problem (e.g. technical competency).
- These conclusions should only be applied to similar situations, populations and constructs.

4 Conclusion and Thoughts on Wider Implications for the Educational Interface Between Industry and Mathematics

In conclusion, we offer this account of our ongoing research as a contribution to wider discussions on the educational interface between industry and mathematics, in the form of numeracy for nursing. Though mindful of the second caveat above, we think that there is at least one lesson to be learned from our research for other fields (including other healthcare fields) in which mathematics plays a key part. This is the prime importance of authentic teaching, learning and assessment of mathematics for industrial purposes. Where mathematics is situated in professional/vocational practice, it should be taught, learned and assessed in relation to that practice, both directly in practice and through authentic and comprehensive simulation of practice; the latter enables individuals to be exposed to the full range of problems associated with the use of mathematics in their professional practice, something which may be impossible to do safely, comprehensively and effectively in real-world, real-time contexts.

References

Black, P., & Wiliam, D. (1998). Assessment and classroom learning. *Assess in Educ, 5*(1), 7–74.

Clochesy, J. M. (2008). Improving medication dosage calculation skills among undergraduate nursing students. *Simul in Healthc, 3*(4), 0.

Coben, D. (2000). Numeracy, mathematics and adult learning. In I. Gal (Ed.), *Adult numeracy development: Theory, research, practice* (pp. 33–50). Cresskill, NJ: Hampton Press.

Coben, D., Hall, C., Hutton, B. M., Rowe, D., Sabin, M., & Weeks, K., et al. (2008). Numeracy for nursing: The case for a benchmark. In T. Maguire, N. Colleran, O. Gill & J. O' Donoghue (Eds.), *The Changing Face of Adults Mathematics Education*: Learning from the past, planning for the future. *Proceedings of ALM-14, the 14th International Conference of Adults Learning Mathematics—A Research Forum (ALM)*, held at the University of Limerick, Ireland, 26–29 June, 2007 (pp. 88–102). Limerick: University of Limerick in association with ALM.

Coben, D., Hall, C., Hutton, B. M., Rowe, D., Sabin, M., Weeks, K., et al. (2010). *Research Report. Benchmark Assessment of Numeracy for Nursing: Medication dosage calculation at point of registration*. Edinburgh: NHS Education for Scotland.

Coben, D., Hodgen, J., Hutton, B. M., & Ogston-Tuck, S. (2008b). High stakes: assessing numeracy for nursing. *Adult Learn, 19*(3–4), 38–41.

Cooke, H. (2009). Theories of risk and safety: what is their relevance to nursing? *J of Nurs Manag, 17*(2), 256–264.

Gulikers, J. T. J., Bastiaens, T. J., & Kirschner, P. A. (2004). A five-dimensional framework for authentic assessment. *Educ Tech Res Dev, 52*(3), 67–85.

Hall, C. (2006) A third of new nurses fail simple English and maths test. Daily Telegraph. Retrieved February 13, 2010 from http://www.telegraph.co.uk/news/main.jhtml?xml=/news/2006/08/05/nhospital05.xml

Hodgen, J., & Wiliam, D. (2006). Mathematics inside the black box: Assessment for learning in the mathematics classroom. London: nferNelson.

Hutton, B. M. (1997). *The acquisition of competency in nursing mathematics*. Birmingham: Unpublished PhD, University of Birmingham.

Hutton, B. M. (2004). *What do we mean by competency in calculation*? chapter presented at the A Calculated Risk: Numeracy needs in Healthcare. Retrieved 20 March, 2009 from http://www.nes.scot.nhs.uk/multi/numeracy/documents/Meriel.ppt.

ISMP (2008). Seven lapses add up to tragedy. [ISMP Medication Safety Alert]. Nurse Advise-ERR, 6(5), 1–2.

NES Numeracy Working Group. (2006). *Identifying and Supporting the Numeracy Needs of Healthcare Staff in Scotland (Consultation)*. Edinburgh: NES.

NMC (2007). Essential Skills Clusters (ESCs) for Pre-registration Nursing Programmes. Annexe 2 to NMC Circular 07/2007. London: Nursing and Midwifery Council.

NPSA. (2003). *A report on the Pilot Data Audit undertaken by the NPSA NPSA Business Plan 2003–04*. London: National Patient Safety Agency.

NPSA (2006). *Safety in Doses: Medication safety incidents in the NHS. The fourth report from the Patient Safety Observatory (No. PSO/4)*. London: National Patient Safety Agency.

NPSA. (2009). *Safety in Doses: Improving the use of medicines in the NHS*. London: NPSA.

OECD. (2005). *The Definition and Selection of Key Competencies*. Paris: Organisation for Economic Cooperation and Development. Executive Summary.

Sabin, M. (2001). *Competence in practice-based calculation: Issues for nursing education. A critical review of the literature*. Occasional chapter 3. London: Learning and Teaching Support Network (LTSN) Centre for Health Sciences and Practice.

Sabin, M. (2006a). *Identifying and supporting the numeracy needs of healthcare staff in Scotland*. A Strategy document Edinburgh: NHS Education for Scotland.

Sabin, M. (2006b). *NHS education for Scotland: Supporting numeracy skills in the Scottish healthcare workforce* Retrieved August 25, 2007 from http://www.nes.scot.nhs.uk/multi/documents/Numeracy_NES_Update.pdf.

Smith, J. (2004). *Building a safer NHS for patients—Improving medication safety*. London: Department of Health Publications.

Weeks, K. W. (2001). Setting a foundation for the development of medication dosage calculation problem solving skills among novice nursing students. The role of constructivist learning approaches and a computer based 'Authentic World' learning environment. Unpublished PhD, University of Glamorgan, Pontypridd.

Weeks, K. W., Lyne, P., Mosely, L., & Torrance, C. (2001). The strive for clinical effectiveness in medication dosage calculation problem solving skills: The role of constructivist theory in the design of a computer-based 'authentic world' learning environment. *Clin Effectiveness in Nurs, 5*(1), 18–25.

Weeks, K. W. & Woolley, N. N. (2007). Innovative and Authentic Assessment of Medication Dosage Calculation Problem Solving Skills, via Engagement and Interaction with an Authentic World Learning Environment. LTO Innovation Grant 2005–06 Report. Pontypridd: University of Glamorgan.

Linking Professional Experiences with Academic Knowledge: The Construction of Statistical Concepts by Sale Manager Apprentices

Corinne Hahn

As it builds a partnership between the school and the firm, the French alternance system helps us to link mathematics to students' professional experience. In this paper we will describe a pedagogical device we experimented with manager apprentices in order to make them confront the different conceptualisations they built through their multiple experiences, at school and at work.

1 To Learn Between School and Workplace

A major well-known difficulty in vocational education is to help students to link professional experience to theory learned at school (see for example Hahn 2000). As the constitutive role of cultural practices on cognition is now widely recognised (Hatano and Wertsch 2001), in order to enhance learning the aim is to confront students with epistemologically rich problems. These problems should be not only inspired by 'real' situations but also familiar, part of their field of experience (Boero and Douek 2008), and related to the community of practice (Lave and Wenger 1991) at work.

However, linking disciplinary knowledge with work practices is not an easy task as work situations are always multidisciplinary. This is a major problem encountered by teachers, especially in higher education where the use of sophisticated technology makes mathematics mostly invisible (Strässer 2000). Although referring to different theoretical frameworks, some authors agree that learning appears through a dialectical process—between conceptualisations in action, embedded in the setting in which they occur and theories or "scientific" concepts—whether they stress the continuity between them (Noss and Hoyles 2000) or the discontinuity (Pastré et al. 2006). A dialectical learning process implies the

C. Hahn (✉)
ESCP Europe, Paris, France
e-mail: hahn@escpeurope.eu

A. Damlamian et al. (eds.), *Educational Interfaces between Mathematics and Industry*, New ICMI Study Series 16, DOI: 10.1007/978-3-319-02270-3_12,

construction of an internal space where knowledge of different levels of generalisation play/work/compete together (Brossard 2008). From our point of view, this implies that the learner should be involved in the construction of the problem. But, on the other hand, how can we be sure that a problem enacted from the learner's personal experience would fit with the school's aims and help the learner to construct the knowledge that she/he is supposed to learn? In vocational and professional school education, curricula often force teachers to follow a prescribed pathway, leaving little room for such activities, when in the field of adult education, the problem differs: the pressure of curriculum is less important and learners usually already have work experience.

2 French Alternance System

In this system, 16 to 26-year-old students sign an apprenticeship contract with a firm and study part time at school in order to prepare a vocational degree. This concerns all types of qualifications from lower levels up to highly skilled jobs such as engineers and managers. As it builds a partnership between the school and the firm, this system encourages us to consider the relationship between them. These two organisations share the same goal of educating the new generation of professionals, although their cultures and the way they consider knowledge are so different that it is a real challenge to help the learner make productive associations between them.

We have been studying the positive effects of the French 'alternance' system on learning for some years now (Hahn 2000; Hahn et al. 2005). As adult education, it offers the opportunity to link school content to students' professional experience, and this system creates a two-way relationship, from school to firm and from firm to school. Alternance may be described as a boundary space, an in-between place where sense making should become easier.

3 Managers and Statistics

Vocational curricula often include statistics because being able to handle data is an important competency in many workplaces and in everyday life. Statistics is supposed to be more easily linked to out-of-school practices. Nevertheless, when we work with our postgraduate business students on problems they have constructed, these problems rarely include a statistical dimension, not even a basic one. In fact, most decisions can be made without considering any statistical methods. This is what usually happens in the field, although statistics certainly provide much insight into many corporate questions and issues (Dassonville and Hahn 2002).

The question is not only to help professionals to improve their understanding of the statistical tools they use in the workplace (see for example Noss et al. 2002; Bakker et al. 2008) but also help them to "see" the statistics that could be useful. Therefore, we need to "enculturate" students into statistical reasoning (Pfannkuch 2005) so that, as managers, they will be able to improve their decision-making processes by using statistical methods.

The decision-making process is not only a question of processing information and finding patterns in observed data. Our rationality is shaped by the values and beliefs that are raised through our participation in different communities and of which we are mostly unaware. Not all of this tacit knowledge can be codified and it shapes not only the means but also the evaluation of ends (Polanyi 1966). We must also consider that scientific rationality as it is developed at school—i.e. explicit logical reasoning—co-exists with other social forms of rationality.

The way the learner solves a problem depends on what the problem means to her/him. The aim of the experiment we will now describe was to find out how these different forms of rationality shaped a statistical decision-making problem and the use of statistical concepts by students.

4 The Experiment

4.1 The Device

We designed a 4-step pedagogical device focussing on the concept of variation. This concept is central in management (e.g., to consider investments' volatility in finance, segmentation in marketing, etc.), as in many other fields, and it is claimed that it is very hard to deal with at any age or level (Garfield and Ben-Zvi 2005).

We had planned to study how the device mediated the construction of statistical concepts and how students' personal experiences shaped their decisions. The device was based on a mini case study about a firm ("T" sells office equipment) that is hiring a sales manager. Students were asked to choose which of the three sales areas they would prefer to manage. They had to make their decision according to information they were given on a group of customers (businesses) located in each area (different group sizes in each area). This information consisted of an Excel File with a set comprised of one qualitative variable (date of first purchase) and five quantitative variables: previous year's amount of sales (with the client), distance (from the client to "T" location), staff (of the client), evaluation of commercial relation (a grade from 0 to 10) and number of different items (sold to the client last year).

First, each student was provided individually with the distribution of one variable from one sales area, and was subsequently asked to write a brief summary of the information he or she received (step 1). Next, we formed groups of three students, with each of them having studied the same variable in a different area, and we asked each group to summarise the information it had received by comparing the three distributions of the same variable in the three different samples (step 2). Therefore, they were able to consider two types of variability, within a group and between groups (Garfield and Ben-Zvi 2005). Then, we built new groups of six students, each of them having different information about one variable (among 6) in all three sales areas[1] (step 3). Last, we asked the students (in groups of 3, as in step 2) to make a final decision about the area they would choose by analysing all available data simultaneously (step 4).

We assumed that step 1 and 2 were closer to school practice and step 3 and 4 closer to professional practice (step 3 was more typically a situation faced by a salesperson, step 4 a situation faced by a manager). We also expected that passing from step 1–2 but also from step 3–4, would lead students to move from a local (data seen as a collection of individuals) to a global point of view and thus to the construction of the concept of distribution (Makar and Confrey 2005).

4.2 The Population

The device was first tested with 36 postgraduate students engaged in a 3-year master's level program including several periods of internship. Most of these students (n = 34) completed a 2-year commerce degree prior to entering the masters program. They all had previously undertaken at least a basic statistic course and had work experience, most of them as a salesperson.

Two questionnaires were given to the students at the beginning of the school year by the teacher in charge of the Business course. The first questionnaire focused on their former experiences and professional project, the second on their statistical knowledge: given a list of statistical concepts, they were asked if they had learned these at school and if they knew how to use them.

The experiment took place during the two first sessions (3 h each) of a compulsory statistics course during the first year. We split the group into two subgroups located in two different classrooms. Each subgroup followed the same procedure. At each step, the students were able to use their personal calculator or computer.

The debates between students during steps 2, 3 and 4 were audio-taped; in addition, we took field notes and collected reports written by the students at each step. Students were told that we wanted to keep track of the discussions in order to help to adapt the course to their needs, what I actually did. They were allowed to stop the recorder if they wanted. Some did occasionally during breaks.

[1] They were not able to use what they had done at step 2.

Table 1 Answers to the questionnaire and observations made at step 1

	Questionnaire					Step 1	
	Learned at school	Met out-of-school	Know how to calculate	Know how to get the result from spreadsheet	Know how to use it out-of-school	Calculated (without mistake)	Calculated (wrongly)
Average	30	16	16	20	18	24	2
Median	18	5	6	7	6	10	5
Standard deviation	24	2	1	4	2	1	3

5 Results and Discussion

We will describe some results concerning the quantitative variables (30 students studied these variables, 6 per variable at the first step). Here, we will mostly focus on the use of the mean, median and standard deviation. Table 1 compares, for 30 students out of 36,[2] answers to the questionnaire (what students claim to know) and observations we made at step 1. What students claimed seems coherent with they were able to do—although they seem to underestimate their capacity to calculate a mean and a median.

Although very few students calculated or used variation indicators, many of them expressed an intuitive conception of variation as evidenced by their references to the shape of distribution. As we suspected, as it is a natural process (Hammerman and Rubin 2004), many students divided the data into subgroups. Nevertheless, we found two different types of strategies. At this stage, 19 students out of 30 built subgroups based on the distribution (use of mean or median, of discontinuities in the data set) and 10 built subgroups referring to a "social norm": the decimal system (hundreds), economic typology (size of firms) or a "school norm" (a good grade must be 5 or above).

Among the hypotheses formulated from our literature review, we forecasted that the students would refer more to school knowledge at step 1 and 2 than at step 3. Indeed, at step 1, many students tried to apply the statistical knowledge learned at school and calculated as many indicators as they could. Nevertheless, many of them already integrated elements of their commercial experience at this stage. Strategies seemed to depend on context and not only on the distribution of numbers: similar strategies were used for the same variable (for example all students who dealt with sales and distance calculated percentages for subgroups). We mostly found references to the context for sales and distance (most important for a salesperson, according to our interviews with professionals).

Our second hypothesis was that steps 2 and 4 would help them to move from a local to a global conception—in particular by using multiplicative strategies

[2] The six remaining students dealt with the qualitative variable.

(as sample sizes were different). That was obvious in step 2: all students who had made lists or ranking of customers abandoned them. This seems to indicate a shift to a global point of view, although they used few indicators: they kept indicators when they were able to agree on a common interpretation. They dropped indicators that they could not make sense of. This is coherent with previous observations that students have difficulties with spontaneous use of indicators (Konold and Pollatsek 2002) and, when they calculate indicators, they do not use common sense in solving the problems (Bakker 2004).

Here is an extract of the discussion in one of the two groups dealing with grade:

Student	You did not calculate the average[3] for your area? For your 20 customers, how many?
S2	I told you that there were 10 (customers who gave a grade under 5) out of 20
S1	Yes, but the total average?
S2	But I told you, it is 10
S1	But the average grade, how much?
S2	I told you
S1	You did not calculate it
S3	The addition of grades divided by the number of grades
S2	Oh this, I did not do it
S3	The average is 6.76 in my area
S3	Is this good or not?
S3	This is not so simple … the average is 6.76 …
S2	But how many have a grade above 5, this I am sure you did not do it?
S3	No, I did not
S1	In my area, there are 31 customers, the general average is 5
S2	Exactly 5?
S1	Yes those whose business relationship is under 5 are 13, that represents 42 %, those whose relationship is above 5 are 18, that represents 58 %, then the end result is positive but not good enough
S3	In my area, 11 customers reach average 5, then we must improve commercial relationship and try to find out during appointments what they really need and adapt commercial policy to improve their satisfaction. (…)
S1	Standard deviation that means the repartition of grades, if they are close or far, is 2,006
S3	That's nice what you did, because me I did not think about standard deviation, no I did not think about it but, standard deviation uh
S2	Better not think about it because there is no use for it

[3] In French there is only one word for 'mean' and 'average', and we decide to translate it by 'average' in the dialogue.

We could claim that the three students were at different stages of understanding but it seems to us that they are not solving the same problem. The problem is shaped by the objective they set themselves: student 1 seems to plan a comparative study in order to understand differences between areas, student 2 is solving a school problem to answer the teacher's request and student 3 wants to answer the question "which is the best area?" at this stage already. Then of course strategies differ. Student 1 referred to what seems to be a scientific rationality: to compare distribution and use statistical concepts. Student 2's actions are based on technical rationality (Schön 1996): he applied techniques learned at school but does not know how he can use the result to answer a question. Student 3's reasoning is pragmatic and he used a simple intuitive strategy.

The second group dealing with grade followed a similar path, although they all calculated the mean at step 1[4]:

S1 For area A, the mean is 4.3
S2 Forme, 5
S3 And for me 6.76
S2 Let me do the calculation again … yes it is 5, Ok
S3 Ok, then what do you want to do?
S1 The number of customers under 5
S2 It is important?
S2 Very important
S1 But you must understand! There are 20 customers in my area, among them, 10 are not satisfied at all
S2 Ok
S3 They do not reach average
S1 Not average = not satisfied
S2 Ok
S1 Then this is global. (…)
S2 (*he gives the standard deviation of the distribution of grades in sector A explains how to calculate it and calculate it for sectors B and C with the spreadsheet as the two others did not do it*)
S1 What is the need finally?
S2 See, we found about the same value
S1 Yes because we do a calculation but we do not know what it is for? If you got a big standard deviation, what does it mean?
S2 (*he kept silent*)

We can see that, finally, the same procedure is chosen by both groups: consider the number of values under 5 and above 5 and do not compare the means. They used an anchoring strategy (Tversky and Kahneman 1974), comparing the values to midrange 5 which they called 'average': it is 'global', as all data can be compared to

[4] I numbered the students from 1 to 3 in both groups but of course they represent different students.

Table 2 Standard deviation for two variables

10 groups	Sales	Staff	Items	Distance	Grade
Average	3	3	4	4	3
Median	1	1	2	2	1
Standard deviation		2	2		
Use of effective commercial context	5			5	5

it when the other means are only 'local'. It is known that students have difficulties with considering the mean as a good representation of a distribution (Konold and Pollatsek 2002). But when both groups dealing with sales compared means, only the groups studying grade used the midrange. This is probably connected with the magnitude of grade values. But we suspect that is also linked to a strong social/school norm in France: 'to obtain average'[5] means 'reach midrange value' which is usually the passing grade. If we refer to Vergnaud's theory of conceptual fields (Vergnaud 1990), it seems that students built a concept-in-action 'average as middle value' linked with the theorem-in-action 'a grade above 5 is good' and activated the scheme 'compare number of data above and under 5'.

In both groups, students who referred to technical rationality switched easily to the strategy of comparison to 5, while students who wanted to study the distribution more deeply (student 1 in group 1 and student 2 in group 2) seemed more reluctant.

At step 3, when groups of six students had to draw a conclusion about one area from their individual study of each of the variables, we noticed that the use of indicators was less frequent and, as in step 2, dependent on the context (see Table 2). They indicated standard deviation for two variables only, those for whom we found no occurrence of commercial comments. It seems that they calculated this indicator for variables that made no sense for them to consider the context. One group went back to a local strategy by numbering customers.

At step 4, when groups of three were supposed to decide on the area they would like to manage, we found almost no occurrence of the use of indicators: only two groups (out of 10) used in their argumentation the average for distance and grade. Their argumentation was based on commercial arguments and mostly built on comparison of percentages within sub-groups. Considering the references to the context, sometimes even at step 1, it seems that students very quickly built a representation of the problem as a commercial problem. As they moved forward in the experiment, they left behind their statistical knowledge. They found it not practical enough for the objectives they had set. Pragmatic rationality prevailed.

The difficulty in moving from a local to a global point of view seems somewhere to reflect the difficulty in moving from a salesperson identity to that of a sales manager, because a salesperson deals with his customers more individually. And this implies the need to mobilise more statistical knowledge.

[5] Translated from the French "avoir la moyenne".

6 Conclusion

We used this experiment as a basis for the statistics course that year. The course was unusually successful. Of course, we could not evaluate on a scientific basis whether the experiment helped the students to change their views about statistics, but many of them mentioned in the evaluation that they became more aware of the utility of statistical methods.

This experiment is going to be extended to a larger population through a Computer-Supported Collaborative Learning system based on the same device (we have added four variables to the file). We plan to use this system with students from our five European campuses.

References

Bakker, A. (2004). Reasoning about shape as a pattern in variability. *Statistics Education Research Journal, 3*(2), 64–83.

Bakker, A., Kent, P., Derry, J., Noss, R., & Hoyles, C. (2008). Statistical inference at work: Statistical process control as an example. *Statistics Education Research Journal, 7*(2), 13–145.

Boero, P., Douek, N. (2008). La didactique des domaines d'expérience. *Carrefour de l'Education 26*:99–114.

Brossard, M. (2008). Concepts quotidiens/concepts scientifiques: réflexions sur une hypothèse de travail. *Carrefours de l'Education, 26*, 67–82.

Dassonville, P., Hahn, C. (2002). Statistics education for future managers: Needs obstacles et possible solutions. *ICOTS 6 Proceedings*.

Garfield, J., & Ben-Zvi, D. (2005). A framework for teaching and assessing reasoning about variability. *Statistics Education Research Journal, 4*(1), 92–99.

Hahn, C. (2000). Teaching mathematics to apprentices. Exploring content and didactical situations. In A. Bessot & J. Ridgeway (Eds.), *Education for mathematics in the workplace*. Netherlands: Kluwer.

Hahn, C., Besson, M., Collin, B., & Geay, A. (2005). *L'alternance dans l'enseignement supérieur, enjeux et perspectives*. Paris: L'Harmattan.

Hammerman, J., & Rubin, A. (2004). Strategies for managing statistical complexity with new software tools. *Statistics Education Research Journal, 3*(2), 17–41.

Hatano, G., & Wertsch, J. (2001). Sociocultural approaches to cognitive development: The constitution of culture in mind. *Human Development, 44*, 77–83.

Konold, C., & Pollatsek, A. (2002). Data analysis as the search for signal in noisy process, *Journal for research in mathematics education, 33*(4), 259–289.

Lave, J., & Wenger, E. (1991). *Situated Learning: Legitimate peripherical participation*. Cambridge: Cambridge University Press.

Makar, K., & Confrey, J. (2005). "Variation-talk": articulating meanings in statistics, *Statistics Education Research Journal 4*(1), 27–54.

Noss, R., Hoyles, C., & Pozzi. (2000). Working knowledge: mathematics in use, In: *Education for Mathematics in the Workplace*, Kluwer, Netherland: Annie Bessot and Jim Ridgeway publishing.

Pastré, P., Mayen, P., & Vergnaud G. (2006). La didactique professionnelle, *Revue Française de Pédagogie, 154*, 145–198.

Pfannkuch, M. (2005). Thinking tools and variation, *Statistics Education Research Journal 4*(1), 83–91.

Polanyi, M. (1966). *The Tacit dimension*. Doubleday, N.Y.
Schön, D.A. (1996). A la recherche d'une nouvelle épistémologie de la pratique et de ce qu'elle implique pour l'éducation des adultes. In J.M. Barbier (Ed.), *Savoirs théoriques et savoirs d'action* (pp.201–222). Paris: PUF.
Strässer, R. (2000). Conclusion. Bessot, A. & Ridgeway, J. (Eds.), *Education for Mathematics in the Workplace* (pp. 241-246). Kluwer.
Tversky & Kahneman. (1974). Judgement under uncertainty, heuristics and biases. *Sciences*, *185*, 1124–1130.
Vergnaud, G. (1990). La théorie des champs conceptuels. *Recherches en Didactique des Mathématiques*, *10*(2-3), 133–170.

Learning Conversation in Mathematics Practice School–Industry Partnerships as Arena for Teacher Education

Gert Monstad Hana, Ragnhild Hansen, Marit Johnsen-Høines, Inger Elin Lilland and Toril Eskeland Rangnes

The on-going research project learning conversation in mathematics practice (LCMP)[1] focuses on communication and learning in the field of mathematics. Its goal is to develop the notion of learning conversation as a didactical concept and tool for describing and facilitating learning processes. The project collects research data from schools that have established partnerships with industrial[2] companies. An aim of these partnerships is for pupils[3] to learn mathematics through experiencing and discussing how mathematics is applied and used at work. This paper gives two examples of research areas within the LCMP-project that are made possible or enhanced by the school–industry partnerships. These examples are connected with several of the themes outlined in the Discussion Document of the

---

[1] LCMP is financed by the Research Council of Norway (NFR) and Bergen University College. LCMP is part of the research consortium Teaching Better Mathematics, which consists of mathematics educators from University of Agder, Bergen University College, Bodø University College, Oslo University College and Sør-Trøndelag University College. LCMP is lead by Marit Johnsen-Høines. Webpage (in Norwegian): Læringssamtalen i matematikkfagets praksis (LIMP), http://www.hib.no/fou/limp/.

[2] Industry is in this paper broadly interpreted to include different types of workplaces where mathematics is used. Different types of companies participate, varying from oil related mechanic industry to shops.

[3] In this paper, children in school are referred to as pupils and student teachers as students.

G. M. Hana (✉) · R. Hansen · M. Johnsen-Høines · I. E. Lilland · T. E. Rangnes
Bergen University College, Bergen, Norway
e-mail: gert.monstad.hana@hib.no

R. Hansen
e-mail: rhan@hib.no

M. Johnsen-Høines
e-mail: marit.johnsen-hoines@hib.no

I. E. Lilland
e-mail: inger.elin.lilland@hib.no

T. E. Rangnes
e-mail: toril.eskeland.rangnes@hib.no

A. Damlamian et al. (eds.), *Educational Interfaces between Mathematics and Industry*, New ICMI Study Series 16, DOI: 10.1007/978-3-319-02270-3_13,

EIMI-study, including teacher training, examples of practice, modeling, and issues related to communication and collaboration. How the school–industry contexts influence students will be a main theme. We want to elaborate on the research context and the questions and discussions that are potential as part of the LCMP-project. We do not aim to give answers to these questions in this paper or at this stage.

A school development initiative "Real-life Education"[4] enables the LCMP-project to study industry as part of the learning environment of pupils and students. In this development initiative, lower-secondary schools have established school–industry partnership agreements. This agreement entails that the learning and teaching of mathematics is organized sequentially in industrial and school environments. Students from Bergen University College participate in the "Real-life Education"-initiative through their practice teaching[5] The experiences gained, by both pupils and students, through the "Reallife Education"-initiative are conjectured to impact their attitude toward mathematics and the way mathematics is communicated. The LCMP-project collects and analyses data from three different layers[6] (Johnsen-Høines in print):

- The school development initiative, where the research focus is on pupils' ability to communicate and learn mathematics.
- The professional development of student teachers engaging in the school development initiative, where the research focus is on the students' communication related to their professional development as mathematics teachers.
- The collaboration between didacticians, school teachers and students, where the research focus is on the communicative learning processes that develop between members of the learning community.

Central to the project is the passage between different spaces of learning. This can be illustrated by learning loops that depict different contexts where pupils and students participate (Johnsen-Høines 2009). Pupils and students participate in the different practices and move between them. Learning this involves an object of study in the project (cf. Lave 1999) and Dreier (1999), who argue that learning is development through participating in and between different practices) (Figs. 1 and 2).

---

[4] In Norwegian: *praksisnær undervisning*. The initiative is administrated by Gode Sirklar AS (http://www.godesirklar.no/).

[5] I. e. practicum (student teaching), which is an integrated part of their study in mathematics/ mathematics education.

[6] Empirical data is collected from teaching and learning sequences in the mathematics classroom and in pre-and post-classroom discussions. Conversations in which mathematical and didactical issues are discussed are recorded and transcribed. The data collected from teaching and learning sequences and from the collaborative communications in which students, teachers and didacticians participate are analyzed. Students and teachers take part in discussing some of these analyses.

Fig. 1 Learning loop—pupils

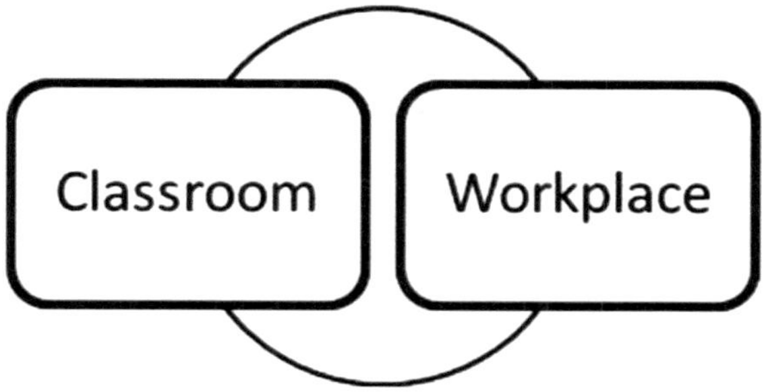

Fig. 2 Learning loop—students

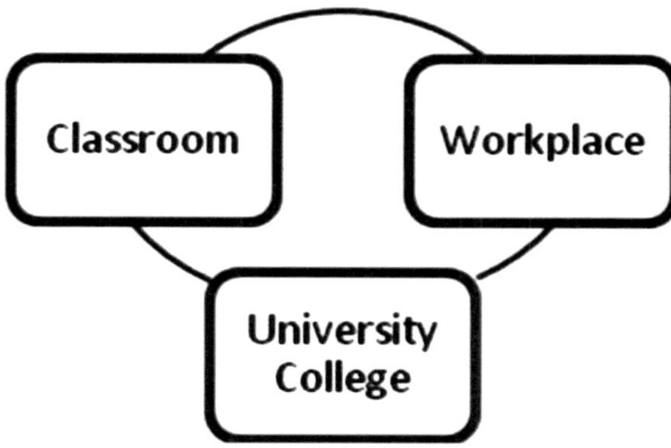

We investigate what kind of meta-learning the pupils' learning loop facilitates. We envision that pupils and students can develop a meta-reflection where the differences between mathematics at a workplace and in the classroom are problematized and reflected on. For example, the tools used by the company for solving mathematical problems (e.g., different technological tools) can be in tension with the school's intentions of what should be learned and how. In other cases, can different use of language and tools be an enriching source for new learning? It is of interest to study how the pupils' knowledge about the company and the company's praxis characterize the conversation on mathematical topics and how the pupils' experiences from school mathematics influence their conversation when visiting the company.

The intentionality inherent in learning and teaching is influenced by the resources of intention available. The workplace context makes resources of intention available that are not necessarily available in the school context. Alrø and Skovsmose (2004, 154163) considers intention as basic to participation in a dialog. As such, the workplace context may enable involvement in dialogs for which the school context, for some pupils, does not give the necessary resources of intention. The intentions-in-learning and intentions-in-teaching of students having their practice teaching connected to a school-industry partnership can also be influenced in a likewise manner. In the first example, it is shown how the context influences the intentions-in-teaching of a task given by students to pupils.

The functionality of knowledge is essential for its use and application. Our consideration of the "functionality of knowledge" is inspired by Mellin-Olsen (1989) and Skovsmose (1994). The contexts given by the school–industry partnerships enable a widening of pupils and students functional understanding.[7] Experiencing mathematics functioning in the industry context will hopefully develop their understanding and view of specific mathematical topics, as well as of mathematics as a whole.

1 Example: Workplace Assignment

The group of four students, a school teacher, and a didactician are preparing for the next teaching period. The school had a partnership agreement with a company producing valves for the oil industry. The students decided on statistics to be the mathematical focus for the pupils. As the students anticipated the necessity of the pupils having some rudimentary knowledge of statistics before they met the people at the company, the students developed a short module containing what they considered to be the most fundamental statistical concepts. However, they did not succeed as desired, and commented that "the pupils did not get engaged, they did not show interest in the learning even though they were told that these concepts would be helpful when working in contact with the industry." The discussion that followed made it clear that the students wanted the pupils to "work on issues they saw as important," they wanted them "to experience knowledge as useful and necessary," and they wanted to organize the activities in a way that "stimulated the pupils' independence and motivation." These discussions are to be seen as a background when the students decided to challenge the company leaders to request the pupils to carry out an assignment. "The pupils should be requested to do a job that the company really needs done," the students said.

The director welcomed the pupils to their first visit at the company by saying: "Statistics is the Alpha and Omega for us, we can just close down if we buy too little metal of the different sorts needed for producing the valves that are ordered or are expected to be ordered. We must have an overview of our production and stock. […] We have expanded so fast and we really have not developed good enough systems." The manager addressed one group of pupils by saying: "We want you to help us!"[8] He gave them a register of unsorted orders.[9]

[7] De Villiers (1994) defines functional understanding as "the role, function or value of a specific mathematical concept or of a particular process".

[8] The director's voice acts here as reference for the pupils on the given tasks. The pupils may interpret the challenge given as referring to statistics as important knowledge and an industrial tool, as referring to the importance of this specific job being done and that the director believes that the pupils are capable to do it, and/or as referring to the importance of being done diligently.

The pupils worked on this assignment in the classroom. The register seemed cryptic and chaotic for the pupils as well as for the students (and for the teacher and the didacticians). The pupils needed to understand the codes used for labeling different metals and products (i.e., valves) made, and contacted the company several times in order to get more information. Additionally, tools for sorting were needed. The students were used as experts when the pupils learned to use Excel to sort. The pupils made new registers that gave a better overview of the register.

The student teachers decided that the pupils were to be viewed as a "consultancy firm".[10] The pupils should ask for information when this was needed, they should present the results when the work was done, and they should challenge the company to discuss consequences if they saw any. The company cooperated on these ideas, as can be seen by them informing the pupils that they would pay for the job—if it was well done. The pupils invited people from the company to a meeting at the school, and presented their result. They told that they had experienced the importance of accuracy, and gave examples of choices they had made in order to obtain an overview they predicted to be sufficient. They discussed the result, and addressed the company by asking critical questions. The presentation and discussion was seen as important in a preperspective: The fact that this was part of the task given gave motivation for and direction to the work. However, it also turned out to be evident that the presentation-session became important for the pupils' and the students' subsequent learning. It became a stage to look back at and make reference to, as well as for bringing about reflections for later work. "What have we learnt" became a meaningful utterance to inquire into.

The teaching and learning sequences referred to above illustrate the context in which the LCMP research is situated, and it gives the background to briefly consider some themes of research interest connected with this example. In this paper, we restrict ourselves to pointing out a few themes that we have found significant in the chosen example: intentionality, functionality, and empowerment (Mündigkeit). The conversations in the initial module on statistics that was given to the pupils appeared to be very different from those in the further continuation. A central issue for the group to inquire into is then: How can the communication be characterized in these situations, and how do the differences emerge? Further discussions deal with investigating the conditions: What influences the conversations and the learning, what are the conditions for developing different sorts of communication in the context of learning?

(Footnote 8 continued)
The director's approach can be seen as influencing the intentions-in-learning and intentions-in-teaching by the way it puts the tasks into the discussion on necessity and functionality.

[9] Different groups of pupils got different challenges. Each student teacher collaborated mainly with one group.

[10] The students defining the consultancy firm role can be seen in the context of note 8, referring to functionality and meaningfulness, and is connected to the students' discussions about ownership and empowerment of pupils using mathematics (intentions-in-teaching).

Intentionality. By studying the learning/teaching sequences and the collaborative communications within the LCMP-group the intentions-in-teaching of the students become evident. They want to[11] organize for the pupils to be enthusiastic learners; they want the pupils to work on issues that they see as "real" and important; they want the pupils to experience mathematics as essential knowledge; they want the pupils to be independent and to see the importance of accuracy. The intentions of the students developed during the process. This development can partly be seen as interplay with the intentions of the manager, the pupils, the teacher, and the researcher. The way we are analyzing students' intention here, it can be characterized as intentions-in-teaching. But we find that the students are describing their intentions as intentions-in-learning as well. They see themselves as learners, of this being part of their professionalization as mathematics teachers.

Functionality. The students have inquired into "functionality of knowledge" as part of their study in mathematics education. The discussions in practice challenge the concept of functionality from different angles. The participants are questioning why and how statistics (and also tools such as Excel) is vital as a tool for the industry and for the students' and the pupils' learning of mathematics. The functionality of theories and didactical models is also discussed in the context of developing the teaching-and-learning sequences, as part of the students' professionalization as mathematics teachers.

Empowerment (Mündigkeit). Communication underpins the project; it defines the research focus and is seen as a tool for developing the collaboration and the teaching and learning interactions. Referring to "inquiry co-operation," as developed by Alrø and Skovsmose (2004), it is vital that participants see contributions by themselves and others as important for the community. Thus, it becomes relevant to organize for and study the students' development as independent, responsible, and collaborative partners. The development of the pupils' empowerment as collaborative learners is seen as important as well, and this is articulated as one of the students' main interests. The collaborative inquiries focused on the empowerment of pupils and of students in the light of one another. The data includes such meta-conversations.

---

[11] Our interpretation "they want to" is based on transcripts from conversations where the students, together with other group members, are inquiring into what they want to try out in their teaching practice and their rationality for making their choices.

2 Example: Mathematical Modeling

Another theme we study is how mathematical modeling can be integrated into education, both in primary schools and in teacher education. Of special interest is to study how students and pupils develop critical democratic competence[12] in interactions with such models.

Mathematical modeling is a theme for the students in their course work at the university college. Together with students, we have tried to include mathematical modeling in the learning loop of the pupils. This also makes mathematical modeling a theme for the students throughout their own learning loop. Questions regarding mathematical modelling are discussed in the LCMP-group consisting of students, school teacher, and didacticians.

A group of students have included semi-authentic models[13] as part of their practice teaching (Hansen 2009). After a study of different mathematical models at the university college, the students decided to let the pupils in the classroom work with trend diagrams and regression to predict company turnover.[14] Two different models were used. These models gave different predictions. This leads to questions about validity and uncertainty; and reflections about how the results given by a particular model may depend on various input data. We regard students' and pupils' abilities to raise such questions as important parts of their critical democratic competence. For such questions to arise, we believe that some models and pedagogical initiatives could be more appropriate than others. By analyzing conversations in the classroom and between didacticians, teachers, and students we hope to get further insight into such issues. This is related to the following research questions:

- How can industrial mathematical models be made relevant for pupils' learning?
- How can some models be more appropriate than others for development of critical democratic competence in the field of mathematics?
- How necessary is the use of authentic models to achieve insight into modelling, and thus to the development of critical democratic competence?
- What criteria do we use when we discuss fruitful ways of working with models?
- Assuming that working with mathematical models increases pupils' critical democratic competence in mathematics, how does classroom communication reflect such skill development?

---

[12] The term "critical democratic competence" is used here in relation to mathematics and refers to peoples' ability to remain critical, considering and analyzing according to use of and results from mathematics in society (Blomhøj 1992, 2003; Skovsmose 1994).

[13] By "authentic model" we mean a model used in real-world applications (i.e. enterprise planning of income, oil recovery, climate forecasts etc.). The word "semi-authentic model" refers to a simplified model compared to daily life industrial models.

[14] This illustrates how the students move in and between different practices in the learning loop.

Our belief regarding the first question is that not all models used for industrial purposes are well-suited for classroom use. Authentic mathematical models often hide complex mathematical structures and advanced technology which excludes many models (cf. the notion of "black box"). For use in primary education these authentic models usually need to be simplified, which leads to the use of semi-authentic models. Thus, it becomes pertinent to investigate how pupils and students in different learning situations can work with semi-authentic models connected to topics in the school curriculum (e.g., concepts such as functions, equations, and linearity). Part of this will focus on how classroom communication may reveal different stages in pupils' development of critical democratic competence.

3 Final Remarks

In this paper, we have included examples where students participate, but the focus is also on the pupils' conversations and learning.[15] What characterizes the pupils' mathematical conversations as they move between classroom and company? Studying pupils' mathematical conversations at the workplace and in the classroom linked to pupils' readiness to apply mathematics in new contexts is believed to give new knowledge about pupils' learning through conversation. We investigate whether the inclusion of outside-school mathematics in mathematics teaching facilitate pupils' participation in problem solving, communication of ideas, and discussions about strategy use and solutions. The pupils' learning loop makes it possible to mathematize and model real-world problems. Mathematization and modeling require discussion; e.g., simplifications must be done and results must be critically evaluated. Through studying mathematical conversations in and outside the classroom, we will search for insight into the impact mathematical conversations have on the pupils' development of mathematical literacy. The OECD describes mathematical literacy as a preparedness to employ mathematics when one needs it, a critical strength to influence through mathematics, and critically consider mathematical results (OECD 2003, 2006). Mathematization, functional understanding, empowerment, and critical democratic competence are all related to mathematical literacy.

We also wish to look for traces of inquiry—whether pupils and students by their own interests together wonder and seek information (Lindfors 1999)—and how they initiate and invite into searching conversations with an inquiring stance in and about mathematics. An inquiring stance will impact the intentionality and empowerment of the participants.

[15] Parts of the LCMP-project are primarily interested in the pupils (this is the first layer described on page 2). A PhD-student, Toril Eskeland Rangnes, is working on pupils mathematical conversations as they participate in the learning loop, with particular emphasis on mathematical literacy. The questions raised in this paragraph will be part of her study.

We have given two examples from the LCMP-project where we see the influence of the school-industry partnership. This illustrates how the context of the project enables us to give insight into the relationship between school and industry. The first example points toward the school–industry context changing the conditions of learning and teaching, and thus changing the intentions of students and pupils. The context gives the knowledge acquired a different functionality and empowers the students and pupils. In the second example, the source of relevant mathematical models in the classroom is enhanced by the partnership with a company. The real-world connection may be helpful in engaging pupils and students in discussions that relate to critical democratic competence.

References

Alrø, H., & Skovsmose, O. (2004). *Dialogue and Learning in Mathematics Education. Intention, Reflection, Critique. Mathematics Education Library 29.* Dordrecht: Kluwer Academic Publishers.

Blomhøj, M. (1992). *Modellering i den elementære matematikundervisning: et didaktisk problemfelt.* Copenhagen: Danmarks lærerhøjskole, Matematisk institut.

Blomhøj, M. (2003). Modellering som undervisningsform. In O. Skovsmose & M. Blomhøj (Eds.), *Kan det virkelig passe?* (pp. 51–71). Copenhagen: L&R Uddannelse.

De Villiers, M. (1994). The role and function of a hierarchical classification of quadrilaterals. *For the Learning of Mathematics, 14*(1), 11–18.

Dreier, O. (1999). Læring som endring av personlig deltakelse i sosiale kontekster. In K. Nielsen & S. Kvale (Eds.), *Mesterære. Læring som sosial praksis* (pp. 196–214). Oslo: Ad Notam Gyldendal.

Hansen, R. (2009). Modellering og kritisk demokratisk kompetanse. *Tangenten, 20*(4), 0.

Johnsen-Høines, M. (2009). Learning Dialogue in Mathematics Practice. In C. Winsløw (Ed.), *Nordic Research in Mathematics Education. Proceedings from NORMA08 in Copenhagen*, April 21–25, 2008. Rotterdam: Sense Publishers.

Johnsen-Høines, M. (in print). Interpretative Research as Collaborative Inquiry. In B. Sriraman et al. (Eds), *The sourcebook on Scandinavian research in mathematics education.* Charlotte NC: Information Age Publishing.

Lave, J. (1999). Læring, mesterlære, sosial praksis. In K. Nielsen & S. Kvale (Eds.), *Mesterlære. Læring som sosial praksis* (pp. 35-52). Oslo: Ad Notam Gyldendal.

Lindfors, J. W. (1999). *Childrens inquiry. Using language to make sense of the world.* New York: Teachers College Press.

Mellin-Olsen, S. (1989). *Kunnskapsformidling: virksomhetsteoretiske perspektiv.* Bergen, Norway: Caspar forlag.

OECD (2003). *The PISA 2003 Assessment Framework.* Organisation for Economic Co-Operation and Development.

OECD (2006). *Assessing scientific, reading and mathematical literacy: a framework for PISA 2006.* Organisation for Economic Co-Operation and Development.

Skovsmose, O. (1994). *Towards a Critical Mathematical Education. Mathematics Education Library 15.* Dordrecht: Kluwer Academic Publishers.

The Threefold Dilemma of Missing Coherence: Bridging the Artificial Reef Between the Mainland and Some Isolated Islands

Guenter Törner, Volker Grotensohn and Bettina Roesken

1 Introduction

1.1 The "DAMPF" Activities of the Mathematicians as a Hotbed for the Educational Interface

The reader should know that the Duisburg region is traditionally an industrial area where the last (very large) German steel factories are located, e.g., the globally operating ThyssenKrupp Steel (TKS)[1] and Hüttenwerke Krupp Mannesmann (HKM).[2] Currently, more than 36,000 persons are working in this industry branch and 15 million tons of steel are produced and processed at Duisburg every year.

This explains that for more than two decades, the mathematical department of the University of Duisburg-Essen has been organizing half-yearly a one-day conference where engineers and scientists working in steel production and university research mathematicians meet. Each conference is structured in the same way: A talk of an industrial mathematician is followed by a talk given by a university professor on actual developments in mathematics. Finally, there is a lunch with intensive personal conversations. These meetings are organized by the so-called DAMPF-group[3] where the German acronym DAMPF means "steam, power and energy."

---

[1] http://www.thyssenkrupp-steel.de/de/.

[2] http://www.hkm.de/english/the-enterprise/the-ironworks.php.

[3] http://www.uni-due.de/mathematik/DAMPF.shtml.

G. Törner (✉) · V. Grotensohn
University of Duisburg-Essen, North Rhine-Westphalia, Germany
e-mail: guenter.toerner@uni-due.de

B. Roesken
Ruhr-University of Bochum, North Rhine-Westphalia, Germany
e-mail: bettina.roesken@rub.de

A. Damlamian et al. (eds.), *Educational Interfaces between Mathematics and Industry*, New ICMI Study Series 16, DOI: 10.1007/978-3-319-02270-3_14,

Meanwhile, cooperative projects have also been initiated and first master and PhD-theses dealing with scientific problem settings were supervised by university researchers. And it was ThyssenKrupp who decided to participate actively in the nationwide Year of Mathematics in 2008 in Germany. That is, meetings with teachers, courses for students, and exhibitions were successfully arranged and financed, emphasizing the importance which the company is attributing to mathematics as a school subject and beyond, last not least, also to the competencies of the company's apprentices.

1.2 Educational Interfaces: An Issue in Deficit

The success of these events served as a hotbed so that the first author was invited to comment on further educational problems since TKS is suffering daily from the low mathematical competencies of their apprentices. The company has becoming aware of these deficiencies when entrance examinations were set up. Further, the problems have been encountered when these young people (16–18 years) enter the firm and attend a 3-year course. Every year, the company is looking for 250 apprentices for technical professions such as electronics technician for industrial engineering, industrial mechanics, mechatronics, and so on. The basic mathematical knowledge needed for these professions is not inconsiderable. This is true for the location of Duisburg as well as for the five other company locations. Thus, the problem which we are faced locally is just the tip of a larger iceberg.

1.3 The Three Dilemmas

1.3.1 Dilemma 1

To become more precise, the first dilemma is constituted by the fact that the ThyssenKrupp company, which is looking for qualified apprentices, is nearly never satisfied with the competencies (Neubrand et al. 2002) of school leavers who are now applying for a position in the company. This deficiency is obvious when we consider that in the year 2009 more than 53 % of all appliers passed the initial test in mathematics with less than 50 %!

The exercises in the initial test have elementary character: changing units of measurement, converting elementary formulae, calculating surfaces (rectangles, trapezoids and sectors of circles), and averaging volumes. Additionally, basic skills in algebra and geometry are tested. Over several years, the company has kept statistics in view of deficiencies, which would be worth analyzing, but this is not the subject of our work that we present here.

For various reasons, it is not possible to reject all applicants who show a lack of mathematical skills. In this respect, compensating for deficiencies—in view of mathematics—is an integral part of the so-called dual education system in Germany. “Dual” means, that the apprentices go to “Berufsschule” (vocational school) twice a week and work three days within the company, where there is theoretical instruction aiming also at minimizing the mathematical deficiencies.

1.3.2 Dilemma 2

It should be remarked—and this is the second dilemma—that mathematics at vocational school (and within the instructional lessons in the company itself) is no longer an independent subject, but part of metallic engineering, electrical engineering, etc. For the most part, our subject is understood as an instruction of “technical mathematics” (e.g. Müller and 2005; Falk and 2007). Thus, mathematics looses its central role and it is not possible to catch up the subject-specific deficiencies. Teachers experience apprentices’ weaknesses regarding the content, without having time to strengthen the mathematical foundations. This makes it difficult to identify deficits in terms of didactical categories. It is well known that learning takes place first of all within a contextual process. The transfer of situative heuristical solutions and rudiments has then to be carried into a different, however, mathematically equivalent context. Nevertheless, such a transfer is not trivial because of many reasons.

1.3.3 Dilemma 3

On the other hand, there is some training at the company’s side, since the company has clear concepts and expectations what should be handled mathematically and “mastered” by these employees. This training is done by the instructors, however, these trainers are seldom qualified (in this area) and mathematically trained experts. This leads us to the third dilemma. Although they try to give their best, the instructors advise their students to manipulate mathematics as they personally think it should be done—in most cases, they are not aware of alternative approaches.

2 The Mathematical Strands

This is not the right place to present the corresponding mathematical contents of the intraplant apprenticeship training in detail, but in the following, we provide some information on typical tasks that apprentices encounter. We will focus on presenting four geometrically based operating tasks (see Fig. 1).

We possess numerous student documents waiting for being analyzed carefully. What in fact seems to be very interesting is the type of exercises (see page 5), which one generally cannot find in standard secondary school books. It reveals that the complexity of the exercises is not mainly based on managing arithmetical

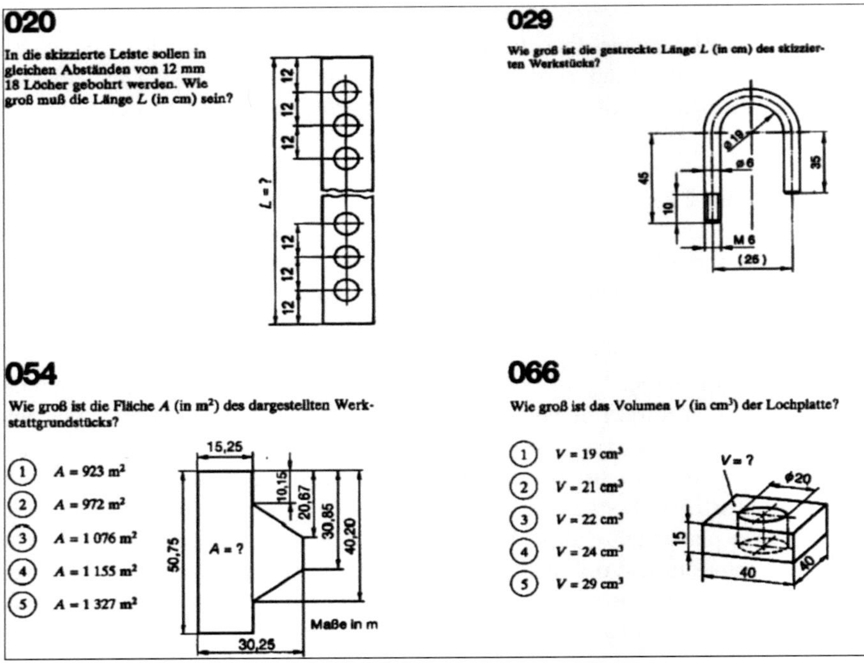

Fig. 1 Geometrically based operating tasks

calculus. It is much rather the amount of information and geometrical decoding that has to be evaluated and is difficult to handle [see the comments in Sträßer (2003) for the school context]. The exercises transport a sense of realism since they are connected to technical contexts and they do not have the nimbus of classical school exercises.

3 The Struggle of the Instructors and Their Dilemmas

3.1 Missing Continuity and Coherence

But still, how can we solve the appearing difficulties? It is a moot point to complain about deficiencies and that minimum standards have not been reached at school [see the discussion in Chap. 4 of Heinze and Grüßing (2009)]. Rather, the problems should be seen in non-existing continuity and coherence [see again Heinze and Grüßing (2009)] for the different fields of school and vocational education. What is missing are communication and cooperation among those partners. However, it is not our issue to discuss and benchmark the approach of the vocational school, nor to suggest isolated solutions [see Sträßer (1982, 1984)].

As aforementioned, the companies' instructors try:

- to work independently and to supplement vocational school, and
- to solve the deficits inside the company.

The instructors are indeed certified and experienced technicians with long-lasting experiences with apprentices, but in turn have limited global didactical and mathematical potential to diagnose the deficiencies in the learning of their pupils. Last year, the first author held a one-day workshop for this clientele. In the following, we refer to that event with a few remarks.

3.2 The "Folklore Estimation" of the Instructors

It is a matter of common knowledge that sustainable advanced training without acknowledging the underlying beliefs of the persons involved is blind [see Rösken (2009)]. Hence, before designing the in-service training course, the first author tried to get in contact via e-mail with the instructors who intended to attend the course. We obtained five replies, far beyond being representative, however, those provided us with some kind of snapshot showing at least some typical convictions. Unfortunately, the statements focus only on the learners' side and the instructors have not reflected their role in the learning process. We summarize some of these quotations and comment them from our point of view:

(a) School is responsible for the deficits that our apprentices are showing.

It would be necessary to identify the deficiencies and to suggest solutions by a third party, because the instructors of the company may just cover operating skills.

(b) Students are not able to set up and manipulate formulae. They try to look them up in books.

Rearranging terms in the elementary algebra is in many cases a necessary skill. Converting formulae is on the other hand everyday business, because it is mastered by "memorizing." In this case, it would be helpful to discuss the approaches of Barnard/Tall (Barnard and Tall 2001), who are using the term of "cognitive units" to describe such phenomena.

(c) Students are unable to qualitatively understand the functional relations in formulae.

With the overhasty change of representation mode, a new problem comes up.

(d) Pocket calculators are used without thinking.
(e) The general attitude toward mathematics can be characterized by sentences like boldness toward knowledge gaps and I am not talented for mathematics.

4 Possible Research Topics

The classic teacher–student relationship is primarily based on teacher's competency in specialized knowledge, and some elaborated pedagogical content knowledge. In front of such a background, the main task of the teacher is to initiate understanding and insight. Although this is the standard philosophy in school, in our context, such an aim almost seems to be very ambiguous from both the teachers' and the pupils' position.

So at first, we are missing:

1. An in itself coherent, intellectual, and simple philosophy that points out how mathematics can be thought on a minimum standard level with a lasting effect.
2. Such a philosophy has to be positively experienced and put into action at the same time by teachers as well as by pupils.

Sträßer et al. (1989) made clear that skill training versus insight is an incorrect dichotomy. Still today, persuasive approaches that balance these two aspects have not been noticed by third parties and the authors have no information regarding the following important aspect:

3. Coping with mathematical deficiencies in a specific context of application, e.g., metal engineering. We do not know how to transfer insights into a quite different context, e.g., electrical engineering and vice versa.

Finally, we cannot offer any straightforward convincing concepts for further education to the instructors at the moment, but we have some visions…

5 Conclusions and Possible Consequences

It seems remarkable to us, that our first steps to strengthen the educational interfaces are based on a fruitful cooperation with mathematics in industry over many years which enforced the belief that university partners can also increase the didactical needs of industry. This estimation results from the cooperation between the university and the company. Without such a cooperation, we—responsible for mathematical education at university—would never have been asked!

Besides, it is important to build a trusting atmosphere between the instructor and the didactician. Research on professional development of teachers—and the situation of instructors is quite similar—can only be sustainable and successful if the teachers are willing to accept any kind of support and empowerment. Therefore, the offers for instructors should not create the impression of compensating for deficiencies on the teachers' side. Thus, we need some specific strategy to address the individual needs of educators that is based on positive messages.

Since the company under discussion has at least five locations spread over a large area, it is not easy to design one measure. With respect to designing a training for the apprentices, one should notice that there are more than 1,000 people starting their job every year and one cannot gather them together to just one course. These organizational difficulties have to be solved. Up to now, the first author has not had any direct contact to the apprentices, and that is why we have to rely on information from the instructors. From our research in the field of beliefs, we know that negative emotions are often linked to the subject math. Because of this, every supporting measure has to start right (here) with the apprentices and has to spark interest, self-confidence, and self-efficiency. So, we conclude with the following question:

How is it possible to establish an intellectual philosophy for mathematics teachers at school that will cover a basic level of competencies and enable a vision of mathematics education for a better performance of students starting their job training?

References

Barnard, T., Tall, D. (2001). A comparative study of cognitive units in mathematical thinking. In M. van den Heuvel-Panhuizen (Ed.), *Proceedings of the 25th international conference of the international group for the psychology of mathematics education (PME)* (Vol. 2, pp. 89–96). Netherlands: Freudenthal Institute (Utrecht) (12–17 July 2001).

Bromme, R. (1994). Kompetenzen, Funktionen und unterrichtliches Handeln des Lehrers. In F. E. Weinert (Hrsg.), *Enzyklopädie der Psychologie*. (Band 3. S. 177–212). Göttingen: Hofgrefe.

Enzensberger, H. M. (1999). *Drawbridge up: mathematics—a cultural anathema*. Natick: A K Peters, Ltd.

Falk, D. et al. (2007). Elektrotechnik. Grundbildung. Technische Mathematik. Braunschweig: Westermann.

Heinze, A., & Grüßing, M. (Eds.). (2009). *Mathematiklernen vom Kindergarten bis zum Studium*. Münster: Waxmann.

Müller, E. et al. (2005). Metalltechnik. Gesamtband. Technische Mathematik. Braunschweig: westermann.

Neubrand, M., Klieme, E., Lüdtke, O., & Neubrand, J. (2002). Kompetenzstufen und Schwierigkeitsmodelle für die PISA-Tests zur mathematischen Grundbildung. *Unterrichtswissenschaft, 30*(2), 100–119.

Rösken, B. (2009). *Hidden dimensions in the professional development of mathematics teachers: In-service education for and with teachers*. Doctoral Dissertation, Duisburg: Universität Duisburg-Essen, FB Mathematik.

Sträßer, R. (Ed.). (1982). *Mathematischer Unterricht in Berufsschulen; Analysen und Daten. Band 28*. Bielefeld: Institut für Didaktik der Mathematik.

Sträßer, R. (1984). Mathematik als Element beruflicher Qualifikation. In H. W. Heymann (Hrsg.), *Mathematikunterricht zwischen Tradition und neuen Impulsen*. (S. 49–79). Köln: Aulis Verlag Deubner & Co KG.

Sträßer, R. (Ed.). (2003). *Darstellenund Interpretieren. Mathematiklehren, Heft 117*. Friedrich Verlag: Seelze.

Sträßer, R., Barr, G., Evans, J., & Wolf, A. (1989). Skills versus understanding. *ZDM Zentralblatt Didaktik Mathematik, 21*(6), 197–202.

Törner, G., & Hoechsmann, K. (2004). Schisms, breaks, and islands—seeking bridges over troubled waters: A subjective view from Germany. In D. Mc-Dougall, D., *Proceedings of the 26th Conference for the International Group for the Psychology of Mathematics Education (PME)—North America Chapter (PME-NA)* (Vol. 3, pp. 993–1001), Toronto, Oct 2004.

The Project "Mathe-Meister": A Mathematical Self Assessment Centre with Diagnostic Feedback for Vocational Trainees

Kathrin Winter

1 Initial Points

In agreement with OECD results (2008) and the educational interfaces between mathematics an industry (EIMI) discussion document (ICMI and ICIAM 2008), there have been many statements of teachers in vocational education that their students may not know

- that they need mathematics in all vocational training,
- how much or which mathematical competences they need to be successful,
- to appreciate their mathematical competences.

One of the main problems in advanced vocational education in Germany is the abolition of mathematic lessons since 2005 for the benefit of situated learning: According to vocational curricula, Mathematical competences shall now be learned embedded in the context, in which it will be needed and will not longer be picked out as a explicit and isolated topic. As a consequence, the students have to check out and train their mathematical competences on their own (Stein and Winter 2009).

Additionally, many trainers miss special mathematical didactical education in the academic studies for vocational teachers. For example study lessons in the field of mathematical diagnostics or bug analysing do not exist. As a consequence they notice missing mathematical competences of the students but they cannot diagnose the mistakes and problems in detail.

Based on these facts Prof. M. Stein (University of Münster, Germany) initiated the project 'Mathe-Meister'.

K. Winter (✉)
Institute for Didactics of Mathematics and of Computer Science, University of Münster, Fliednerstraße 21, 48149 Münster, Germany
e-mail: kathrin.winter@uni-muenster.de

A. Damlamian et al. (eds.), *Educational Interfaces between Mathematics and Industry*, New ICMI Study Series 16, DOI: 10.1007/978-3-319-02270-3_15,

2 Project 'Mathe-Meister'—Conditions and Aims

The self-assessment-centre is meant to help future students of advanced vocational courses to

- appreciate their own mathematical competences in view of their chosen business,
- detect their individual mathematical deficits by using the tests and
- realize the possible negative effects of their deficits to their future master education.

There is even more than analysing of mathematical deficits on this project. Attendants with deficits in mathematical knowledge or competences get

- information and hints for getting the correct answers to each item,
- advice to helpful textbooks, educational computer programmes and lectures at vocational training places, especially at the "Industrie- und Handelskammer" (chamber of industry and commerce) or "Handwerkskammer" (chamber of handicrafts) and
- an individual diagnostic report to each individual wrong chosen answer to the test items.

3 Way of Looking at the Problem

Using new technology is state of the art in industrial as well as in handicraft jobs. So using the World Wide Web for realising the project suggests itself. Realising such a Web-based test including a real-time analysis report with individual diagnostic information to every user represents an actual and yet not solved problem in an interrelated field of computer science, didactics of information technologies, didactics of mathematics and vocational training.

As a consequence, it is necessary to check out the basic mathematic themes and competences for each embedded occupational area which are required to work successfully. With regard to every mathematical theme and competence we developed diagnostically significant test items. The individual diagnostic report is based on the chosen answers for every test item in a multiple choice test design. To realize this idea we had to design and implement concepts how to collect, analyse and select special answers for a mathematic Web-based self-assessment centre for vocational training.

4 Basic Mathematical Competences in Several Vocational Fields

To identify the required basic mathematical themes and competences, we analysed training books and interviewed teachers. The disappearance of mathematic lessons in advanced vocational training also implies a lack of mathematical textbooks in this field. So we had to analyse training books of each job field to extract the mathematical items. We give an example of an exercise for Electrical Engineers. In this exercise the students have to calculate different data in an engineering drawing. Figure 1 shows the engineering drawing and the given data. Figure 2 shows one of the calculations based on this exercise.

To check out the needed basic mathematical competences to solve this exercise we analysed the exercises of different jobs. Figure 2 demonstrates the extraction of basic mathematical competences and themes in one example of calculation (Stein et al. 2010; Winter 2011):

Based on the identification of basic mathematical themes and competences, a pool of 228 test items was designed. For evaluating the test items and get a compact test, 454 students of the target group were tested. In addition, the test was presented to approximately 1,500 pupils of secondary schools. Statistical methods and diagnostic knowledge have been combined to determine the final pool of test items (Stein et al. 2010).

5 Test Design and Diagnostic Feedback

In the self-assessment centre "Mathe-Meister", the user can choose his intended occupation. He will get the special test for this occupation. Special examples for his occupation help to understand the use of the mathematical competences behind

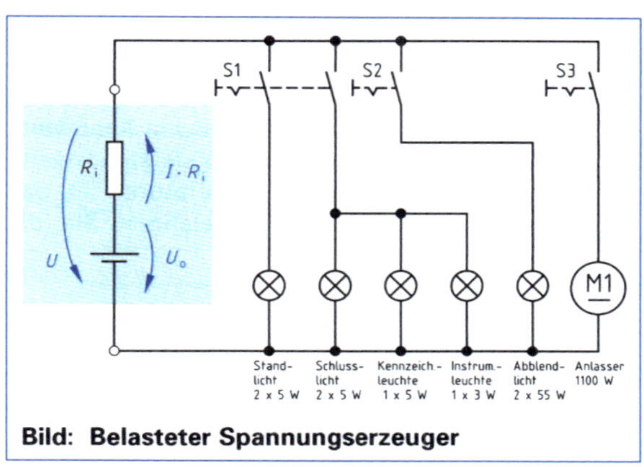

Fig. 1 Introductory example for masters in electronic engineering (Bastian et al. 2005, p. 5). Exercise: calculation of different data in an engineering drawing

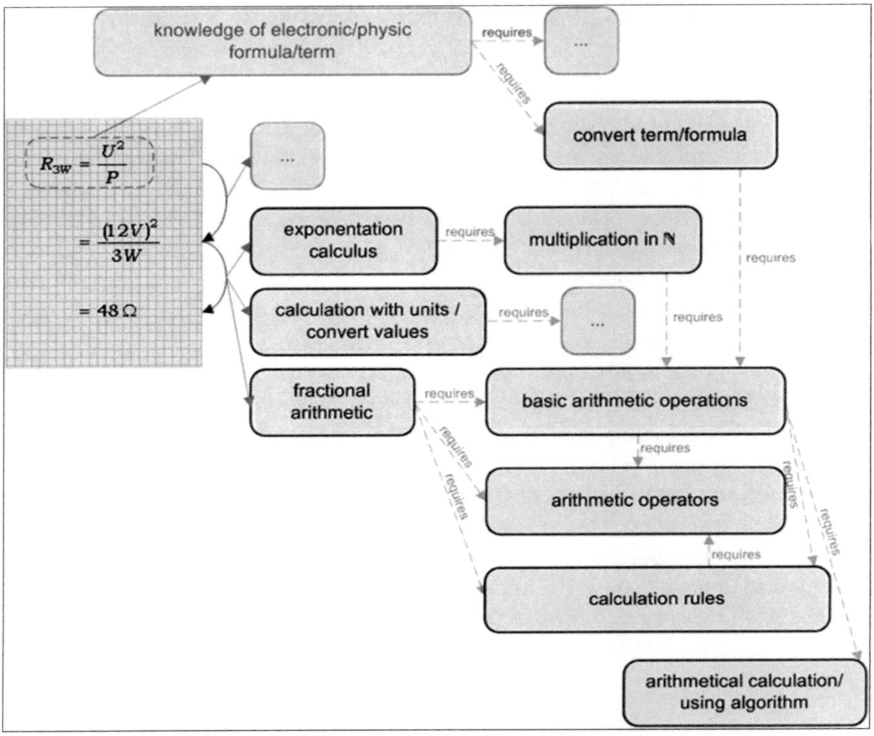

Fig. 2 Extraction of mathematical themes and competences to solve an exercise in a school book for electronic engineers

the test items. For example, calculation with fractions is shown in individual example exercises for future carpenters, for coiffeurs and for brick layers.

To get specific tests for different occupations is one of the main aims. Based on the book analysis and interviews with teaching persons, lists of needed basic mathematical themes and competences were generated. For example a carpenter needs to know geometrical figures and how to calculate different data for them. A coiffeur does not need geometrical knowledge. But both had to know the calculation of percentage. Using those theme listings and the evaluated pool of test items specific tests for every occupation were composed.

The feedback of the “Mathe-Meister” test shall show the individual insufficient mathematical competences in detail. Based on the test results, the respective mathematical themes will be presented in the feedback. For every theme, different books, computer programmes and special lessons for mathematical training in the respective occupation will be identified.

To realise a real-time feedback including individual diagnostic information, we decided to use a multiple choice test design. An analysis of mathematical self-assessment in the Web reveals three main problems with regard to diagnostic potential and feedback (Winter 2011):

- multiple choice answers can be chosen at random
- only a few multiple choice answers have diagnostic potential[1]
- there is no diagnostic feedback—even information about the mathematical topics are rarely to be found

The multiple choice design of the "Mathe-Meister" test has been modified in contrast to existing tests (Winter 2011):

1. The multiple choice answers for a test item are not directly shown. At first the user shall try to find an answer by himself. After a defined time slot the multiple choice answers are shown and the user can select one.
2. All given multiple choice answers for every item are developed from the knowledge and the results of bug theories.
3. For every incorrect answer there is a separate declaration about the probable mistakes the user may have made.

So the report to the user includes detailed information about the bugs he probably made. This detailed and complex feedback of themes, tips and bug report will be usable as a helpful basis for an individual mathematical training.

6 Prospects

The project will be closed at June 2011. Actual information can be found here: http://www.mathe-meister.de.

After closing the project add-ons for more occupations will be developed. Additional possible extensions for other vocational trainings and fields are under consideration.

7 Positioning Within the EIMI-Study

For the aims of "EIMI" Study (ICMI and ICIAM 2008), we relate our project to the following aims and questions of the EIMI study:

- to make people aware of the importance of mathematics in vocation and vocational training
- to show hidden mathematics in the workplace
- to give mathematical or rather diagnostic information to vocational teachers.

[1] Here, diagnostic potential means reproducible knowledge of bugs (Winter and Wittmann 2009).

References

Bastian, P., Eichler, W., Riefler, S., Rinn, H., Spielvogel, O., Tkotz, K., Winter, U. (2005). Prüfungsvorbereitung fachrechnen elektrotechnik. (preperating for examination in calculation for electronic engineers) Verlag Europa-Lehrmittel, Haan-Gruiten.

ICMI, ICIAM. (2008). Educational interfaces between mathematics and industry—discussion document. http://www.iciam.org/EIMI/eimi_dd-09.pdf.

OECD Organisation for Economic Co-operation and Development Global Science Forum. (2008). Report on mathematics in industry, July 2008. http://www.oecd.org/dataoecd/47/1/41019441.pdf.

Stein, M., Winter, K. (2009). Das Projekt Mathe-Meister: Strukturen und Konzeption. (Project "Mathe-Meister": Structure and conception), in: Beiträge zum Mathematikunterricht: Vorträge auf der 43. Tagung für Didaktik der Mathematik vom 02. März bis 06. März 2009 in Oldenburg, WTM-Verlag, Münster (pp. 887–89).

Stein, M., Winter, K., Podlogar, H., Jordan, R. (2010). Homepage of the project "Mathe-Meister". http://www.mathe-meister.de.

Stein, M., Winter, K., Jordan, R., Podlogar, H. (2010). Das projekt mathe-meister: Stand der Dinge. (Project "Mathe-Meister": As things are now) in: Vorträge auf der 44. Tagung für Didaktik der Mathematik vom 08. März bis 12. März 2009 in München, WTM-Verlag, Münster, in print.

Winter, K., Wittmann, G. (2009). Wo liegt der Fehler?—Schülerinnen und Schüler analysieren fehlerhafte Lösungswege beim Rechnen mit Brüchen und Dezimalzahlen (6–8. Klasse). [What's the bug?—Students analysis of bugs in calculations with fractions] in: PM Praxis der Mathematik in der Schule—Sekundarstufen 1 und 2 Heft 27, Aulis Verlag, Hallbergsmoos (pp. 15–21).

Winter, K. (2011). Design and development of multiple choice test items with diagnostic potential for individual deficit and bug analysis for using in a web based mathematical self assessment centre (WTM-Verlag, Münster).

Part IV
WG University and Academic Technical/ Vocational Education

University and Academic Technical/ Vocational Education

Nilima Nigam and José Francisco Rodrigues

1 Introduction

In this chapter we report on the university-level teaching and learning of topics at the interface of mathematics and industry. We focus on questions from several topics from the EIMI document: the Teaching and Learning of Industrial Mathematics, the Teaching and Learning for communication and collaboration, Curriculum and Syllabus issues, and on the Teacher training. We shall see that there is a diverse range of pedagogical and training practices surrounding industrial mathematics internationally. After reviewing the specific questions that the Working Group 5 (WG5) addressed at the 2010 EIMI conference in Lisbon, and which were partially complemented and discussed at the Macau 2011 Workshop, we present some reported experiences, discussions, and recommendations. We also present brief introductions to some selected papers from the Conference Proceedings.

The WG5 met in two sessions and brought together experts with two different perspectives. One group consisted of experts specializing in mathematics education at the university level and in preparation of engineers and applied mathematicians. The other group consisted of researchers and practitioners with experience in industrial mathematics. The works included four short presentations by William Lee (Ireland), Adérito Araújo (Portugal), Avenilde Romo (Mexico), and Edwige Godlewski (France). In addition, there were two Skype presentations by Lou Rossi and Bogdan Vernescu, both from the USA and who were unable to attend the Lisbon meeting in person. The WG5 included, in addition to the presenters, the two coordinators, Pilar Romero, Juan Tejada (Spain), Jaime Carvalho

N. Nigam (✉)
Department of Mathematics, Simon Fraser University, Burnaby, Canada
e-mail: nigam@math.sfu.ca

J. F. Rodrigues
Universidade de Lisboa, Lisbon, Portugal
e-mail: rodrigue@ptmat.fc.ul.pt

A. Damlamian et al. (eds.), *Educational Interfaces between Mathematics and Industry*, New ICMI Study Series 16, DOI: 10.1007/978-3-319-02270-3_16,

e Silva (Portugal), Ajit Kumar (India), Rudolf Strässer (Germany), Jinxing Xie (China). The questions in front of the WG5 and the presentations were discussed by all the participants. We found the range of international experiences and perspectives to be particularly illuminating. Participants were able to share their experiences and recommendations with each other. Equally importantly, we learned what university-level initiatives did not work in particular contexts; this enabled participants from other countries to reflect upon whether such initiatives could, with modification, succeed in their universities. The discussions and debates were summarized and presented to the plenary by Sarah Peterson (Canada).

The specific questions the WG5 addressed were: (1) How should one organize university-level teaching and learning in order to make industrial mathematics visible? Is this increased visibility of mathematics wanted/necessary? (2) What is an appropriate level of detail for educational purposes in order which can serve to generate interest and excitement in an industrial mathematics problem, without overwhelming the learner? (3) Are there specific skills for use in relation to industrial mathematics, and how do we teach mathematics as a second language? (4) What is the role of mathematical contests and competitions in developing and assessing communications skills in mathematics? (5) What are the advantages and disadvantages of creating specific courses on mathematics for industry versus including the topic in the standard mathematical courses at various levels? (vocational training? role of internship?); (6) What are the appropriate levels of understanding, good practices, and guidelines for implementation of new measures in teacher training related to industrial and applied mathematics?

Question 1 *How should one organize university-level teaching and learning in order to make industrial mathematics visible? Is this increased visibility of mathematics wanted/necessary*?

Several participants gave examples and reported their experiences and a first observation raised by this question is, naturally, to whom industrial mathematics should be visible, not only for the students but also for the society, since mathematics is everywhere and it is necessary to make it more visible. On one hand, this depends on the level, and taking the Master's level as example, mathematics should be visible to students as well as their future employers. On the other hand, teachers-in-training often are not exposed to industrial mathematics, but they could be disseminators to students and to the society, as for instance, the case of the interesting Georgia Tech GIFT program. As a difficulty, it was observed that this requires and will put pressure to provide more real-world examples within "service" courses, what is hard for the math professor because often they do not fully know the industrial context of the mathematics they teach. Among the recommendations, we report the following: (1) collaborate with teachers and experts in other disciplines to expose mathematics in context; (2) open some black boxes at every level (e.g., at undergraduate level, write your own Differential Equations solver to understand the software black box); (3) Share and disseminate good examples, e.g., on the Web; (4) Stimulate, encourage, and reward curiosity.

Question 2 *What is an appropriate level of detail for educational purposes in order which can serve to generate interest and excitement in an industrial mathematics problem, without overwhelming the learner?*

First, it is necessary to notice that problems from industry are often not formulated as mathematical problems, and therefore there is a need to learn a lot about the specific context of the problem and to distill essential information for the specific question from the mass of existing details in the application. Then it is necessary to formulate the problem and study the problem mathematically. This requires a student to draw on existing mathematical training and search the literature for relevant ideas. Often, they may need to learn new mathematical concepts. Since many mathematical problems do not have closed-form solutions, the student may have to select appropriate software tools or develop new tools to solve the problem. Finally they have to 'close the loop' by relating their mathematical results back to the specific application of interest.

Several international experiences were described and discussed. A remarkable range of novel pedagogical approached have been implemented globally to enhance the training of students in the use of mathematics to solve industrial problems. Some initiatives are short-term (week-long courses or workshops); others can be long-term and immersive (e.g. internships). In Europe and USA, week-long modeling courses produce very successful results for industry. These are modeled on the successful mathematical study groups with industry (UK, North America, Ireland, Portugal, Australia, China, etc.). In these settings, the problems from industry are not clearly formulated as mathematical questions. Participants have one week to produce newmodels, new perspectives, and insights. When using such a format to train students, it was noted that these short time frames may require students to get help navigating prior knowledge pertaining to the application. Existing week-long industrial graduate-level problem-solving workshops involve a mixture of distilled prior knowledge, and guided problem formulation.

There are also successful examples of longer-term courses. For example, in a University of Delaware final year undergraduate modeling course, students work on real-world problems and develop habits of mind for effective practice of mathematics in applied and industrial contexts. In another successful example in France, a financial mathematics program was developed and taught by academics with actual banking experience.

In mathematical modeling course with large enrollments, as in some institutions in China, it is not practical to have inclass open-ended modeling activities. Instead, students are shown case studies, then encouraged to work on more realistic problems outside class.

We discussed models of industrial internships (e.g., WPI in USA, MITACS in Canada, UPMC, and UPPA in France), which provide immersive (long term) learning experiences. These are ultimately very valuable, but finding suitable industrial internships for a large number of students requires an intensive commitment of resources.

Given this range of pedagogical approaches in use globally, the question of an appropriate level of detail to present must necessarily depend on the level, length and format of the course. Several recommendations on how much detail to present were proposed: obviously, always taking into account that the level of detail depends on level of course: (1) Base activity or course on existing resources, like books, or other materials which are already carefully structured, as in the US Consortium for Mathematics and its Applications (COMAP) or in the Contemporary Chinese Undergraduate Mathematical Contest in Modeling; (2) Present the general problem statement and provide 3–4 relevant articles or sources; (3) Present case studies in class, assign open-ended problems for team-based assignments outside class; (4) Present only the problem statement; (5) Point students to important concepts/courses in other disciplines. The WG5 consensus recommendation was to encourage independent and actual industrial problem-solving, not just passive learning, as well as to. There were divergent opinions on role of books, but there was a consensus that shared resources for industrial problems would be needed; (6) Point students to important concepts/courses in other disciplines. There were divergent opinions on role of books, but a consensus that shared resources for industrial problems would be needed was easily obtained.

Question 3 *Are there specific skills for use in relation to industrial mathematics, and how do we teach mathematics as a second language*?

The discussion around this issue started with general and essential non-technical skills, like teamwork, cooperation, communication in and with multi-disciplinary teams, different and specific communication skills needed for academia and industry, presentation skills for different audiences, management of information resources, effective and efficient time management, ability to assess and manage risks, initiative and creativity.

It was observed that any curricula in industrial mathematics should encourage the acquisition of both transversal and specific competences. The *transversal competences* (instrumental, personal and systematic) should not be distinguished from a general Subject Benchmark statement for engineering skills, attributes and qualities, as referred in (Alberti et al. 2010). These are reproduced here by their own intrinsic interest:

- Knowledge and Understanding
  - Specialist (Discipline) knowledge.
  - Understanding of external constraints.
  - Business and Management techniques.
  - Understanding of professional and ethical responsibilities.
  - Understanding of the impact of engineering solutions on society.
  - Awareness of relevant contemporary issues.
- Intellectual Abilities
  - The ability to solve engineering problems, design systems, etc., through creative and innovative thinking.
  - The ability to apply mathematical, scientific, and technological tools.
  - The ability to analyze and interpret data and, when necessary, design experiments to gain new data.

  - The ability to maintain a sound theoretical approach in enabling the introduction of new technology.
  - The ability to apply professional judgment, balancing issues of costs, benefits, safety, quality, etc.
  - The ability to assess and manage risks.
- Practical Skills
  - Use a wide range of tools, techniques, and equipment (including software) appropriate to their specific discipline.
  - Use laboratory and workshop equipment to generate valuable data.
  - Develop, promote, and apply safe systems of work.
- General Transferable Skills
  - Communicate effectively, using both written and oral methods.
  - Use Information Technology effectively.
  - Manage resources and time.
  - Work in a multidisciplinary team.
  - Undertake lifelong learning for continuing professional development.
- Qualities
  - Creative, particularly in the design process.
  - Analytical in the formulation and solutions of problems.
  - Innovative, in the solution of engineering problems.
  - Self-motivated.
  - Independent of mind, with intellectual integrity, particularly in respect of ethical issues.
  - Enthusiastic, in the application of their knowledge, understanding and skills in pursuit of the practice of engineering.

The specific competences depend on each "industry". It can be readily seen that mathematics also has a transversal nature. Transversal mathematical competences should be incentivized in specific courses and can be used to obtain specific competences in many different situations. The *Mathematical and technical skills* were summarized as follows:

- Modeling and analytical skills, knowledge of numerical methods;
- Simulations, experience with mathematical models in industry;
- Ability to handle huge amounts of data by integrating mathematical, numerical, and statistical methods—Ability to identify and select key information;
- Software skills—Students need sophisticated understanding of mathematical software—Programming skills—gain better understanding of "black box" tools, build own tools when needed;
- Ability to apply mathematical, scientific, and technical tools;
- Analytical skills—formulation of problems.

Question 4 *What is the role of mathematical contests and competitions in developing and assessing communications skills in mathematics?*

Challenging in Mathematics is a classical issue that goes beyond the traditional school and has a well-established tradition in elementary mathematics, in particular in the Olympiads, that encompasses millions of students, teachers, mathematicians, and hundreds of associations all over the world. The topic was even the subject of the 16th ICMI Study (Barbeau and Taylor 2009). The less known first Mathematical Contest in Modeling, was initiated in the USA in 1985 by COMAP, and was followed in 1999 by another series on Interdisciplinary Contest in Modeling. The benefit to the students and the utility for mathematics education at tertiary level was soon recognized in China, where a huge and successful series exists and was reported at the WG5 by J. Xie.

Finally, it was observed that contests and modeling weeks may simulate the real-world workplace and provide a good environment for students learn to deal with time pressure, different time constraints in industry, along with the development of skills like teamwork, presentation of results, etc. It was observed that the inclusive spirit of these contests implies no one fails in competitions. Even if some competitors naturally get prizes, everyone learns and wins with the participation. Students learn to tackle problems with no single correct answer, to identify and use which assumptions should be reasonable. Models should be creative and innovative and the solutions should be justified, efficient, and useful.

Question 5 *What are the advantages and disadvantages of creating specific courses on mathematics for industry versus including the topic in the standard mathematical courses at various levels*? (*vocational training*? *role of internship*?)

First it was observed and commented that the reputation of applied/industrial math versus pure math is an issue in current academic culture. Diverse approaches to overcome the situation were reported. For instance, in the USA, the WPI model integrates modeling and applied problems throughout the curriculum from the ground up and the case of the University of Delaware the students take a modeling capstone in their final year. In China, modeling courses were introduced with great success, and participation in national and international competitions stimulated those courses. In Europe, the situation in several countries was discussed. In particular, the European Consortium of Mathematics in Industry (ECMI) is promoting courses at the master's level, specific to industrial mathematics, in several European universities. Several commentators observed that it is a big challenge to find educational opportunities in industry, but that the Study groups with industry (faculty and stu- 206 dents) are a good tool. Also, modeling weeks for students can prepare them for internships that should be incorporated in their training, as is already often the practice in several universities. It was also observed that the European Science Foundation (ESF 2010) has recently published a Report on a Forward Look with, in particular, the recommendation to academia that ''mathematical societies an academic institutions must harmonize the curriculum and educational programs in 212 industrial mathematics at the European level'', with two ''road map implementations'': the creation of a common industrial mathematics curriculum and the setting up of a pool of industrial mathematics engineers, on one hand, and the development of new criteria to assess and recognize careers in industrial mathematics, on the other hand.

Trying to answer question 5, the WG5 discussions recognized the advantages of a distinct course where students will:

- See mathematics applied to substantial long-term problems.
- Improve their employment opportunities.
- Learn mathematics within a specific context.
- Focus on additional skills (communication, teamwork, interdisciplinary collaboration, etc.).

On the other hand, if there was a consensus that resources should be shared at national/regional level, some disadvantages and difficulties were pointed out:

- Start up cost is high—making industry connections is really hard.
- Students may not have appropriate mathematical expertise.
- Perception problem in some countries –need to convince students and faculty colleagues.

Question 6 *What are the appropriate levels of understanding, good practices, and guidelines for implementation of new measures in teacher training related to industrial and applied mathematics*?

A first observation dominated this point: the majority of mathematics teachers are trained in pure math only, even if some exposure to applied/industrial math already exists in some countries. It was also observed that there are some tools and examples for teachers provided by associations, as COMAP the USA, and by mathematics societies in some countries.

From the contributions to the study and the WG5 discussions the following recommendations were suggested:

- Applied and industrial mathematics societies should provide more materials for teachers.
- Include more applied mathematics examples in existing teacher-training programs.
- Universities should provide lifelong learning/in-service education for teachers (e.g., Georgia Tech GIFT program summer internships, which place teachers in industrial "real world" projects).
- Enable secondary school teachers to highlight mathematical models arising in other subjects (e.g., physics, biology).

2 Complementary Selected Articles

These conclusions and recommendations are based on WG5 discussions at the EIMI conference and on several contributions included in the original proceedings (Araújo et al. 2010). These proceedings contain different articles and reports on developing curriculum options that prepare the students for a career at the interface of mathematics and industry, for strategies to create opportunities for research participation at the undergraduate and graduate level, including industrial internships, modeling camps, and summer schools, as well as on setting up opportunities for secondary school teachers to participate in academic-industrial interactions.

Due to the large number of contributions, it was not possible to include all those that are related to the four Sections of the Discussing Document, namely the Sect. 5, on Teaching and Learning of Industrial Mathematics, the Sect. 7, on Teaching and Learning for communication and collaboration, the Sect. 8, on Curriculum and Syllabus issues, and the Sect. 9, on Teacher training. Nevertheless the following nine articles, edited from the 2010 Proceedings or added after the Macau Workshop in 2011, are part of this WG5 report.

1. *Mathematics for Engineering and Engineering for Mathematics*, by M. Albertí, S. Amat, S. Busquier, P. Romero and J. Tejada, is a report on behalf of the Spanish Committee for Education of the CEMat, which plays the role of ICMI in Spain, the country that hosted the International Congress of Mathematicians in 2006, congregates three types of complementary contributions to the Study. First, a report on mathematics and "engineering" in secondary education, based on research developed in a city in Catalonia that confronts the mathematics is used and how it is used at work, as well as the several characteristics of the industrial field that are absent in secondary education and in mathematics education as well: competitiveness, production, and valorization method of results. Second, an analysis of the mathematical competences in the engineering curricula at university level versus industrial demands, describing the case of the agro-industry of the region of Murcia, with which the Polytechnic University of Cartagena has projects. Third, contains description of some Spanish master programs under the label of "Mathematical Engineering". This part analyzes the examples of two Universities in Madrid (Complutense and Carlos III), and the University of Santiago de Compostela in the European framework.
2. *Laboratory of Computational Mathematics: an interface between academia and industry*, by S. Barbeiro, A. Araújo and J. A. Ferreira, is short report that illustrates how a university laboratory, associated to a mathematical research center in Portugal, can act as an interface between academia an industry, with concrete industrial oriented and sponsored research projects, providing support to an educational program and to the integration of students in the job market, as well as integrating the national efforts of mathematics relationship with industry, in particular, hosting an European Study Group with Industry.
3. *Improving the industrial/mathematics interface*, by J. P. F. Charpin and S. B. G. O'Brien. In the context of experience with a network of applied mathematical modelers across Ireland, centered at the University of Limerick, the authors discuss how necessary is the refocusing of our mathematical resources in mathematical modeling, which is central to the practical applications of industrial and applied mathematics, for the regeneration of economic success. Questioning why are mathematicians who participate in industrial mathematics

in the minority and what can be done to improve the situation, the Irish answer states that the solution must lie in the training: mathematics curriculums at all levels need to be redesigned to reflect the ever-growing interest in mathematical modeling and provide the students with the basic skills necessary to become real modelers. As an important tool, the article describes the study groups with industry, which started in Europe in 1968 and is now a well-established institution and a leading workshop model for improving the industrial/mathematics interface all over the world.

4. *Two masters on 'Mathematics for Industry' at the Universities of Paris and of Pau*, by Godlewski et al. (2010). Within the dual French system of higher education, the evolution of the university diploma in applied mathematics at the master's level is described in two different contexts, one in Paris, one in the small provincial town of Pau. These programs have evolved according to the European implementation of the Bologna process; they have taken into account the new range of applications, for instance in Finance, and profited from the spectacular development of computer science. Emphasis on contacts with industry is given, in particular, concerning the computational mathematics and in the professional programs, where the internships are immersed in industry, most often in a private company or in some public research institution. However this contribution does not cover the French recent experience on the educational interfaces between mathematics and finance. In fact, since the beginning of the nineties, Mathematics, and more particularly the theory of probability, have taken an increasing role in the banking and insurance industries, which interactions and their consequences at the level of training in France in these domains as well as at the level of research were also discussed in the WG5 on the basis of the Proceedings contribution (Godlewski and Pagés 2010).
5. *Mathematics in Industry and Teachers' Training*, by M. Heilio from Finland, sustains that for education and mathematics teachers' training a change should be visible in curriculum development, up-to-date contents, novel teaching methods and that the development of mathematics modeling education is a crucial part of this endeavor and university pedagogy of applied mathematics. Referring to case examples of industrial math projects illuminating the educational challenge, Heilio suggests and analyzes some implications for teachers training programs and practices at school level, sustaining that a modern view of mathematics should be reflected in curricula and educational practices.
6. *Interfaces between Mathematics and Industry and the Use of Technology in Mathematics Education in India*, by A. Kumar, describes the current situation on interaction between Mathematics and Industry, which is still almost non-existent in contrast with the rapid economic growth of India, and suggests some of the steps that must be taken in order to improve that interface. A second part deals with the use of computer aided technology in mathematics education at various levels in India and mentions a future project proposed that, in the view of the author, can significantly revolutionize mathematics teaching using computer-aided tools.

7. *Modeling*: *Developing Habits of Mathematical Minds*, by J. A. Pelesko, J. Cai Louis, and F. Rossi. To develop undergraduate mathematics majors, specific habits of mind associated with the effective practice of mathematics in applied and industrial settings is the central aim of the experience described in report from the Department of Mathematical Sciences of the University of Delaware, USA. Their reform has led to the creation of a new mathematical modeling capstone course possessing many novel features. Over the past years, this course has developed the confidence and ability in their students to see the intimate connections between mathematics and real world processes. This chapter, the authors identify seven key habits of mind associated with effective mathematicians and present best practices for reinforcing these habits in our students.
8. *The evolution of graduate applied math courses in the Institute of Mathematics, University of the Philippines*, by C. P. C. Pilar-Arceo and J. M. L. Escaner IV. This short article describes the recent development of the applied mathematics education in a developing country in Asia by looking into the history and development of graduate mathematics programs in the University of the Philippines Institute of Mathematics. Particular focus is be given to the role industry has played in this evolution and relates lessons learned from past programs, various efforts to promote certain applied math areas and to sustain others which are already strong, valuable collaborations, recent achievements, and future directions. As a late contribution presented at the Macau EIMI workshop, it is a good example of global trends within the educational interfaces between mathematics and of currents interactions between industry in developing countries and academic experiences.
9. *The Vertical Integration of Industrial Mathematics the WPI Experience*, by B. Vernescu from USA, describes several initiatives, including a project oriented education curriculum, introduced almost 40 years ago, at the Worcester Polytechnic Institute (WPI), has facilitated a major change in the mathematics education. Since the mid-1990s the WPI faculty have developed a successful model that introduces real-world, industrial, projects in the mathematics training, at all levels from middle and high school to professional masters, the Ph.D. program and faculty research. The faculty and students affiliated with the Center for Industrial Mathematics and Statistics have developed project collaborations with over 35 companies, businesses and government labs. These projects serve to motivate students to study mathematics and prepare them for interdisciplinary work in their careers. The development of several vertically integrated educational programs are referred and has had funding from NSF, SIAM, the GE Foundation, the Alfred P. Sloan Foundation and Intel.

Two other contributions were also selected and discussed, but since they are also considered in other Working Groups these two contributions appear in the general section of selected chapters in this book:

A Meta-analysis of Mathematics Teachers of the GIFT Program Using Success Case Methodology, R. Millman, M. Alemdar and B. Harris, from USA, reports on the interesting North American experience of the Georgia intern-fellowships for teachers (GIFT) project initiated in 1991. This successful initiative aims to providing first hand connections between classroom activities and real-world applications for teachers in grades six through twelve (students between 12 and 17-years old) with "real life" experiences in the applications of the school disciplines of mathematics, science, and technology.

An Introduction to CUMCM: China/Contemporary Undergraduate Mathematical Contest in Modeling, by Jinxing Xie. In the past thirty years, among the most remarkable changes in the educational interfaces between industry and mathematics, the introduction of mathematical modeling courses and a high successful and competitive mathematical contest in modeling are the most important. China university mathematical education in the past thirty years has significantly changed, and will continue to change, in particular the contents and forms of mathematical education. This chapter summarizes the history and current status, including the organizing, training and judging systems, of the mathematical contest in modeling in China universities. The aim of the contest is to expose students to the real-world challenges inherent to mathematical modeling and applications, and provide educational (creativity, challenge, etc.) experience unique to problem bare learning. The contest has been influenced by and has also significantly influenced the teaching of mathematical modeling and applications in China. According to Chinese practice, the contest provides a good educational interface between mathematics and industry.

References

Alberti, M., Amat, S., Busquier, S., Romero, P., & Tejada, J. (2010). Mathematics for engineering and engineering for mathematics: *Proceedings of the EIMI 2010 Conference*, Lisbon, Centro Internacional de Matemática, pp. 11–40.

Barbeau, E. J., & Taylor, P. J. (Eds.). (2009). *Challenging Mathematics in and Beyond the Classroom—The 16th ICMI Study*, Berlin: Springer.

ESF (2010). Mathematics and industry, An European science foundation forward look. http://www.esf.org/flooks.

Friedman, A., & Lavery, J. (1993). *How to start an industrial mathematics program in the university*. Society for Industrial and Applied Mathematics (Philadelphia).

Godlewski, E., & Pagès, G. (2010). Educational interfaces between Mathematics and Finance: the French context: *Proceedings of the EIMI 2010 Conference*, Lisbon, Centro Internacional de Matemática, pp. 227–233.

Godlewski, E., & Madaune-Tort, M., Dossou-Gbete, S. (2010). Two examples of a program 'Mathematics for Industry' at the master's level in a French University: Université Pierre et Marie Curie-Paris 6 and Université de Pau et des Pays de l'Adour: *Proceedings of the EIMI 2010 Conference*, Lisbon, Centro Internacional de Matemática, pp. 211–226.

Mathematics for Engineering and Engineering for Mathematics

Miquel Albertí Palmer, Sergio Amat, Sonia Busquier, Pilar Romero and Juan Tejada

1 Introduction

The Spanish Committee for Mathematics (CEMat) and its Committee for Education decided to participate at this International ICMI Study 20 with a report about the current situation and perspectives of the educational relations and connections between mathematics and industry in Spain.

ICMI-Spain has realized that there are several activities that are being developed in this field in the country, both at university and secondary education levels. Some Spanish universities and research centers are active members of the European Consortium for Mathematics and Industry (ECMI) and some of their directors are members of CEMat and its Committee for Education. There are instructors and innovation teams in secondary education involved in the development of studies about the relations between industrial activities and mathematics in secondary education. The "Dirección General de Formación Profesional" of the

M. Albertí Palmer (✉)
Department of Mathematics, Institut Vallès, 08206 Sabadell, Catalonia, Spain
e-mail: malbert8@xtec.cat

S. Amat · S. Busquier
Department de Matemática Aplicada y Estadística, University of Politécnica de Cartagena, Cartagena, Spain
e-mail: sergio.amat@upct.es

S. Busquier
e-mail: sonia.busquier@upct.es

P. Romero
Department de Astronomía y Geodesia, University of Complutense de Madrid, Madrid, Spain
e-mail: romero.pilar@upct.es

J. Tejada
Department de Estadística e Investigación Operativa, University of Complutense de Madrid, Madrid, Spain
e-mail: juan.tejada@upct.es

A. Damlamian et al. (eds.), *Educational Interfaces between Mathematics and Industry*, New ICMI Study Series 16, DOI: 10.1007/978-3-319-02270-3_17,

"Secretaria de Estado de Educación y Formación" of the Ministry of Education (MEC) has expressed its will to deepen into this field from the point of view of a professional education reform.

The present contribution brings together three works that consider, at secondary and university levels, the educational relationships between engineering, in a very broad sense, and mathematics. Such works take as its starting reference some experiences developed in Spain. The first part, developed by M. Albertí, is devoted to explore certain questions related to the previous issues in the context of secondary education. The second work, due to S. Amat and S. Busquier, analyzes, in a particular case, what type of math is more suitable to be taught to engineers. The third and last one deals with the development in several Spanish formative proposals of the concept of Mathematical Engineer, work developed by P. Romero and Tejada (coordinator).

The common framework of the contribution is exploring (from the formative viewpoint) the two-way road between mathematics and engineering. On one way, it is interesting to determine and include in the secondary education and university syllabuses the math that play a role in the real world when it comes to problem solving and applications. The other way is related to value the contribution to mathematical education of the methods, techniques, procedures, and experiences from the real world and engineering.

2 Mathematics and "Engineering" in Secondary Education

Recently, a research concerning mathematical activity at work contexts has been developed in the town of Sabadell (Catalonia, Spain). As a result of this study we have reached several conclusions related to certain statements presented in the ICMI-ICIAM's *Discussion Document* of the *International Study on Mathematics and Industry*. From our viewpoint, these conclusions are basic for a didactic approach of the problem in the secondary school. The main questions are the following.

First, it has been always said that Mathematics are used almost everywhere, but the contexts of its use and the way it is used may have changed through the last years. What several decades ago had to be solved with paper and pencil was solved later with calculator and it can be solved now with computer. Software often appears incorporated to machinery. Which kind of Mathematics is used today at work?

Second, the presence of Mathematics in real-life contexts has usually been seen, by professional mathematicians as well as by mathematics educators, as applications of Mathematics. However, it is not often like this. History of Mathematics shows that before becoming academic, many elements of Mathematics were grounded on practical problems. Axioms, theorems, and proofs are not the beginning, but the final stage of a process through which academic mathematics are settled. The use of Mathematics at work, is limited to applications of academic Mathematics? Do workers solve a problem in the same way than it does a mathematics educator?

Third, almost all mathematics educators come from an academic context and never have worked in another job different from teaching. There are three reasons for which practice work mathematical aspects should be considered in a truly competence mathematics education. First, educate citizens to be mathematically competent, and not only at work, but culturally. Second, to prepare students for the work world anticipating at class some of the activities they will have to face once finished the secondary school. And third, make clear the relationship between Mathematics and social and cultural environments. Then a question arises: how can a teacher promote mathematical competency coherent with a non-academic work or industrial environments if his or her mathematical education was developed outside these environments?

We suggest solutions to these three points of interest considering both teacher and student interests from an educational point of view. Teachers need to educate people mathematically competent. This means real life competency. To achieve this goal they need to put their students in real life situations of which work situations are a part of utmost importance. Students must learn mathematics and realize that the mathematical knowledge they have acquired is really useful to answer or to create answers to the problems they will face in their lives, also as workers, after secondary education.

Concerning to secondary education, the approach could be determined by three main questions related to the role to be developed by the different educational stages towards Industry:

a. *Industry and secondary education curriculum* Which are nowadays the general and actual frameworks between Education and Industry? What is studied concerning to Industry at the secondary school? Which relation or presence has Industry in the mathematics curriculum? Which mathematics is involved in those topics related to Industry that are studied in another matters?
b. *Mathematics for Industry* To know directly after interviews to workers and industrial business men which mathematics is used and/or needed at industrial field. Such a field is so huge that research should be limited to the most outstanding industrial sectors in the aim to precise a fundamental mathematical kernel.
c. *Industry and competency education* Which mathematical aspects and skills should be studied deeply at secondary school in the benefit of people's competency? It's desirable that someone who has finished secondary education can face his or her incorporation to the industrial work field, vocational training, or university technical studies.

The following considerations are concerned with our secondary school perspective. Industrial view should be obtained interviewing industrial professionals. Our starting point is the identification, development and valorization of a curriculum, including innovative applications of mathematics and highlighting industrial mathematical problems. But then the point is: Which are the 'industrial mathematical problems'?

To specify an adequate level for workers means an agreement between educators and professionals. It can be achieved through a mathematical education by competences. After passing secondary school stage one should be mathematically competent to face a professional training industrially oriented. But, do the teacher and the professional share the same idea of competency? Maybe they complement each other? Anyhow, it is essential to know differences and intersections to know which ones must enter into mathematical secondary education.

Student activities at secondary school should fit three aspects. First, they should not be isolated from reality. Second, extra academic activities for students should be addressed towards the knowledge of industrial world, visiting in situ some industrial enterprises according to their relevance, availability or spatial proximity. Third, education authorities should promote, as it is the case of Catalonia (Spain) the realization of work practices out of school in industrial enterprises.

But not only students must be the focus. Teachers must know what happens outside of their academic context. Teacher training by nonacademic professionals can lead to mathematical developing, both in its knowledge and educative aspects.

Several characteristics of the industrial field are absent in secondary education and in mathematics education as well: competitiveness, production, and valorization method of results. Lack of competitiveness, often existing between pupils, is usually vanished by teachers and educators, probably as a consequence of the traditional lack of it between public education professionals. Nevertheless, some competitiveness between educators is now fomented by authorities.

There are nowadays reasons for introducing and intensifying the academic weight of realistic and productive activities to be developed individually or in group, in collaboration and cooperation, the aim of which transcends teacher numerical qualification. Education in search of the 10, and not only the teacher's one, but also the student's and social environment's one. This means to introduce a bit of *industrial culture* into the secondary school.

2.1 Mathematics at Work: Not Only Applications of Mathematics

To know which Mathematics is used and how it is used at work it has been elaborated a *Guide of Work Mathematical Activity* (Albertí 2009, pp. 100–102). Given the wide extension of the context, the study was limited to actually accessible jobs for a people after having finished secondary school in the town of Sabadell (Catalonia, Spain). This guide relates mathematical activities developed at work with the five content groups of the Spanish secondary school curriculum. All these work mathematical activities were identified after interpellation to professional workers of every job. It has been confirmed, according to the ICMI document, that workers still think that only something truly hard deserves to be considered as mathematical. For instance, workers do not see as mathematics the calculation of the area of a room or calculated estimations of the measure of a magnitude.

Jobs studied came from different labor sectors. But all of them were closely related to Industry in one of two senses: first, because the job belonged to an Industrial sector, as it was the case car industry, and second, because nowadays most of the tasks are delegated to special technology and such a technology usually comes from Industry. As mentioned above, the main mathematical activities at each of these works were related to the five contents of secondary education: (1) Numeration and calculation; (2) Change and relations; (3) Space and shape; (4) Measure; and (5) Chance and statistics.

After all we got a series of mathematical activities rooted on work contexts. Five activities shared by almost every job were identified as UWA (universal work activities), where often one find mathematical activity: (1) To follow hygienic and security rules at work; (2) To follow chief work orders and instructions; (3) Budget estimations, stocktaking and cashing up; (4) Label, map and operating manual interpretations (including software); (5) Creation and interpretation of graphics and maps.

Teaching–learning capabilities were checked to bring them to academic contexts in the aim that students experiment at classroom several of the real-world mathematical activities they will need to overcome once finished their secondary school education. Is Mathematics actually applied in practice? If so, does its applications agree with those applications referred by mathematics educators in the classroom? In this sense, we have found several discrepancies.

In *Building* right angles are traced on the basis of the so called by workers as 'Egyptian triangle,' i.e., a triangle of sides 3, 4, and 5 m or a smaller proportional one of sides 30, 40 and 50 cm. Workers know that such a figure produce a right angle, although they don't know a proof of it. In fact, they don't need to know such a proof. Nevertheless, Mathematics educators not only should know this proof, but the proof should belong to the mathematical education they offer to their students. Pythagorean theorem is made of two implications. The first one is the most popular and usually presented in secondary education classrooms in Spain. But people working in building make right angles thanks to the other implication, which never is proved, studied, or showed in the classroom.

In *Multimedia Design Production* is interesting to save money inserting the maximum number of visiting cards (just to put an example) into one sheet of paper. Experts do the task experimentally, but it could be done easily preparing an EXCEL page to determine their maximum possible number and the way they should be placed into the sheet of paper. So we did it. It was a spatial optimization problem solved with EXCEL which solution was truly welcomed by the professional.

Car industry workers have to know how to interpret graphics. But often they need to do in a way not usually studied in academic contexts. Academic graphic interpretations too often look at the function involve too closely, the teacher asking for images and anti images, growing and decaying intervals. But in reality, sometimes, these are not the most relevant factors. The mathematician and the mathematical educator must be careful and work together with professionals when developing mathematical models to fit real practice. Let us not forget that theoretic solutions to practical problems often are not the best practical solutions of the problem.

Work and Industry is a source of real mathematical activities. Math activities from work contexts are real, practical, and contextualized problems where teachers can situate students. They are authentic situations not invented nor supposedly real according to teacher's imagination. Demidovich (1976): 91 contextualizes in Building a popular problem of constructing an optimal fence from a wall. *Pisa* 2003 *Study* (2005) declares as work context its problem 22, consisting in calculating how many different pizzas can be prepared combining the available ingredients. During our research we had the opportunity to ask about these questions to workers. No one answered affirmatively, pizza makers looked at us astonishingly.

Mathematics coming from work contexts cannot and should not determine the scope of Mathematics secondary school curriculum, but part of it should be incorporated at it. If not, on which aspects can we base the relation of our matter with reality? Such a relation should not be left to teacher's imagination and creativity. This would not be honest and it could transmit the perception that such a relationship is arbitrary, depending on the teacher hobbies. To prevent it, it is necessary that teachers of Mathematics know what happens outside their academic context. So they will see Industry as a source for mathematical problem solving and creation from where to get authentic and real mathematical activities to be developed at classroom.

3 Mathematical Competences in the Engineering Curricula Versus Industrial Demands

The relation between university, research institutes and industry is now more active. As mathematics provides the language for communication and discovery in many scientific disciplines and modern industry, it is very important to incorporate the use of mathematics in the new engineering curricula in a satisfactory way. However, this task is not trivial and many problems have to be solved. Mathematics, including statistics and computing, has to be one of the most important parts in almost any science or engineering education. Mathematicians provide a logical thinking and a language that explains many complicated processes. It is crucial in the technological development.

The link between pure mathematics and its applications in the industrial progress is a crucial task. In particular, the engineer has to obtain the necessary competences to develop this collaboration. In the last decade the collaboration between the university and the industry has increased considerably in the Region of Murcia (Spain). Many research groups in the Polytechnic University of

Cartagena (UPCT) have industrial projects. This fact has pointed out the need to review the mathematical competences in the new curricula of the engineers.

This study is organized into four parts. The first will present the wished competences. The new curricula elaborated (to start in 2010) are analyzed in the next part. As a particular example, that is very important in our region, we detail the mathematical competences in the Agricultural Engineering. The following part explains the type of collaboration with the Industry of the Region of Murcia in Spain. Finally, we present some discussions and conclusions by analyzing the acquired mathematical competences in the university in comparison with the demanded in the agricultural industry.

The new curricula encourage the acquisition of both transversal and specific competences. The transversal competences can be divided into three subgroups:

a. *Instrumental transversal competences* Ability to analyze and synthesize; Ability to organize and plan; To communicate suitably both orally and written in Spanish; To communicate suitably both orally and written in English; Basic computational abilities; Ability to use the information; Resolution of problems; Decision making;
b. *Personal transversal competences* Critic and auto critic ability; Teamwork; People relations; Ability to work in a team; Ability to communicate with other experts; To show knowledge and understanding of the diversity; Ability to work in an international context; Ethical approach in the different fields of the profession;
c. *Systematic transversal competences* Ability to apply the knowledge in practice; Ability to knowledge; Adaptation in new situations; Creativity; Ability to become a leader; To show knowledge and understanding of others′ cultures; Ability to work on your own Initiative*;* *Qualité*; Motivation.

The specific competences depend on each degree. The mathematics has a transversal nature and the above competences can be incentivized in our courses. Moreover, it can be used to obtain specific competences in many different situations. The specific competences are related with the type of work that the student can be done. The UPCT is a polytechnic. There are several engineering degrees with different specific competences, however only one department of Applied Mathematics and Statistics. The material prepared by our department has several particularities depending on the specific characteristics of each degree. Basically, in order to obtain its specific competences, the real applications are always related with the degree.

The specific mathematical competences are: Ability for the resolution of mathematical problems in engineering. Applications of mathematical tools related with: linear algebra; geometry; differential geometry; differential and integral calculus; ordinary differential equations and partial differential equations; numerical analysis, statistics and optimization. The use of MATLAB program in engineers problems.

Traditionally mathematics and engineering have been learning in different separate courses. In particular, there is usually too big a gap between the different fields during the learning process. In the last two years we have worked in decreasing the distance between the different areas in our university. The collaboration between researchers, teachers, and industry is very important.

A particular situation can be found in the Agricultural sector. Many groups in our university work with the industry of this sector and many deficiencies in the actual curricula have been detected. With the new curricula in the Agricultural Engineering at UPCT and with the mathematical courses on Algebra and Calculus, Differentials Equations, Applied Statistics, Numerical Analysis and Informatics, we aim to respond to new challenge by taking into account working life requirements. Moreover, it is motivated both the own work and the teamwork. The learning process is active, the use of computer tools has been increased and the applications and the relation of the mathematics with real problems emerge as the key point.

But, is it enough for the industrial applications? To answer this question, we will analyze the acquired mathematical competences in comparison with the demanded in the agricultural industry.

The agro food sectors in the region include: Horticulture, Fruit culture, Canned, Citric, Almonds and Cereals, Olive and Olive Oil, Vineyard and Wine, Fishers and Aquaculture, Farming, Meat Industry. The contribution of the agro food sector at the region economy is the most important. The implication of the government is also very important, as the public expenditure in Euros for I+D+I in the region is increasing every year.

On the other hand, in our university, the Higher Technical School of Agricultural Engineering has a very nice international reputation. This School has some leader research groups that have active collaboration with the agro food companies. In most of these projects high mathematical competences are necessary. Actually, our department works with these groups in order to improve these competences. The necessary mathematical tools include: basic statistic, nonlinear least-square, ANOVA, mathematical models using statistic or differential equations and computer applications.

It is clear that Agricultural industry needs professionals with high mathematical competences. As we have described, the different sectors use all the mathematical tools that compose the new curricula. Thus, the goal will be to obtain an engineer that can work fluently with a mathematician (or any scientific) and give clever solutions to the industrial problems. Of course, we need also a compromise from the other side: the mathematicians. In fact, the need for a strong collaboration between the different parts: schools, universities, businesses and governments, concerning the promotion of mathematics, science and technology, should be unanimously recognized. The main reason for such an initiative is to overcome the dissatisfaction of industry with the quality of specialists emerging from universities.

The pedagogical proposal in our department could be classified halfway between the classical teaching techniques and the new pedagogical proposals introduced for adapting teaching methods to the new European Framework for Higher Education. The innovation is focused on the use of teaching videos developed by the teachers themselves as a complement for classroom lessons. The proposal was divided into three stages. The first stage was developed during the months of September and October (the first two months of the first semester in Spain). At this stage, the contents of the subject were introduced with lectures. Then workgroups of 2–3 people were formed, and personalized meetings were held with all the groups. Next, during the second stage in November, the videos made by the teachers were given to the students. The videos explained both the theory and the problems derived from that theory. The exams from previous years were provided. During this month, we had only computer-practice lessons. In the last stage, during the months of December and January, questions raised by the students were resolved, and the students were evaluated. The results were strongly positive.

In conclusion, we would like to point out that mathematical technologies have the potential to bring real world applications to life in the mathematics classroom. Collaboration between researchers, teachers and industry seems crucial. Moreover, this task should be iterative introducing the necessary changes during the process. Many people are working in order to improve the curricula of the engineers and the collaboration between schools, universities, businesses, and governments seems also very important.

The new curricula seem to go the correct direction but we would appreciate the real change in the near future. But, it is a very difficult task. The gap between the different areas and between the university and the industry has to decrease. Moreover, most research takes place at departments of universities. Professors and are involved teaching of basic courses and at many universities their research is considered a personal hobby. Moreover, the majority of first-year students we receive at our universities are poorly prepared. In particular, the industry is unable to find enough engineers who are sufficiently educated by our universities. We hope that with the new curricula and the new methodologies this situation will change. In our opinion, we have performed the first step in the good direction.

4 Mathematicians as Engineers

Finally, in this third part, we shall consider the main characteristics of some master programs developed in Spain under the common title of Mathematical Engineering. These kinds of programs have been developed in Spain in the last years although they have a longstanding tradition in, for instance, United Kingdom, France or Italy.

In the previous section, an analysis of mathematics contents and skills has been made, that, in a particular case, must be a part of the engineers training, i.e., it is about mathematics for the engineering, about mathematics for industry. From another point of view, in this section we examine some proposals that could develop skills or contents that may contribute to the formation of a specific type of mathematician who has the skills and attitudes that could be considered typical of an engineer training. Taking into account the training proposals that are being offered in this sense, two typical denominations can be found in Europe: Industrial Mathematics and Mathematical Engineering.

The first one is a usual denomination within the titles offered by universities and colleges belonging to the ECMI, which are consistent with the name of the Consortium itself. There are also training proposals (degrees, masters and doctorates) named Mathematical Engineering in countries like France, United Kingdom, Italy or Spain, which in some cases, are offered by universities that also belong to the ECMI.

The professional performance of a mathematician in the industry, business, or management can vary from one country to another depending on the economical structure of each of them. That is, the development opportunities as an industrial mathematician, in a country with a rich and autonomous industry such as France, may be very different from those one could find in a country such as Spain, whose economy is focused on services and highly dependent on the foreign know-how. This situation may partly explain why the denomination of Mathematics Engineering has been adopted in Spain against Industrial Mathematics so far.

The first question that we want to address is the following one: Does such training proposals agree with the same professional profile? As we will see the answer is that, even though they have many common features overall (the usage of mathematics in solving real problems), they differ in essential aspects with regard to the professional performance that may be expected from the students of the different training proposals. In such case, the second question we propose is if it is worthy to define a common professional profile of the mathematical engineer.

One previous issue that cannot be ignored is that of denomination, that is the extent to which the denominations agree with the contents and how they communicate their objectives. According to our opinion, the name of Industrial Mathematics is an evolution of that of Mathematics for Engineering original concept. Such designation would be appropriate at some times and in some countries where the industrial weight was very important. However, the ECMI itself was forced at one point to explicitly extend the field of its activity beyond the pure industrial environment, in the classical sense, to the more general field of business and management as it is showed with the introduction of the Techno-Mathematics and Econo-Mathematics terms. As referred in the Discussion Document of this study, "the term "industry" obviously refers to a diverse range of activities, producing goods and services. Under constraints such as time and money, these activities generally attempt to optimize limited—sometimes scarce-resources, both material and intellectual. The overarching goal is to maximize benefits for certain groups of people while, ideally, minimizing harm to other groups and the natural environment." In this sense it can be said that although the denomination of Mathematical Engineering fits better the proposed professional competences, it represents a strong anomaly as it is the only Engineering discipline that references the instruments used, instead of the specific area of developed activity.

However, it can be good reasons to keep, in some countries at least, the designation of Mathematical Engineering. In the economic activity sectors, the applied mathematician competes with disadvantage with classical engineers limiting them to a role of subordination and support to those engineers. This unfavorable competition occurs just from the entrance of the students in degree studies. At least in Spain, there is a very strong social pressure for qualified students in mathematics to choose classical engineering degrees. There are indications that Mathematical Engineering can change somewhat this situation as it is observed with the experience of the Mathematical Engineering degree offered, for instance, by the Polytechnic of Milano and the more recent one of the Complutense University of Madrid.

Leaving aside, for the moment, the issue of the denomination, the question is if these proposals are directly focused to the development of a professional profile with features clearly differentiated from other previously existing. As it has been already said, the answer is not clearly affirmative, even under the same denomination. We analyzed the content and main characteristics of three Spanish Master Degree Programs in Mathematical Engineering offered at present by the Carlos III University of Madrid (UCIIIM), the Santiago de Compostela University in cooperation with the Coruña and Vigo Universities (USC-UDC-UVIGO) and the Complutense University of Madrid (UCM), respectively. Although the general aim of these three Masters is to develop mathematical, statistical, and computational skills to solve technologic o scientific problems in business and industry and other related task with focus on competencies as modeling, programming, and simulation as well as ability to handle huge amounts of data by integrating mathematical, numerical, and statistical methods, we have found different results when trying to reach the declared goals.

First, the durations of the Master Programs are different; ranging from the UCIIIM Master, which is spread over two years and covers a total of 120 European Credit Transfer System (ECTS) to the UCM Master with 60 ETCS in total (and till 60 ECTS as previous complementary formation depending of the background of the entering students), corresponding to a year study. On the other hand, in the USC Master are required courses up to 90 ETCS. A more detailed analysis of the course content offered us deeper information about the differences in the knowledge of the mathematical contents and practical competences that these Masters provide to students, and make us to easily identify the core mathematics curriculum needed to achieve the goal of providing qualified students with extensive, detailed knowledge of the theory and methods combined with a high level of practical competence in Mathematical Engineering. It must be said that all the used information about the Masters on Mathematical Engineering was obtained from the web pages. It was assumed that all the essential characteristics, in order to interest to potential students, are reflected on them.

After exploring the details of these Programs of 2009–2010, we made a comparative analysis, concluding that curricular issues are different in some aspects described in what follows. First at all, a general distinction can be made into Master Programs orientated to one specific field, let us say, for example to Econo-

Mathematics or Techno-Mathematics, as UCIIIM, and, on the other hand, with a more global orientation as the UCM Master Program. Although it is not excluded an academic or professional goal, the UCIIIM and USC- UC-UCVIGO Masters were considered as Research Masters and, therefore, linked to a Ph.D. program. On the other hand, the UCM Master program is seen, as a whole, as a professional Master program. This distinction is crucial since these goals could mark differences between the contents and the skills required in order to obtain them.

The structure and contents of the Masters also established important differences between them. On one hand, the MUSC-UC-UVIGO Master presented a classical offer in Industrial Mathematics that is supported by an excellent research group in the field. On the other hand, the UCIIIM Master, since it lacks a common part, was actually a group of four Masters. Finally, the UCM Master contains an important common part, which helps to ensure almost identical competences between the students. It deepens less in a particular area, but it tries to get the students adapted easily to different professional challenges.

The contents do not completely determine the competences attained in a Master's degree, since the methodology is especially relevant when we try professional training. Only the UCM Master refers explicitly to a learning methodology based on case studies and on experience modeling and solving real problems but some activities developed in the other masters could be based in the same methodology.

The duration of the bachelor degrees in Spain affects in an important way the Master degree offers. After a four-year degree, to complete a training period equally in ECTS to those offered in other European countries, it is only necessary a year Master's degree of 60 ECTS. However, this approach makes difficult the harmonization between Spanish Masters and European Masters without penalizing the Spanish students with an extra year or without declining the possibility to join the Spanish masters to other European students. The disparity of proposals regarding the duration of the three studied masters faithfully reflects this situation. Particularly complicated could be adapting them to the general guidelines that are being discussed within the ECMI orientated to a proposal of European Master on Industrial Mathematics. For instance, the duration of the Master Thesis of the UCM Master is very far away from the mentioned guidelines.

About the employment factor, there are also differences between them. Due to the current situation in the Spain industrial activities, it is hard to obtain permanent positions for the students with the mentioned profiles. Actually, for certain singular problems a consultant is hired, whose survival will depend on the continuous realization of projects. In this type of consulting highly specialized professionals could be needed, professional with an excellent mathematical background to tackle highly complex problems. On the other hand, if what we want is to train people to make them have a strong quantitative profile in several different areas and with experience in adaptation to others, it might be needed a more transversal orientation towards problem solving. It is predictable that in the near future we will be having a higher number of companies interested in this last type of professional, a kind of quantitative engineer since they are used to work on different areas.

Finally, would be worthy to define at European level a versatile professional profile in the field of mathematics with a broad orientation to employment in industry, business, and management having engineer's own competences and requirements? That is would it be worthy to formally create the figure of Mathematical Engineer? We think that any proposal on this figure must take into account the main characteristics of engineer training.

As a conclusion, we think that is worth to propose the professional profile of a mathematical engineer. The training process of such a professional must take into account an appropriate balance between Mathematics, modeling activities coupled with another discipline such as Statistics or Physics with the aim to develop knowledge, understanding and experience of the theory, practice, and application of selected areas of Mathematics but also including Statistics and Operations Research and learning mathematics not only in mathematics courses but including context-based mathematics courses in other disciplines that employ significant mathematical methods so that students will be able to use this skills and techniques of these areas to solve problems both of a routine and of a less obvious nature arising in industry, commerce, and the public sector and they could function comfortably in work situations that require quantitative and analytical skills to solve more general mathematical problem. But we think that in order to define the professional profile of the mathematical engineer, these specific requirements must be developed under a learning methodology that guaranties the acquisition of the skills, attributes, and qualities of an engineer.

References

Albertí, M. (2009). *Activitat matemàtica a l'àmbit laboral a l'inici del segle XXI. Implicacions per al currículum de l'ESO.* Departament d'Educació de la Generalitat de Catalunya. Servei d'Innovació i Recerca Educativa. http://phobos.xtec.es/sgfprp/entrada.php.

Coe, R. (1999). *Changes in examination grades over time: Is the same worth less?*

Croft, A. (2000). A guide to the establishment of a successful mathematics learning support centre. *International Journal of Mathematical Education in Science and Technology, 31*(3), 431–446.

Demidovich, B. (1976). *Problemas y ejercicios de análisis matemático.* Madrid: Editorial Paraninfo.

Estudio Pisa 2003 (2005). *Pruebas de matemáticas y de solución de problemas.* INECSE. Ministeriode Educación y Ciencia. Monografía SUMA 03. FESPM. Madrid.

Gil-Albert, F. (1999). Tratado de arboricultura frutal. Volumen III: Técnicas de plantación de especies frutales. Ministerio de Agricultura, Pesca y Alimentación. Ediciones Mundi-Prensa. Madrid.

Lawson, D. A., Halpin, M., & Croft, A. C. (2003). *Good practice in the provision of mathematics support centres* (2nd ed.). LTSN MSOR Occasional Publications Series, 3/01. ISSN 1476 1378.

LMS. (1995). *Tackling the Mathematics Problem.* London: London Mathematical Society, Institute of Mathematics and its Applications, Royal Statistical Society.

Luk, H. S. (2005). The gap between secondary school and university mathematics. *International Journal of Mathematical Education in Science and Technology, 36*(2, 3), 161–174.

MacGillivray, J. L., & Cuthbert, R., (2003). Investigating weaknesses in the underpinning mathematical confidence of first year engineering students: *Proceedings of Australasian Engineering Education Conference* (pp. 358–368). The Institution of Engineers: Australia.

Maddocks A. P., Dickens J. G., & Crawford A. R., (2002): *Encouraging lifelong learning by means of a web-based personal and professional development tool*. Manchester: ICEE 2002, UMIST, 18–22.

Perkin, G., Croft, A. C. (2004). Mathematics Support Centres- the extent of current provision. *MSOR Connections, the Newsletter of the LTSN Maths, Stats and OR Network*, 4(2), pp 14–18. Available http://mathstore.ac.uk/newsletter/may2004/pdf/supportcentres.pdf.

Proceedings of the 1999 British Educational Research Association Annual Conference. Brighton, UK Sept (1999). Available from http://www.cemcentre.org/research/examchanges/BERA2.html.

Robinson, C. L., & Croft, A. C. (2003). Engineering students—diagnostic testing and follow-up. *Teaching Mathematics and its Applications, 24*(4), 177–181.

Sala, M. (2002). *Análisis sectorial de la industria catalana. Proceedings of VIII Jornadas de Economía Crítica*, ISBN: 84-8448-146-8.

Smith, A. (2004) Making Mathematics Count—The report of Professor Adrian Smith's Inquiry into Post-14 Mathematics Education. February 2004. The Stationery Office, 2/04937764.

Tariq, V., & Cochrane, A. C. (2003). Reflections on key skills: Implementing change in a traditional university. *Journal of Education Policy, 18*(5), 481–498.

Laboratory for Computational Mathematics: An Interface Between Academia and Industry

A. Araújo, S. Barbeiro and J. A. Ferreira

"If Europe is to achieve its goal of becoming the leading knowledge-based economy in the world, mathematics has a vital role to play" (MACSI-net 2004). In spite of the fact that in many industrial sectors the value of mathematics is already proven, there is a need for positive action to promote the use of mathematics by European industry. A dynamic mathematical community interacting actively with industry and commerce, on the one hand and the science base, on the other, has been pointed out as important ingredients for competing in the global market of the future where innovation will be the key to success.

It is widely perceived that graduate education in mathematics focuses almost exclusively on preparation for traditional academic research careers. Also, because of the interdisciplinary and diversity which non-academic employers typically demand, the knowledge of technical areas outside mathematics is of utmost importance in non-academic positions. In Portugal, the number of PhD and Master graduates with a degree in mathematics is small and the quantity of those having non-academic careers is almost imperceptible. Nevertheless, some indicators show the tendency for a slight change. With this scenario, there is the perception that the creation of institutional high level connections between academic mathematics groups and industry could produce a favorable impact (Vicente 2006).

The activity in Laboratory for Computational Mathematics. The laboratory for computational mathematics (LCM) was created in April 2005, integrated in Centre for Mathematics of the University of Coimbra (CMUC), a research center that comprises most of the research-active members of the Department of Mathematics of

A. Araújo (✉) · S. Barbeiro · J. A. Ferreira
CMUC-LCM, Department of Mathematics, University of Coimbra,
3001-501 Coimbra, Portugal
e-mail: alma@mat.uc.pt

S. Barbeiro
e-mail: silvia@mat.uc.pt

J. A. Ferreira
e-mail: ferreira@mat.uc.pt

A. Damlamian et al. (eds.), *Educational Interfaces between Mathematics and Industry*, New ICMI Study Series 16, DOI: 10.1007/978-3-319-02270-3_18,

the University of Coimbra (DMUC). Currently, the Centre has 76 members holding a PhD and 14 research students. The Centre makes a significant contribution to fundamental mathematics and includes research in Algebra, Analysis, Numerical Analysis, Optimization, Probability and Statistics, Geometry, History and Methodology of Mathematics.

The foundation of LCM followed the recommendations regarding the CMUC activities proposed by the research unit evaluation panel in 2002 (FCT 2002): "Computational Mathematics and Numerical Analysis are important subjects on which Portugal is somewhat lagging behind in spite of isolated pockets of expertise. The panel believes that this is well positioned and has the capacity needed to lead a nationwide initiative and to provide a solid foundation on which to build a major Center of Excellence in Scientific Computing/Computational Mathematics in Portugal." During its 8 years of existence, LCM has been promoting research in computational mathematics and scientific computing as techniques for the solution of challenging problems arising in biological sciences, engineering, finance, and management. The activity of LCM includes interdisciplinary research, high-performance computing and the development of numerical software, in collaboration with industry and also promotes PhD and Master's level education.

The relationship with industry. The term "industry" has been used to denote business and commercial firms, research and development laboratories, commercial and not-for-profit research and production facilities, i.e., activities outside the sphere of education and purely academic research. As an interface between the University and industry, one of LCM's major tasks is to identify relevant social and industrial problems that should be tackled. The Portuguese Science Foundation, which supports most of the projects in LCM, has been considering the relevance of the project toward obtaining comparative advantages for Portugal core criteria for funding, in accordance with the objectives stated in its strategic plan. To identify such problems, contacts outside the mathematics community are important. To pursue this goal, LCM organized several workshops (www.uc.pt/uid/lcm) and an European Consortium for Mathematics in Industry (ECMI) study group (miis.maths.ox.ac.uk), where industry was invited to present problems that could be undertaken by mathematicians.

The project portfolio of the LMC includes 17 application-oriented projects, some having industrial partners, namely hospitals, banks and bank holding companies, data and software companies and industrial firms. As an example, we highlight the project "Simulation of a Moving Bed Reactor used in the Pulp and Paper Industry". An important mill of the major Portuguese firm Portucel, which is one of the world's biggest producers of bleached eucalyptus Kraft pulp for the packaging industry and one of Europe's top five producers of uncoated wood-free paper, is located near Coimbra. In order to optimize the quality of the pulp, this industry has a real need for tools that enable the simulation of experiments that cannot be afforded or that might be risky in a real industrial context. The most critical piece of equipment in a Kraft pulp and paper plant is the digester, a complex heterogeneous reactor where a moving bed of wood chips contacts and reacts with a combination of chemical products, in order to dissolve lignin and therefore to release the fibers of cellulose. The incidence of the work developed at LCM has twofold: from an engineering point of view, to give a description of the transient behavior of the digester; from a mathematical point of view, the project gave the possibility to study a new kind of numerical methods, specially tailored to the phenomena that take place in the digester. "Reaction–diffusion in porous media" is another research project with many relevant real applications, namely the contaminant transport in groundwater, which, in Portugal, has a big strategically interest. The final goal of this project is to predict the evolution of this kind of pollution and to define strategies to reduce its environmental impact. The models and the software package developed can also be used to study biological filters.

We must point out that in most of the projects at LCM, the degree of commitment of the companies involved is still less than it would be desirable since, in general, they do not contribute with funding. In addition, many of the on-going projects are engineering real problems but they do not target any particular industrial application. However, the laboratory has strong interactions with engineers, physicists, physicians, and finance specialists from Portuguese universities and also with computational mathematicians from elsewhere in Europe and in the US. The cooperation is formalized in a number of organizational frameworks, apart from other numerous contacts with colleagues all over the world.

There has been a strategical effort to develop a variety of mechanisms to facilitate a constructive relationship between mathematics and industry. To achieve this objective, contacts and collaboration with industrial partners are vital. Since 2007, LCM is strongly committed with the organization of the ECMI European Study Groups with Industry in Portugal.

European Study Groups with Industry (ESGI) were originated in the UK in 1968 under the name Oxford Study Groups with Industry. The concept has been adopted by other countries, and study groups have become a well-established institution and the leading workshop for interaction between mathematics and industry in Europe and in the world (Freitas 2009). The purpose of these one-week meetings is to strengthen the links between academia and industry by using mathematics to tackle industrial problems that are proposed by industrial partners.

The first study group to be hosted in Portugal was the 60th in the ESGI series. It took place, one year later the 2006 Meeting of the Portuguese Mathematical Society, at the Instituto Superior de Engenharia de Lisboa and counted with the collaboration of several British specialists including John Ockendon FRS. Two problems were presented, one on traffic flow monitoring, proposed by BRISA, and another on a Stewart platform simulator, proposed by FunZone Villages. The second Study Group in Portugal was held at the University of Porto in 2008, and it is being held regularly on a year basis.

In April 2009, Coimbra hosted the 69th edition (www.mat.uc.pt/esgi69). The work in Coimbra was focused on five problems: "Optimizing a complex hydroelectric cascade in electricity market", "Management of stock surplus", "Estimating the price elasticity of water", "Fraud detection in plastic card operations", and "Reliability of a customer relationship management", proposed by the

Portuguese firms REN, SONAE, Águas de Portugal, SIBS and Critical Software, respectively. The experience was very fruitful, both for the University and the industrial partners. At the end, the firms were asked to answer a questionnaire and the answers reveal the desire of the industry scientists to stay in contact with current research being carried out at the universities.

Besides the study groups, another way to strengthen the links with the outside community and to create institutional connections with local industry is to offer a number of short-courses, with topics of interest to both industry and academia, open to members of the University as well other professionals and industry.

The educational programmes. An important part of the mission of LCM and DMUC is the transition and integration into the job market of its students. For this reason DMUC runs a career service, giving students a first working experience, preparing them for a better integration in the job market. The cooperation between the University and industry from all over the country has made this career service a success.

Several of DMUC's former students are now working in industry. When they were asked about their academic preparation, they all tended to agree, according to what it was also pointed out in the SIAM report (SIAM 1998), "that they were well educated for several important aspects of non-academic jobs: thinking analytically, dealing with complexity, conceptualizing, developing models, formulating and solving problems. However, many felt inadequately prepared to attack diverse problems from different subject areas, to use computation effectively, to communicate at a variety of levels and to work in teams." Taking into account this scenario, there has been an effort to incorporate modifications in undergraduate mathematics curriculum in order to overcome these drawbacks. The number of really problem-oriented courses is increasing and every student must have contact with courses that link mathematics and computing. The students are also encouraged to organize a regular interdisciplinary seminar focusing on a large variety of themes. Sometimes, non-academic mathematicians are invited to talk about their work.

The topics covered by LCM have a prominent place in the educational programmes at DMUC, especially at the Master level. The applied Master programs of DMUC are divided into several areas: Applied Analysis and Computational Mathematics, Computation, Statistics, Optimization, and Financial Mathematics. In these Master specialties, the students are in contact with real problems. Some companies, like Reuters, Critical Software and Mercer, support these Masters. Students have the possibility to develop their MSc thesis in these companies, being supervised by a member of the mathematical faculty and a member of the company staff. Quantitative Methods and Financial Mathematics is another applied Master program, shared by DMUC and the Faculty of Economics of the University of Coimbra and involves companies like Bloomberg, Goldman Sachs, and the Portuguese banks Millennium BCP and Montepio Geral.

One of the goals of the LCM is to incorporate students in its projects. Since its foundation, LCM gives visibility to the work developed in CMUC in applied areas and as consequence, the number of MSc and PhD thesis made under LCM projects has increased. We give some examples. One thesis where the major goal was to develop a platform to help the users of public transportation in Coimbra to obtain the best path to travel between two points was done in the framework of a European Project named Civitas, a partnership between Critical Software and the local public transport firm. In other thesis, software to treat the clinical data of the Portuguese Society of Cardiology using both parametric and non-parametric variance analysis was developed. There are also several students that developed their PhD thesis or have on-going work in the scope of LCM projects. As an example, we mention the thesis "Memory in diffusion phenomena", developed within the project "Non-fickian diffusion in polymers and medical applications" which studies mathematical models to simulate diffusion phenomena in materials with memory like polymers. Another PhD thesis, entitled "Controlled drug release" develops a mathematical model and a software package to simulate the drug delivery from contact lens loaded with drug and containing particles, also loaded with drug, dispersed in the polymeric matrix. Both thesis have interdisciplinary character involving chemical engineers from the Chemical Engineer Department and a medical doctor from the Faculty of Medicine.

Since "applications have been the driving force in the science and Mathematics" (Friedman and Littman 1994), LCM strongly supports the idea that applications are extremely useful to motivate the teaching in mathematics. But, apart from the great effort to introduce real-world applications, we believe, agreeing with OECD report (OECD 2008), that "curriculum should not become a light version of the accepted curriculum for future researchers." The students that want to study industrial mathematics should "be familiar with the standard canon of the discipline." In spite of the work that has been done, there is an urgent need for more training in the area of industrial mathematics. It is essential to attract bright students to this area and to convey the challenge and the excitement of solving practical problems.

Conclusions and strategy for the future. The success stories indicate that there has been an increasing interest in strengthening the relation between academia and industry and we feel that mathematics can provide a competitive edge for Portuguese industrial organizations. Based on our own findings and on the experience of other similar laboratories, we are lead in two directions: building better relations with non-academic organizations promoting the role of mathematics; developing strategies that might be useful to encourage shifts in the curricula with the objective of promoting closer ties with industry.

References

Freitas, P. (2009). *European study groups with industry at 40 years*, SIAM News, 42, n 2, March http://www.siam.org/news/news.php?id=1537.

Friedman, A., & Littman, W. (1994). *Industrial mathematics: a course in solving real-world problems*. Phyladelphia, SIAM.

FCT (2002). Comments and recommendations regarding the unit activities, research orientation and application of funds. *FCT report for CMUC.*

MACSI-net (2004). A roadmap for mathematics in European industry. *MACSI-net Newsletter*. Retrieved from www.macsinet.org/newsletter.htm.

SIAM (1998). The SIAM report on mathematics in industry. *SIAM*. Retrieved from www.siam.org/about/mii/.

Vicente, L. N. (2006). Matemática industrial em Portugal. Análise e perspectivas. *Fundação Calouste Gulbenkian, Tema Ciência e Sociedade*. Retrieved from www.mat.uc.pt/lnv/papers/industrial.doc.

Improving the Industrial/Mathematics Interface

Jean P. F. Charpin and Stephen B. G. O'Brien

Mathematics is ubiquitous in the sciences and engineering. Arguably a science is considered to have come of age when it has become sufficiently mathematical as illustrated by the burgeoning areas of mathematical biology and mathematical finance. Despite all the potential applications, some mathematicians have moved away from industry and from real applications. From the late 1960s under the influence of the Oxford group [Alan Tayler and Leslie Fox, see Ockendon and Ockendon (1998), Tayler (1990)], interest in modeling real industrial problems has steadily grown. Industrial mathematics and its near synonyms is problem-driven mathematics for the sake of the sciences while pure mathematics may be regarded as being mathematics for its own sake. In this context 'industry' is interpreted in a very broad sense: the remit of these groups includes more than collaboration with industry (problems may come from anywhere in the sciences, e.g. mathematical biology, mathematical finance, the environment). Collaborations between 'industry' and mathematicians can prove extremely successful; study groups with industry are an excellent example. These are typically weeklong intensive sessions involving mathematicians and industrialists/scientists who propose the problems. But in order to fully exploit industrial activities such as these, with the intention of improving the interface between industry and mathematicians, we must introduce modifications in the training of mathematicians. We suggest that there should be an element of industrial mathematics in every third level mathematics group. We will discuss the development of such activities in the context of experience with the 'Mathematics Consortium for Science and Industry (MACSI)', a network of applied mathematical modellers across Ireland.

J. P. F. Charpin · S. B. G. O'Brien (✉)
MACSI, Department of Mathematics and Statistics, University of Limerick, Limerick, Ireland
e-mail: stephen.obrien@ul.ie

J. P. F. Charpin
e-mail: jean.charpin@ul.ie

A. Damlamian et al. (eds.), *Educational Interfaces between Mathematics and Industry*, New ICMI Study Series 16, DOI: 10.1007/978-3-319-02270-3_19,

The terms 'industrial mathematics', 'applied mathematics' and 'mathematical modelling' are often used as near synonyms to make the distinction with pure mathematics whose central tenet is formal proof and which is not generally concerned with real problems arising outside of mathematics. An applied mathematician is a kind of lapsed pure mathematician in the sense that (s)he would like to prove every result formally but is sometimes unable to do so and must make intuitive leaps in the search for understanding. Mathematical modellers excel at paring problems down to their essence and using mathematical tools to discover the essential mechanisms, which govern processes. Otherwise, for example, the flow of air over airplane wings would not be understood. Classical applied mathematics is associated with names such as Archimedes, Newton, G. I Taylor, Stokes, Reynolds, Kelvin all of whom regularly used mathematics to understand phenomena in the physical world, essentially operating as mathematical modellers.

The Mathematics Subject Area Group (2002) was unanimous in identifying three skills which it believes every mathematics graduate should acquire: the ability to conceive a proof, to solve problems using mathematical tools and to *model a situation.* In point of fact, many groups do not recognize the latter as a key skill. The concept of modelling is central to the practical applications of industrial and applied mathematics. The UK Smith Institute (2004) identifies mathematical modelling and simulation as the core outlet for applied mathematics in Europe. Furthermore this document states that: 'Mathematics now has the opportunity more than ever before to underpin quantitative understanding of industrial strategy and processes across all sectors of business. Companies that take best advantage of this opportunity will gain a significant competitive advantage: mathematics truly gives industry the edge'.

1 Evolution of the Training

Many mathematics courses do not include a real modelling component and this neglect continues through to postgraduate level and beyond. Of course, many courses claim to have a significant modelling component but this is often achieved by relabelling an old course (typically a differential equations course). Unfortunately, some mathematicians pretend to be modellers and to solve real problems in order to obtain research funding.

In fact, they pick out problems which they can deal with but are typically of no interest to the industrial collaborator. The reality is that employers seeking mathematical graduates would like them to have genuine modelling skills, to really understand the discipline which they are modelling, to care about the significance of the results which they obtain and hence to have a broad scientific background. It is important that students gain exposure to suitable texts, e.g. Fowler (1997), Ockendon et al. (1999), Mattheij et al. (2005), which espouse the real philosophy of modelling. Modelling starts at the application level and a key part of the process is the formulation of the mathematical problem. Very few undergraduate third-level courses even touch on this extremely difficult topic. Most mathematics students are used to being presented with nicely formulated mathematical problems, which have equally elegant solutions. In reality, the hardest part of the process often involves asking the right question! Once the problem has been formulated and solved (hopefully), just as important is the interpretation of the solutions for the benefit of the industrial/scientific partner. What do the solutions really mean? How can the knowledge gained be put to use?

Genuine applied mathematical modelling deals almost exclusively with problems which arise outside of mathematics: in physics, chemistry, biology, finance, economics and industry in general. In many traditional pure mathematics undergraduate courses the emphasis is placed almost completely on mathematical rigour and technique with no genuine attempt to develop physical intuition and a feeling for real problems. In effect, the current generation of pure mathematicians is training the next generation to be like itself, to be logically rigorous and to prove theorems. In this philosophy, mathematics is closer to logic than science or engineering. There is nothing wrong with this: such mathematics involves significant intellectual activity. However, what industry needs is mathematicians who are genuine scientists and who are interested in solving real problems. In the controversial words of Dr. Bernard Beauzamy (2002), trained to doctoral level in pure mathematics, and later to set up his own mathematical consultancy company, 'it (pure mathematics) brings solutions which nobody understands to questions nobody asked'. As already mentioned, applied mathematics is organically linked with the sciences, engineering and industry. Strengthening links with industry requires a well-designed training: while providing students with a fundamental mathematical education (logic, rigour etc.), courses can also provide them with a set of practical skills which are useful in the real world. This means, naturally, that courses must provide them with more genuine mathematical modelling skills and at least introductory courses in the connected disciplines. The degrees should involve industrial internships and even exposure to experimental work. Students must learn that the genuine application of mathematics involves starting with a real problem (physical, chemical…) and will usually mean communicating with engineers, physicists or any other scientist with practical knowledge of the problem, translating that problem into mathematics, solving the mathematical problem and interpreting the solutions in such a way that is meaningful for the practitioners at the source of the problem. Educating present and future mathematicians about the close links between applied mathematics and industry is not limited to the classroom. This important aspect can take many forms. A natural route is to set up an industrial mathematics group in every mathematics department (Friedman and Lavery 1993). This has been the approach at the university of Limerick, where MACSI (www. macsi.ul.i.e.) has developed a range of activities to stimulate links between present or future mathematicians and industry. These include: *Outreach Lectures* for the younger students and the general public, to stimulate interest in industrial applications of mathematics, to show the importance and ubiquity of mathematics in everyday life and that mathematics is the underpinning technology for modern society; a *Summer School*, organised every year to introduce secondary school students and first year university students to modelling concepts, including during the week always a site visit to one of our industrial partners; a summer *Internship Programme* in MACSI, where undergraduate students are confronted with the reality of applied mathematics and its industrial applications and work in conjunction with an experienced researcher on a topic suitable for their background; *Modelling Workshops*, which are organised on a regular basis with a typical session lasting up to a day and involving an industry representative, postgraduate students and research staff; and finally, *Modelling Classes* given by experienced practitioners occur several times a year.

The course topics can range from standard modelling techniques like asymptotic and non-dimensionalisation to the presentation of study group reports and examination of real scientific problems. Coupled with an adequate and varied under-graduate curriculum, these activities familiarise mathematicians of all levels with real-life problems and promote long-term links with industry.

2 Study Groups with Industry

Study groups with industry are organised on a regular basis on all continents. In the European context, study groups with industry (as initiated in Oxford University in the 1960s and continued under the umbrella of the European Consortium for Mathematics in Industry (ECMI)) are week long meetings where groups of industrialists, mathematicians and other scientists work intensively on problems proposed by the industrialists.

The study group format is standard. Mathematicians and other scientists gather for approximately a week with industrial collaborators to find solutions to a set of problems proposed by the latter. The first morning, industry representatives present the problems. One must appreciate that the problems presented are usually not mathematical problems to begin with. Typically, they are descriptions of a complicated industrial process, which is not well understood from a scientific point of view. Usually, there is a specific question of the type 'How might we prevent this happening?' Sometimes, the request is vaguer to the effect that if we can help to model the situation, something useful may come from the mathematical solutions. When all problems have been presented, the academic/scientific participants select the problem(s) they would like to work on. The first afternoon, subgroups of the scientific participants meet with each industry representative and ask far more detailed questions. Ideally, at the end of the day, the team should have defined in broad terms the approximate goals for the week. It is important to realise that in some cases, a successful outcome at the end of the week may be a properly formulated mathematical problem (i.e. the correct mathematical question).

During the rest of the week, the group works on the problems and progresses towards a solution. Participants may choose two strategies. Either they focus on a single problem all week or they may decide to hover between different problems. It is really a matter of taste. Some people like to work intensively on one problem:

other prefer to make smaller contributions to a number of problems. The industrial partner may or may not attend all sessions. (S)he also should be easy to reach if more information is required. A mid-week progress report may be required when the groups present their results. On the last day, all groups present their results to the industry representatives and the other academics. A report describing the work of the group is written in the weeks following the study group and given to the industrial partner.

In a study group, one is confronted with real applications of mathematics and must focus on getting significant results quickly. However, the first study group one attends can be an overwhelming experience, particularly for students. For this reason, specially designed students sessions are often organised. The goal is to familiarise students with the concept of a study group and with some of the standard techniques and ideas, which commonly occur. To achieve this, before the study group, an experienced mathematician runs a session where (s)he uses one or several past (and generally simplified) study group problems with the students. (S)he plays the part of the industrial representative and presents the problem to the students. Using the solution previously obtained, (s)he can guide the students and help them rediscover the results. Working in groups, the students all contribute to the final result and this constitutes an excellent first contact with industrial problems.

MACSI organised two study groups, ESGI62 and ESGI70 in January 2008 and June 2009 respectively. Topics included *Fluid mechanics*: spin coating and self-assembly of di-block copolymers, initiating Guinness; *Chemistry*: improvement of energy efficiency for wastewater treatment; *Biology:* lubrication of an artificial knee; *Electronic*: blowing up of polysilicon fuses, the effect of mechanical loading on the frequency of an oscillator circuit; *Engineering:* polymer laser welding, polishing lead crystal glass, solar reflector design; *Environment:* designing a green roof for Ireland; *Finance:* on the estimation of the distribution of power generated at a wind farm using forecast data; uplift quadratic programming in Irish electricity price setting.

These examples reflect traditional subjects arising in study groups. Optimisation, population modelling and medical problems are other traditional topics. One of the most unusual problems was submitted to mathematicians in Bristol in 2003. Participants were asked to study the artificial incubation of penguin eggs. (In the end, this involved modelling the fluid mechanics inside the eggs.) The problems submitted to study groups are extremely varied but they reflect the skills that are expected from a mathematician doing modelling: although daunting at first, they may be split in a series of subtasks and interim models which are much easier to tackle.

A successful study group is extremely beneficial to all the parties involved:

- Mathematicians and scientists are introduced to new and original problems. A single experience at a study group is enough to convince most mathematicians that industry is an extremely fertile source of interesting problems. They get the

opportunity to apply their skills to new exciting problems and get the opportunity to see if the mathematics really works;
- Industrial collaborators obtain direct access to academic expertise. Universities, among other things, are repositories of knowledge and excellence but too often these resources are not easily available to the people who need them most. In the short term, a problem is usually at least partially solved for the industry (or at least some insight has been gained) and this often encourages them to look at their problems from a new perspective and to look for newer and innovative solutions. In the long run, the research carried out during a study group can deliver real solutions for the industrial partners and may lead to patents and genuine financial gains. Potential clients of the industrial partner are often impressed by the improved scientific approach;
- Study groups are unique scientific occasions and are more productive than traditional conferences. Excellent work relationships often develop leading to long-term collaborations.

Study groups are a very important way of improving the industrial/mathematics interface. If study group participants are completely alien to most of the concepts necessary and do not have the appropriate scientific training to solve industrial problems, this sort of initiative is bound to fail. The interface between industry and mathematics can only improve if appropriate training is offered to young (and older) mathematicians.

3 Conclusion

Developing a successful interface between mathematicians and industry requires much effort but the rewards are tangible. Industry provides new and interesting scientific problems; mathematicians provide insights, which allow the industrialists to improve their products. We note in passing that many scientific councils are now including economic relevance among the key requirements in proposals for research funding. Mathematical modellers hold a unique position in the scientific world with their ability to interact with practitioners in so many different areas in industry, the sciences and engineering. In our opinion, to date this advantage has not been exploited to its fullest extent, partly because of the divide in the mathematical world between pure and applied mathematics. It is our thesis that the way forward involves changing mathematics curriculums and placing more emphasis on real mathematical modelling. Study groups with industry are clearly one of the most significant ways of strengthening the industry/mathematics interface but such interactions can only further develop if the mathematicians develop a set of skills more suited for real industrial problems.

4 Acknowledgements

The authors are supported by the Mathematics Application Consortium for Science and Industry (MACSI) funded by the Science Foundation Ireland Mathematics Initiative Grant 06/MI/005 and IvA award 12/1A/1683.

References

Beauzamy, B. (2002). Real life mathematics. *Irish Mathematical Society Bulletin, 48*, 43–46.

Friedman, A., & Lavery, J. (1993). *How to start an industrial mathematics program in the university*. Philadelphia: SIAM.

Fowler, A. C. (1997). *Mathematical models in the applied sciences*. Cambridge: Cambridge University Press.

Mathematics Subject Area Group. (2002). Tuning educational structures in Europe. In towards a common framework for mathematics degrees in Europe? *Newsletter of the European Mathematical Society,* Sept, 26–28; http://www.emis.de/newsletter/newsletter45.pdf.

Mattheij, R. M. M., Rienstra, S. W., ten Thije Boonkkamp, J. H. M. (2005). *Partial differential equations: modeling, analysis, computation.* SIAM, Philadelphia.

Ockendon, H., & Ockendon, J. R. (1998). Alan Taylor (Obituary). *Bulletin of the London Mathematical Society, 30*, 429–431.

Ockendon, J., Howison, S., Lacey, A., & Movchan, A. (1999). *Applied partial differential equations*. Oxford: Oxford University Press.

Smith Institute. (2004). Mathematics: giving industry the edge. http://www.smithinst.ac.uk/News/Roadmap/Roadmap.

Tayler, A. B. (1990). Mathematics: an industrial resource. *Physics World 3,* 23–24.

Two Masters on 'Mathematics for Industry' at the Universities of Paris and of Pau

Edwige Godlewski, M. Madaune-Tort and S. Dossou-Gbete

Over the years, the range of applications of mathematics has been progressively enhanced and applied mathematics, modeling, and simulation have taken an increasing role. This motivated the French Department of Education to launch a Diploma in applied mathematics at the master's level 30 years ago. The two first examples of such programs were created in different contexts, one in Paris, one in Pau, a smaller provincial town. These programs have evolved according to the European implementation of the Bologna process; they have taken into account the new range of applications, for instance in Finance, and profited from the spectacular development of computer science.

The DESS (*Diplôme d'Etudes Supérieures Spécialisées*) in applied mathematics were created in French Universities in the late 1970 [see CNE (2002)]: they correspond to the fifth year of higher education, the graduate program is chosen only in the fifth year, but coherent with the four previous years. The first diploma DESS in applied mathematics was created in Université Paris 6 (the largest scientific French university), now called *Université Pierre et Marie Curie-Paris 6* (or UPMC), in 1977. It was then called *DESS de mathématiques appliquées* and was backed up by the competence of the French greatest laboratories in applied mathematics (numerical analysis, probability and statistics) and continuum mechanics. It appears that the second one was created in 1979 in Université de Pau, now *Université de Pau et des Pays de l'Adour* (UPPA), a smaller multidisciplinary University in the southwest of France. The name of this last program changed in 1998 in *DESS Ingénierie Mathématique et Outils Informatiques* in order to enhance the close link between mathematics and computer sciences at industry level. This University benefits from a favorable industrial environment

E. Godlewski (✉)
Université Pierre et Marie Curie, Paris, France
e-mail: edwige.godlewski@upmc.fr

M. Madaune-Tort · S. Dossou-Gbete
Université de Pau et des Pays de l'Adour, Pau, France
e-mail: simplice.dossou-gbete@univ-pau.fr

A. Damlamian et al. (eds.), *Educational Interfaces between Mathematics and Industry*, New ICMI Study Series 16, DOI: 10.1007/978-3-319-02270-3_20,

thanks to the presence of important industries: Elf Aquitaine (now Total) in oil exploration and Turbomeca (now in Safran group) in aeronautics (helicopter engines).

The French organization of higher education, following the Bologna declaration on the European space for higher education in 1998 has undergone a change to implement the Bologna Process and passed to the so-called LMD frame (for *Licence—Master—Doctorat*). All courses are organized in compliance with the European Credit Transfer System (ECTS) of credits accumulation. Hence the *DESS de mathématiques appliquées* in UPMC has become in 2004 a Master's degree level program *Master Ingénierie mathématique, Mathématiques Pour l'Entreprise* (MPE) which exactly means 'Mathematics for Industry' in the sense given in this study. The MPE program is specialized in a second year (M2) of a master's program, which lasts 2 years, the 1st year (M1) being in common with other programs. In the same way, the *DESS de mathématiques appliquées* in UPPA has become in 2004 a Master's degree level program in Applied Mathematics with two possible curricula: Mathematics, Modelling and Simulation (MMS) and Stochastic and Computational Methods for Decision making, *Méthodes Stochastiques et Informatiques pour la Décision* (MSID). This change of name reflects a greater awareness in the teaching teams of the importance of identifying a 'core curriculum of mathematics for industry' or, at least, the teams wanted to make the program 'more appealing and exciting to students and the professionals in industry'.

The situation in France is specific because of the coexistence of two separate tracks for the training of students: Universities and Schools of Engineering (*Écoles d'ingénieurs*). While in Schools of Engineering, less and less mathematics are being taught (outside Mathematical Finance) since more time is given to management and economy, in French Universities the traditional high level of training in mathematics has been more or less preserved and is appreciated in industry, so that French students following this path still find good job opportunities thanks to their specific skills. However France is presently facing a debate on education at all levels (school, training of teachers, universities). Several reform projects may have a deep impact on our programs.

The UPMC Master 'Mathematics for industry'—The professional Masters degree *Ingénierie mathématique* in Paris offers presently two possible curricula: 'Mathematics for industry' and 'Financial engineering and random models'. In these two curricula, students have little choice and benefit from a supervised training, which does not allow to having a job outside. However, some students need to earn their living and face difficulties at being present all week long. For some other students, the supervision is too heavy, and they need more freedom. There was for 4 years a third curriculum *Outils mathématiques et logiciels* (OML), which was meant for these students. It was stopped because too few of them were able to manage the two activities at once. For the few who succeeded in carrying on this program, it was a good program, and we still think that if more students were interested, we could open the curriculum again, with little additional forces. The present consequence is that a few low income students, those who do not benefit from a grant, do not succeed in getting their master's degree in engineering mathematics, which we deplore. The first curriculum exists since 2004, and truly since 1977 since few changes were brought with the transformation from a DESS to a master's degree. This does not mean that it did not evolve for 30 years, but the first principles that lead to its creation are still believed to be active. For what concerns the second curriculum, *Ingénierie financière et modèles aléatoires*—IFMA, it was open later, in 2006, to face the increasing demand of both students and the banking and insurance industries (more precisely capital markets) for high level formations in applied mathematics including stochastic modeling and simulation using numerical probability methods. We will not analyze this program, which is the object of another study on Mathematics and Finance (Godlewski and Pagès 2010). Note that in the past few years, more students enrolled in mathematics at the master's level attracted by the positions and salaries offered by the later though unemployment does not affect the former program. But traditional manufacturing industries did not communicate enough on the need of well-trained engineers or of hiring mathematical scientists.

The study will focus on the first curriculum, the *Mathématiques Pour l'Entreprise* or MPE program. The program covers a wide range of applied Mathematics: numerical analysis, partial differential equations, scientific computing, and either probability, statistics and mathematical finance, or mechanics. Graduates of this program will have a dual profile of mathematician and engineer, able to model various phenomena and to develop new methods of numerical simulation. Careers are possible in a number of sectors using scientific computing, mechanics, stochastic modeling, or statistics: R&D departments in public organizations or industry, software developers and service providers, studies and estimates departments in banking and insurance. Two curricula are possible with a common core syllabus in numerical analysis and scientific computing (whose courses are in numerical analysis and computation, mathematical software, advanced programing and algorithms): probability–statistics, with an initiation to quantitative methods in finance; mechanics of continuous media (fluids and solids), jointly with the Mechanics and Energetic specialization of the Engineering Science master (concerns few students).

Each curriculum comprises elective courses in the first semester and ends with a practical training period, an internship in industry (usually 6-month long), completed by a formal report and an oral defense. Conferences introducing companies and career guidance workshops are organized. Some training in English is also provided, with the possibility of passing the Test of English for International Communication (Toeic) to certify language skills and enhance international competitiveness. The precise list of courses can be found on the website www.ljll.math.upmc.fr/MPE. The list has evolved with time, for example courses on matlab, C++, MPI, Java, VBA, Fluent, GPU,… and new optimization algorithms (genetic) were gradually introduced, maintaining the highest level in mathematics, with reinforced ability in computer science together with good control of programming language and common software.

Comparing data from 2004 to 2010, one can observe that the number of students coming from a first year of the master's degree at UPMC is increasing and also indicate a very good success ratio (90 %) thanks to a selective entry and a close training, following individually each student. There is also a percentage of 11 % of students undergoing a doctoral program. The doctoral positions offered after a professional program are often funded, thanks to a collaboration between an enterprise and the French ministry of research (so called *CIFRE conventions industrielles de formation par la recherche*). In the 40 % 'found job' column, we have all kind of (sometimes short time) positions found by our students in the year after the program. Employment concerns mostly service industry: finance, insurance, CAD, aircraft design, automotive engineering, civil engineering, petroleum engineering (oil exploration), energy, image and signal processing, telecom, networks, communication industry, possibly pharmaceutical industry... There are broad possibilities, in both low- and high-technology industries. It is difficult to have a precise employment rate on a longer period because of the lack of data, students requested to stay in touch with the faculty forget to do so; after some time, few of them answer our request. The two statistics/mechanics curricula yield similar rates, the first more naturally in studies and estimates departments in banking and insurance.

The UPAA Master 'Mathematiques et applications'—The organization of the former DESS and present Master in Pau has been conceived taking into account the already mentioned industrial environment in petroleum engineering (Total) and aeronautics (Turbomeca). Specific required skills have been identified, leading to the definition of appropriate courses. However, a broader range of applied mathematics is covered so as to ensure more professional prospects to the students. There are two possible curricula: *Mathématiques, Modélisation et Simulation* (MMS), and MSID.

We do not give details on the MMS curriculum since the content does not differ much from that of the above MPE diploma. Note, however, that it may also lead to a doctoral program, so that the training periods may take place in a University laboratory, which is not possible for a professional program. Moreover in collaboration with the University of Zaragoza (Spain), the program may be joint to a Spanish one (*Licenciatura de Matemáticas*) and become a Spanish-French diploma for students spending part of their training in each university.

The MSID curriculum trains the student in the statistical and computing methods of treatment and data analysis, in the advanced programming and in the management of the computer systems. The students stemming from this program will have acquired the skills required to exercise the professions where the administration of databases and the statistical methods generally are indispensable to the production of syntheses of relevant data for the decision making. Main topics are: statistical modeling, data analysis, experimental design, use, development and adaptation of software for the IT and statistical processing of data, conception and administration of databases and information systems, conception of Internet applications for the management, the administration and the data processing, running and management of projects in research and development.

Professional prospects are in industry and services (development, planning for local government, environment, biology and medicine).

The theoretical and professional courses are gathered as follows:

- computing: Computer systems, Databases and advanced Programming
- geomatics: geographical Information system—stochastic methods: applications of the stochastic simulation, statistical methods, data mining, and image processing.

The curriculum in statistics at graduate level has been set up at UPPA as a response to the combination of the following facts. In the early 1980s, the number of graduates was low relative to the need of statistical well-trained collaborators in the companies; many mathematical sciences departments in the French universities had a Ph.D. program in statistics but there were no programs specifically devoted to the training in applied statistics (statistics for business and industry) at graduate level. The specific situation at the university in Pau where most of the faculties of the mathematics department maintained close relations with companies at local as well as national level in the field of industrial mathematics; the statistics group of this department was not strong enough to support a Ph.D. program in statistics but its skills enabled it to consider setting up training in applied statistics at graduate levels. Besides, the debate within the French professional societies for statistics had shown the necessity to strengthen the teaching of statistics at a high education level and to create professional-oriented training in statistics; moreover there were government incentives that lead to the creation of the DESS.

The student flows in UPPA since the beginning of the Masters level program show a regular growth of the number of graduates from 2005 to 2008 and then a sudden drop. Such a fall is a common phenomenon for science studies in French universities. However, the number of students interested in our programs is quite good compared with programs in other domains of mathematics in France, due to the teaching of mathematics for industry in our curricula. Indeed, the number of students preparing for the Master degree is greater in 2010 than in 2009. Altogether, there are 27 (14 for the MMS, resp. 13 for the MSID curriculum). As the success ratio is close to 100 % thanks to a selective entry, it is expected more graduates in the future and therefore a stability of the flows.

The main industrial sectors receiving students for training periods are: petroleum engineering with industries such as Total (21 %) or research organisms such as IFPen, aeronautics (15 %) with industries such as Turbomeca or research organisms (51 %) such as CNES, computer engineering, banks (10 %) and insurance companies, food processing, administrations (regional and local authorities). Students generally do their period of practical training in the region; due to an additional financial cost only 2 or 3 % do it in the Paris region. From 2005 to 2009 for the two curricula, the graduates finding a job in Industry (31 %) work in the industrial sectors listed above; some of them stay in the company where they did their training period. Other graduates (22 %) study for a Ph.D. doctorate, among them 30 % are sponsored by an industrial company.

Important features for both Paris and Pau programs—Let us emphasize some of the main lines that guide our teaching projects: All the faculty involved in teaching courses are specialists of their domain (applied mathematics, computer and information science, statistics or mechanics) and have contact with research in industry using applied mathematics. Some of them even work in a company but have a high level training (for instance PhD) in applied mathematics; Emphasis is laid both on analysis (theory with axioms, definitions, theorems, programming languages, algorithms…) and some use of 'black boxes' (students may learn to use them by just knowing the great line of inner workings). So, the program lies somewhere between edutech (teaching and learning with technology) and indutech (using technology in industry). Since it is impossible to provide a relevant program with depth in all areas using applied mathematics, a few of them only have been chosen, all of them for their relevance (they are used in the applications, though not always at this level). For instance modeling with pde (partial differential equations), numerical analysis of some methods and simulation are in the core of the MPE-UPMC or MMS-UPPA programs; this is linked to the talents of the local teaching team.

The idea is that students having been well trained on these chosen topics ought to: learn quickly many other topics in applied mathematics; understand a problem modeled with a meaningful mathematical formulation; have ability in new computational implementations; know that black boxes are built with something inside, which they could understand and be able to analyze results obtained even with black boxes.

At least that is what is expected, and what is indeed observed for a large part of our students. To be honest, the program prepares them more at applying existing analytical tools and computational techniques than at innovating or discovering new tools and techniques. However, in a world of growing complexity and economic competitiveness you can hardly innovate from a low scientific level, having learnt just the first basic rudiments. We do think it is useful to have (and hope our best students are left with) sufficient and broad enough scientific background to be able to benefit from a specific professional formation in the industry and then enhance innovation if their company aims at being innovative. This does not concern all our students; some of them find jobs which do not need such competencies, but the latter are appreciated, besides their computational skills, for instance for their good organization skills, reliability, analytical minds, qualities often associated to a training in mathematical science.

We are also conscious that more qualifications are needed for a job in industry, for instance communication skills and ability to work in a team (Friedman & Lavery 1993). Not minimizing the importance of these, our belief is that our competence lies in high-level applied mathematics and that this training can help them discover latter new tools and techniques. Thus, the way we choose to prepare them involves a lot of up-to-date high-level classrooms. Besides the long training period gives our students a good opportunity to get some of these skills. Eventually, our main justification is that the program is professionally recognized. The interest of companies for the students from these programs is their complementary training with a solid grasp of theory and excellent computing skills.

Contacts with industry are crucial for both students and faculty members. *For the students*: during the first semester in university, career guidance workshops are organized which help them to writing curriculum vitae and cover letters, preparing to ask for positions. Some meetings or professional seminars with alumni, research engineers, professionals are held about once a week. If not numerous, these first contacts are important and valuable and make a difference with classical academic programs. Besides, a few courses are taught by professionals: for UPMC, a research engineer introduces to parallel computing, another initiates students on Code_Aster (which is an Open Source software package for numerical simulation in structural mechanics). The program is completed by a long training period. For the professional programs all the internships are immersed in industry, most often in a private company or in some public research institution. The student receives some allowance for his work. An advisor from the company has defined some assignment and supervises the student's work. The student must write a master's 'thesis' or memoir, in which he or she describes his activity in the company, the tasks he has fulfilled and possibly adds some material concerning the mathematical modeling involved. At the end of the period, he gives a formal talk to present his work to the faculty committee at the university, in presence of his company advisor. The evaluation is done in concert, taking into account several features among which not only the academic skills but the students involvement and his willingness to understand the company's requirement.

For the professors, every year, contacts with all the companies where the students have done their training period improve their knowledge of the employment market and the needs from industry. For instance a course on software (Fluent, Python) used in the companies receiving students has been introduced in the MMS program.

Critical evaluation and long-term analysis. The present situation is indeed specific in several respects, due to the reforms of the school system and the debate linked to the coexistence of two different programs for the training of students, Universities and Schools of Engineering (in French *Ecoles d'ingénieurs*) and among them, the elite colleges called *Grandes Écoles*.

Ingénieur is a specific diploma delivered only by the *Ecoles d'ingénieurs* together with the grade of master. It is better valued by industry than a corresponding university diploma and considered as a key to well paid jobs. The training of engineers may be provided as part of a component of a university, as in the Polytech network; even in this case, there are few connections between these specific components of a university and the classical curriculum (master's degree) taught at university. The training supposes the acquisition of generic knowledge of the kind that permits later changes of career or the pursuit of studies 'throughout life' more easily than the more specialized master's degree in science and technology. Besides, if passing the competitive entrance exams of *Grandes Écoles* requires years of preparation and knowledge of the French education system, they are seen as the country's premier path to prosperity and power. Students from working class or immigrant families often lack this knowledge and thus follow more frequently university programs. Indeed, the democratization of the French university even if partial and limited, is far greater. In France, university is open to nearly every one, with little filter at entrance besides *baccalauréat* and offers very low cost programs for students, while giving them the opportunity of achieving a high level training as our Master's program do. If more low-income students can attend *Grandes Écoles*, they will be taken among the universities' possibly best students. It is already partly true since *Grandes Écoles* try to attract the students with good potential and allow them to enter their program in their second year of training. It will be even truer if the government forces its most prestigious schools to be less elitist about the students they let in, which is the object of an actual debate. The debate might also accelerate the merging between the two different programs.

The French Universities are affected by the recent LRU, i.e., the Liberties and Responsibilities of Universities. Aimed at radically renewing French universities, the LRU act passed in 2007 and established the right of universities to become autonomous in budgetary matters within 5 years and its implementation began in earnest only in 2009. It already affects deeply the French educational system and this will be even more so in the forthcoming years. Moreover, many initiatives came lately trying to gather University and *Grandes Écoles*, in particular because research is much more active in the university laboratories than elsewhere. This is true also in Applied Mathematics. In order to favor the creation of 'poles of excellence' that are aimed at improving the ranking of French universities, the LRU reform encourages competition between public institutions of education and research. The two universities in Paris and Pau may thus evolve quite differently.

The French school system is again the subject of reforms that may have a deep impact on the level of our students. For instance, the number of hours taught in mathematics in *Lycée* could still decrease; the last year before *baccalauréat* it has already decreased from 9 h in the 1980's to 6 h in the 1990's, and is presently 5 h and a half (plus 2 h optional which few students take). In primary school too, the number of teaching hours in school has lately decreased by 2 h. The ministry of education is also modifying deeply the structure of degree programs that prepare secondary school teachers and the content of competitive exams required for teaching in primary and secondary education; the training the teachers would receive in their original area of interest (thus in mathematics as far as we are concerned) could suffer. Last but not least concerning higher education, the reform of *agrégation*, which many students aiming at an academic career used to prepare, by postponing by 1 year the date of the competitive exam could dissuade the Ph.D. students from preparing the *agrégation de mathématiques* which is a very competitive exam as well as leading the future teachers away from Ph.D. programs in mathematics. Some fear that these reforms will have a negative impact on the French mathematical school and more generally for the school system. No doubt that the long range consequences on our programs will be important.

A further critical evaluation including an evaluation of the Strengths, Weaknesses, Opportunities, and Threats (SWOT analysis) of our educational projects in a context of relation with industry can be found in (Godlewski et al. 2010).

We hope the numerous initiatives to promote applied mathematics at all levels in industry will converge in particular to an increasing number of students enrolling. Concerning our programs, we think they have been successful programs, and hope they will progress. While they have evolved continuously in two different locations for 30 years, taking into account local constraints, they still look much alike and are coherent. However, the above-mentioned reforms in France make difficult a prospective analysis. Let us take a date and compare them in 10 years!

References

CNE. (2002). Les formations supérieures en mathématiques orientées vers les applications, rapport du Comité National d'Evaluation, juillet 2002, France ISSN: 0983–8740.

Friedman, A., & Lavery, J. (1993). *How to start an industrial mathematics program in the university*. Philadelphia: Society for Industrial and Applied Mathematics.

Godlewski, E., & Madaune-Tort, M., Dossou-Gbete, S. (2010). Two examples of a program 'Mathematics for Industry' at the master's level in a French University: Université Pierre et Marie Curie-Paris 6 and Université de Pau et des Pays de l'Adour. In Proceedings of the EIMI 2010 Conference, Lisbon, Centro Internacional de Matemática, pp. 211–226.

Godlewski, E., & Pagès, G. (2010). *Educational interfaces between Mathematics and Finance: the French context.* In Proceedings of the EIMI 2010 Conference, Lisbon, Centro Internacional de Matemática, pp. 227–233.

Mathematics in Industry and Teachers' Training

Matti Heilio

Mathematics, modelling and simulation, so-called mathematical technology, is emerging as a vital resource to achieve competitive edge in knowledge based industries and development of society. This vision about the role of mathematics has inspired efforts to enhance knowledge transfer between universities and industry. Especially, it means a challenge for university education. A modern view of mathematics should be reflected in curricula and educational practices. This has implication for the way how mathematical modelling should be inserted to the curricula at various levels. The main focus of this article is on undergraduate teaching at tertiary level. However, some important implications are suggested concerning the schools preparing students for universities.

The development of technology has modified in many ways the expectations facing the mathematics education and practices of applied research. Today's industry is typically high tech production. Sophisticated methods are involved at all levels. Computationally intensive methods are also used in ordinary production chains, from brick factories to bakeries and laundry machines. The increased supply of computing power has made it possible to implement and apply computational methods. Terms like mathematical technology, industrial mathematics, computational modelling or mathematical simulation are used to describe this active contact zone between technology, computing and mathematics. A mathematical model is assumed to represent the structure and the laws governing the time evolution of the system or phenomenon that it was set out to mimic. Once we are able to produce a satisfactory model, we have a powerful tool to study the behaviour and hence to understand the nature of the system.

M. Heilio (✉)
Lappeenranta University of Technology, Lappeenranta, Finland
e-mail: matti.heilio@lut.fi

A. Damlamian et al. (eds.), *Educational Interfaces between Mathematics and Industry*, New ICMI Study Series 16, DOI: 10.1007/978-3-319-02270-3_21,

1 Industrial Mathematics, Educational Challenge

The computer age has generated a need and a window of opportunity for a new kind of expertise. This presents a challenge to the educational programmes and curriculum development. Some universities already offer specialised MS-programs oriented towards the industrial needs. A good educational package would contain a selection of mathematics, computing skills and basic knowledge of physics, engineering or other professional sector. It would be very important to train oneself to work in a project team, where the interpersonal communication is continuously present. To become a successful applied mathematician ready to tackle the fascinating tasks and challenges, development questions in modern industry, the student need a solid and sufficiently broad theoretical education and operational skills in the methods of applied mathematics.

There is a special need to revise the university pedagogy of applied mathematics. Reflection to the preparatory levels of high school should be elaborated. Regarding the undergraduate programmes at universities the following means should be considered: revision of syllabi and curricula, use of computing experiments, data tools and novel teaching methods. The curriculum should contain knowledge of theoretical mathematics and a collection of applied courses. The students should have also knowledge of some applied science or application fields, knowledge in physics, engineering or other "client discipline" of mathematics. The quality of classroom examples is important. Problem based learning, topical fresh exercises are called for. Mathematics teachers should have interest for different areas of modern professional life. Modelling examples can be found from work places, hobbies, talking with people from various professions, sometimes reading newspapers with mathematically curious mind.

However, the most important single skill is the experience in projects. For successful transfer of mathematical knowledge to client disciplines the theme of mathematical modelling is a crucial educational challenge. The lectures, books and laboratory exercises are necessary, but the actual maturing into an expert can only be achieved by "treating real patients".

2 Modelling as a Course Subject for Industrial Mathematics as Profession

The computer age has generated a quest for new special expertise. Many universities offer specialised MS-programs that equip the students with the skills that are needed in the mathematical projects in the R&D-sections in industry. The job title in industry is very seldom that of a mathematician. It can be a researcher, an engineer, a research engineer, systems specialist, development manager. Industrial mathematics is teamwork. Success stories are born when a group of specialists can join their expertise and visions together in a synergic manner. The teamwork makes communications skills a necessary matter. It would be very important to train oneself to work in a project team, where the interpersonal communication is continuously present.

To become a good applied mathematician one should be curious about other areas as well, to be interested and learn basic facts from a few neighbouring areas outside mathematics. Examples from industrial math projects illuminating the educational challenge come from Economics and management, Flow phenomena, Systems design and control, Measurement technology, Signals and image analysis, Media and entertainment industry.

Many departments have introduced modelling courses in the curriculum in recent years. A course in modelling may contain study of case examples, reading texts and solving exercises. The actual challenge and fascination is the students' exposure to open problems, addressing questions arising from real context. The real world questions may be found from the student's own fields of activity, hobbies, summer jobs, from the profession of their parents, etc. Reading newspapers and professional magazines with a mathematically curious eye may find an idea for a modelling exercise. A good modelling course should

a. contain an interesting collection of case examples, which stir students' curiosity
b. give an indication of the diversity of model types and purposes
c. show the development from simple models to more sophisticated ones
d. stress the interdisciplinary nature, teamwork aspect, communication skills
e. tell about the open nature of the problems and non-existence of "right" solutions
f. bring home the understanding of practical benefits, the usage of the model
g. tie together mathematical ideas from different earlier courses.

The modelling courses have been run in different forms. Traditional lecture course with weekly exercise session is a possibility. It would be important to implement group work mode and PC-lab activities in the course. The most rewarding form of activity might be projects and weekly session where the student report and discuss about their work and progress on the problems.

One of the innovative educational practices introduced in the recent decades is a *Modelling Week*, as an intensive mathematical problem solving workshop which simulates the real life R&D procedures. Students from all over Europe come together and work in teams of five or six on modelling real world problems. The cases originate from industry, different organisations or branches of society. The problems are brought in and presented by the problem owners. The teams are guided by a group of academic staff members. The "instructors" usually play the role of the problem owners. The students are allocated to problem-teams on the basis of their areas of interest and mathematical expertise.

The week starts with the problem owner giving a brief outline of the problem, the industrial context and the relevance of the problem for his/her company. The team questions the problem owner about the problem and the expectations. To identify and understand the "real" problem may take some time. The students must formulate a model and recognise the typically non-unique mathematical problem. The analysis follows leading to analytical studies and efforts to find techniques for numerical solutions.

Typically, the group arrives at an approximate solution. At the end of the week the student groups have to present their findings in public. Further, they are assumed to produce a decent written report, often a short article that will be published in the proceedings of the Modelling Week.

3 Modelling Education, How Much and When?

The presence of mathematical modelling on today's technology agenda and in the challenge of knowledge-based society indicates the importance of the questions of teaching of mathematical modelling in curricula at various levels of education. Does this support the idea that modelling education should be promoted for all children at all school levels? How much, in what way and when? These questions should be discussed seriously in the math education community. The skills of mathematical modelling mean an essential competence which is needed in many science related fields, technology, engineering, economics, biomedical professions, etc.

Modelling means a set of specialised science based skills that can be compared with the expert skills of, let us say, airline pilots and brain surgeons. Our society really needs these skills but we do not arrange mini-courses for airline pilots and surgery on primary school levels. This somewhat provocative statement is meant to emphasise the important question of how mathematical modelling should be inserted to the educational system.

The overall understanding of mathematics in today's world should be explained in mathematics classroom. Examples from applications should be used as pedagogical fermentation of the learning process. The joy of problem solving and the use of mathematics for real life situations is an ideal way to build interest and enthusiasm in the classroom. However, the pedagogical challenge must not be overlooked. The phrase of professor Helmut Neunzert at ICTMA14 "… modelling can be learnt but not taught in a usual way", contains an important message. One should also remember that often educational fashions or New Math reforms tend to drift to overdrive or hype level.

The supply of good classroom examples and case studies from different application areas is a key factor for the development of attractive and inspiring educational modules. We would need a flow of fresh problems. It would be important to maintain contacts to different special sectors, professions, diverse pockets of innovative processes. Good case-histories and exercise material would draw the attention of the students give motivation and make the theoretical ideas transparent.

What makes a good classroom problem is not trivial. The examples may be topical or artificial, fresh or odd relics, witty or dull/mundane, curiosity stirring or perfect/closed, transparent or wizard tricks, realistic or toy-like/concocted, understandable or intricate/complicated. The challenge is how to carry over the idea of mathematics as a useful environment of problem solving. How to shape the image of a modern brand of expertise, an emerging profession of the future?

A set of bad examples is presented below which describe the typical symptoms of poor or questionable ways to create application examples. The problems may hinge upon unrealistic contextual assumptions, model equations dropped from the blue space without a hint of the origin, pure mathematical tricks vested in an application-sounding dress: *Hanging a rope for drying laundry* (Derive the equation of the hanging rope. Find the equation in a case of non-homogenous gravity?); *Cost-function optimisation* (The total cost over the life-cycle of a building element is estimated to be given by a certain formula $P(x)$, where x is the insulator thickness. Find the optimal insulator thickness); *Reliability distribution* (The following model is used to describe the time-to-failure distribution of an electronic component $f(a,\ t) = (1 - at)\ (-0.02t)$; following data is given [...]. Estimate the model-parameter a); *North Sea fish population* (Model for the size of the fish population is given as a certain simple differential equation. Study the growth of the population).

The goal of modern applied mathematics instruction is to produce genuinely useful mathematical aptitudes—not just to do mathematics. The lecture room presentations and exercise labs should avoid the temptation of well-posed problems in well trodden areas—elegant but unrealistic manoeuvring of computational acrobatics. The aim is to see mathematics in action—to produce benefits.

4 FP7 Programme Science and Society

European Union launched within the 7th Framework programme a call with title Science and Society. Natural Sciences, Mathematics, Informatics and innovation are vital to increasing competitiveness, enhancing and expanding the economy and improving the quality of life and promote international competitiveness. Europe needs to generate more trained scientists, engineers and researchers to meet the challenges of global competition. This is essentially the content of the Lisbon Agenda and a key question is the interest of youth in science and engineering education. In the last 20 years, an alarming decrease in the number of youths studying Science and Engineering has been observed ("Science Education Now: A Renewed Pedagogy for the Future of Europe", Michel Rocard group report on Science Education, 2007). The trend seems to continue and the phenomenon is known worldwide.

The Universities, as initial and continuous teacher trainers, need to change their approaches to train new and in-service Science teachers. Inquiry based and problem based learning strategies are known to develop student attitudes to their studies of the Science such as raised interest, curiosity, commitment, autonomy in study and interdisciplinary and awareness of the context and real life. This is a challenge for varied local stakeholders: schools, universities, governmental and non-governmental organisations, local authorities, education research laboratories, science centres, libraries, citizens' associations, local media, enterprises and technological parks.

I suggest that a theme *Mathematics and society—real life perspective in mathematics education* should gain attention in this discussion. The attitudes of young people for math and science teaching and learning are affected by

- teaching methods, syllabi and educational content
- teacher's knowledge about their subject and the world around the subject (other sciences, society, industry, technology, R&D activity, including development challenge in public governance)
- organisation and administration of education as a national enterprise
- public awareness about the role and significance of science, technology and the science education within general public
- competition in the minds of the students between our subject and many other fascinating topics (entertainment, fitness, environment, peace, democracy, global responsibility, self expression, fashions and tribal cohesion, family values and friendship).

A big question is how the demanding science education can be promoted in the challenging, complex, multifaceted cultural arena. No doubt main objective should be fermentation of interest, curiosity, commitment and diligence in the minds of students. Teacher training is a key question. The new generation of math teachers should have a sound overall understanding about the important role of mathematics in today's world so that he/she can bring to the classroom a flavour of the fascinating special skills of mathematics, modelling, simulation and computing. The children hopefully will become aware of several new professional career opportunities in the field of science based professions where mathematical models are the modern space-age toolkit.

The idea of computational technology and modelling as a component in creativity and innovation in all science and science based professions should be brought to the attention of 15–18 year old boys and girls by their science teachers backed up by media and other stake holders.

This goal can be achieved only if the curricula in undergraduate and graduate level in universities and teacher training programs are developed. Universities should offer courses, seminars, project experience for student about "Mathematics applied to real world". Mathematics teachers should have been exposed to such material at undergraduate level and in graduate studies, such as:

- general lectures about industrial mathematics
- courses in mathematical modelling
- problem seminars, project work about small modelling cases
- possibility to participate in Contests of Mathematical Problem Solving, etc.

We should apply this viewpoint to the teacher training and re-training programmes, curriculum development at undergraduate and graduate level, development of teaching culture at universities and programmes preparing teachers for high schools and colleges.

Interfaces Between Mathematics and Industry and the Use of Technology in Mathematics Education in India

Ajit Kumar

This chapter consists of two parts. The first part deals with interaction between Mathematics and Industry in India. The interaction between the two is almost nonexistent. We look into some of the reasons behind this and mention some of the initiatives taken in the past to develop some interactions between the two communities. We also suggest some of the steps that must be taken in order to improve the interface between mathematics and industry in India. The second part deals with the use of computer-aided technology in mathematics education at various levels in India. We also mention a future project proposed by us that can significantly revolutionize mathematics teaching using computer-aided tools.

1 Initiatives for Industrial Collaboration

While in many countries, the interaction between industries and mathematics is fairly established, in India, there is hardly any interaction between the two. In this chapter, we look at some of the reasons behind this nonexistent interaction. In spite some initiatives of the Indian Society of Industrial and Applied Mathematics (ISIAM) and industrial mathematics group (IMG), in order to develop better relations with industries, these initiatives have not lived up to their expectations, and seem to have not made any visible impact. There have been also some recently initiated academic programs in mathematics, which are designed to cater to industrial needs. However, these are not sufficient for a vast country like India with its fast-growing economy. Mumbai being the financial and industrial hub of India, one could expect far better interaction between industry and the mathematics community. However, in reality, it is practically nonexistent. The number of industry-supported projects in some of the very good mathematics departments and institutes is almost negligible.

A. Kumar (✉)
Department of Mathematics, Institute of Chemical Technology, Mumbai, India
e-mail: a.kumar@ictmumbai.edu.in

A. Damlamian et al. (eds.), *Educational Interfaces between Mathematics and Industry*, New ICMI Study Series 16, DOI: 10.1007/978-3-319-02270-3_22,

ISIAM was proposed on the occasion of a symposium on "Differential Equations and Industrial Applications" in September, 1990 at the Department of Mathematics, Aligarh Muslim University, India, and was formally approved by its members in 1991. This society was affiliated to the International Council of Industrial and Applied Mathematics (ICIAM) in May 1999. Several national and international conferences have been organized by ISIAM. The society also publishes its proceedings in its Journal (Indian Journal of Industrial and Applied Mathematics) and provides valuable information for teaching and research of industrial and applied mathematics. It has also constituted two awards for mathematicians with significant contributions to industrial problems. However, the activities of the Society have been confined to organizing workshops and symposia, and not much emphasis is given to industry–mathematics collaborations. The society has not come out with any concrete proposal for bridging the gap and developing strong interaction between the industry and mathematics. ISIAM has organized a satellite conference of International Congress of Mathematicians 2010 on Mathematics in Science and Technology from August 15–17, 2010, with the aim to improve the mathematics–industry interaction.

IMG was started in 1991 by a group of people from the Mathematics Department and the Chemical Engineering Department at the Indian Institute of Technology (IIT), Mumbai. The main aim of IMG was to provide necessary inputs for Mathematical Technology in terms of (1) Research and Development Projects, (2) Consultancy Projects, (3) In-House and In-Campus Workshops, (4) Curriculum Development Activities, and (5) Skill Development Activities. Initially, they had discussions with a few industries and research organizations in and around Mumbai and organized a few workshops on Industrial Mathematics at the IIT Mumbai and at the MS University, Baroda. The main objective of this workshop was to provide an insight into some important mathematical and statistical techniques used in industrial applications, covering the following four modules: Systems Analysis, Optimization Techniques, Finite Element Methods, and Industrial Statistics. The group also received substantial funding from the Department of Science and Technology (DST) in the late 1990s, which was later discontinued. This group does not seem to be active currently.

If we look at industrial-supported projects in last 10 years in some of the finest mathematics departments in Mumbai, Bangalore, and Hyderabad, we conclude that there have been no industrial-sponsored research projects in those departments and only a few consultancies. These consultancies are mostly related to statistics and financial mathematics. However, most of these departments do have research projects in Mathematics that are funded by the government agencies. In fact, it is easier to convince the government funding agencies for research projects, rather than Industry. This just shows that the interaction between Mathematics and Industry, even in the financial hub of the country and in some of the very good mathematics departments is almost nonexistent. However, the fault lies with academicians as much as it with the Industry. Rarely mathematicians approach industries with research projects. Industries too do not have enough confidence in academicians for deliverables. However, there are a few industries that do give financial support for organizing workshops and conferences.

On the other hand, the undergraduate and postgraduate mathematics courses in India mostly include topics from pure mathematics and not much emphasis is given to practical and industrial applications. The use of computer-aided tools and software in these courses is also essentially nonexistent. A few exceptions are some master courses having industrial applications, most of which have been started recently in Madras, in Roorkee, in North Gujrat, in Patan, in Mumbai, or in Ujjain, Madhya Pradesh. Apart from those courses, some universities offer courses in applied mathematics, but its number is very limited for a vast country like India. Although, Mumbai is the financial capital of India, the number of courses in mathematics offered in Mumbai which are relevant to industries is negligible. One can hope that many more such courses will be initiated (especially in Mumbai) and that it will have significant impact in future collaboration between industries and academia once these students join the industries. The poor interaction between mathematics and industry is a consequence of the very small number of mathematics graduates, post-graduates, and Ph.D. students presently working in industries and in a severe scarcity of applied mathematicians in the country. Other reasons for this situation are: the minimal use of computer-aided mathematical tools and software in mathematics education at all levels; the difficulty for mathematicians to convince industries about their work and its relevance to real world problem; poor R&D departments in most of the industries, and when they exist they are not willing to share their research activities; industries are not forthcoming to support mathematics projects although they see its applications and relevance and, in general, do not have enough confidence in the academic world for the deliverables of the project; on the other hand, in general, academicians tend to approach only government agencies for research projects.

Indian universities have poor research culture in mathematics and the elite research institutes have enough research funds from the government agencies, therefore, they hardly bother about the industrial supports. The Institute of Chemical Technology (ICT), Mumbai is having one of the strongest links with industries, so much so that it is often cited as the role model example in India. Most often the industrial projects at ICT are written by faculty members of technology branches and submitted to the industry for funding. Rarely does the industry approach them with their problems. This shows poor R&D on part of industries. Thus, if we wish to develop stronger interaction between industry and mathematics, the academia has to take a lead.

We have tried to gather industrial industry perceptions about mathematics, by talking to various people in the industries and also by taking their views using a questionnaire. Almost all of them do acknowledge the importance of mathematics in industrial innovations, analysis, optimization, and control of industrial processes etc. Most of the industries use ready-made (black-box) software for simulation, analysis, and solving mathematical problems. Industry faces problems that extend well beyond the horizon of classical topics in mathematics. Many of these problems have a significant mathematical components, and intellectual challenges. Industries acknowledge that: students having good mathematical background do much better compared to other students, however, mathematics courses offered at the UG and PG level in most of the universities need to be reviewed. Mathematical background of students to handle real-world problems is not adequate, visible use of mathematical software in mathematics education is the need of the hour. Hardly, any industry has given support for a mathematics research project, but most of them are open to this idea provided the projects have some merits. Many industries give financial support for organizing workshops and conferences. Industries are also willing to give supports for infrastructures to the departments of mathematics that would like to offer courses that are relevant to industry. Industry do not see a need for any training programs in Mathematics for their staffs and people working in R&D, primarily because the mathematics which they use, most often is very simple or ready-made codes are available. Most of the industries are not willing to invest money on fundamental research as they do not see any immediate gain out of it, whereas from the academician point of view fundamental research is a vital component of innovations. Most of the industrial projects and consultancies at the academic institutions like ICT and IIT are through personal contacts and reputations.

Some steps should be taken for developing a closer interaction with industries: curriculum must be redesigned in order to include industrially relevant topics across the board; real-world problems and its solutions must be included in the curriculum; many more masters programs specifically catering to topics of industrial importance must be started, especially in major universities and institutes in India; mathematics community has to take lead and approach the industries with projects having real-world applications; Industry should fund postdoctoral programs at academic institutions; the academia along with the government should take a leaf out of some of the established international centers in order to bring the two communities closer for better coordination.

Although, mathematics is considered to be the queen of sciences, and is at the heart of any scientific and industrial innovations, the interface between the Industry and Mathematics in India is almost nonexistent. The onus lies with both the communities to work together to strengthen coordination and cooperation, which can lead to technology development and meet the societal requirements and challenges in India. Stronger interaction between mathematics and industry will be beneficial to both to the partners and to national economies and this in turn will inspire new mathematics and enhance the competitive advantage of industries as well. Industries must show greater confidence in mathematicians to tackle their problems. They must consider it as a long-term investment and also as part of their obligations to society and must not have a myopic view of immediate gain.

2 Use of Technology in Mathematics Education in India

Few working in mathematics education today would be unaware of the emergence of computer technologies, and related mathematical software in recent years for teaching, learning, and research in mathematics. The use of Information communication Technology (ICT) in teaching mathematics can make the teaching process more effective as well as enhance the student capabilities in understanding basic concepts. Calculating technology in mathematics has evolved from four-function calculators to scientific calculators to graphing calculators and now to computers with computer algebra system (CAS). Introduction of computer algebra systems, advantages and disadvantages of using CAS have been explained in (Kumar and Kumaresan 2008). Challenges of implementing CAS-based mathematics teaching and overcoming these challenges at the undergraduate and postgraduate level mathematics in India have been explained in (Kumar and Kumaresan 2008).

While use of computer-aided tools in many countries in teaching and learning mathematics have made a significant impact at all levels, in India the progress and awareness of this technology has been really very slow. Mostly, it has been confined among researchers and a handful of university and college teachers in well-established research institutes, IIT's, and university departments. At the undergraduate and postgraduate level, in recent years, many people have taken the lead to create awareness about ICT tools among mathematics teachers, by conducting workshops and conferences and the University Grant Commission (UGC) runs refresher courses for college teachers. However, many more consolidated efforts are needed in order to create a significant impact. As a result of some of these efforts, use of some software (especially free and open source software like Sage, SciLab, GeoGebra, KASH, MAXIMA, WinPlot etc.) at the undergraduate level in universities like University of Mumbai, University of Pune have been initiated. However, effective use of software in mathematics teaching has to be assessed in due course, as there are many concerns and challenges. But most college teachers in rural areas are not even aware of the existence of mathematical software and its impact in mathematics teaching. The use of ICT in mathematics teaching is mainly limited to the postgraduate levels. At the undergraduate level in most of the universities, it has been almost nonexistent. Main reasons behind this are infrastructure constrains and unawareness of the mathematics teachers. In recent years, the government funding agencies are providing generous funding for creating computer laboratories, and related infrastructures in colleges and institutes. But what we need is to create awareness and provide teachers with innovative teaching modules. The author, along with Professor S. Kumaresan of University of Hyderabad, India is planning to organize a series of workshops across India in order to create awareness. We give a brief description of this proposal later. Even in the streams like Engineering and Technology, use of mathematical software in teaching mathematics have been very limited in most of the colleges and institutes. At best most of them are teaching some programming languages, spreadsheet programs like Excel. Use of software, like MatLab, MathCAD, Mathematica, statistical software are confined to individual preferences. At school level, use of ICT is even less visible. There have been some efforts to create awareness about use of technology in mathematics teaching at school levels. One such effort is the National Conference and Workshop on Technology and Innovations in Mathematics Education (TIME) conducted at the Department of Mathematics, IIT in 2005, 2007, and 2009. The main aim of this conference is to provide mathematics teachers of schools, a platform to discuss about innovative ways of mathematics teaching at the same time create an awareness of effective use of ICT, and the kind of innovations that it can bring in mathematics teaching. One can clearly see during these workshops that many school teachers have started using tools like, GeoGebra, Geometer's Sketch Pad, Excel spreadsheets, power point Presentations etc. However, it is mostly at personal level and officially it has not been implemented in the curriculum. The Homi Bhabha Centre for Science Education (HBCSE), Mumbai, the center devoted to research and development in science and mathematics education has also taken some initiatives recently to create awareness by conducting some workshops at schools. However, this is not enough and it must take further initiative in order to create awareness, develop innovative teaching modules for schools using various teaching tools as they have got necessary infrastructure and funding. There have been some initiatives to set up virtual GeoGebra Institutes in India, which can provide on-line support system for teachers not only using GeoGebra but also using other innovative tools. There have been few other conferences also dealing with ICT use in mathematics education in India. The above initiatives are not enough, and we need more consolidated efforts by concerned mathematics teachers at local levels to organize many more workshops and discussions to bring in technology in the mathematics curriculum and to revolutionize mathematics teaching. The different educational boards should also seriously looks into the various aspects of implementing technology in mathematics educations.

The author, along with Professor S. Kumaresan of the University of Hyderabad, has submitted a proposal to government agencies under which, we plan to conduct several workshops across the country to make the mathematics teachers aware of computer-aided technologies and how to make use of such tools effectively in their teaching. We would develop teaching modules for teachers based on which they can design their own teaching modules or they can adopt modules developed by others. We first plan to send questionnaires to university and college teachers, teaching mathematics across the country to make assessments of their knowledge in mathematical software and their opinions about whether these tools can have a significant role in improving the teaching standard of mathematics. Based on these data, we may be able to create an awareness of the importance of the use of mathematical software in mathematics teaching. We also wish to create a pool of experts who can help in creating awareness, develop teaching materials, training others, and look into the various aspects of implementation of these tools in mathematics teaching at college and university levels. We would like to give stress on the use of free open source mathematical software so that it will be easily accessible to teachers and students. We may also explore the possibilities of developing mathematical software in other Indian languages, which can then be made freely available.

Acknowledgments I sincerely thank Professor S. Kumaresan for introducing me to this project and for his invaluable inputs. I also thank all my colleagues who have shared their experiences in industries. Last but not the least I thank, people from the industries for their views on this topic.

References

ICIAM. International Council for Industrial and Applied Mathematics, (http://www.iciam.org/).
IMG. Industrial Mathematics Group, Indian Institute of Technology, Mumbai, India, (http://www.math.iitb.ac.in/rd/img/).
ISIAM. Indian Society of Industrial and Applied Mathematics, (http://www.siam-india.org/).
Kumar, A., & Kumaresan, S. (2008). Use of mathematical software for teaching and learning mathematics. In *Proceedings of ICME 2008* (forthcoming).
OECED (2008). A report on mathematics in industry, (http://www.oecd.org/dataoecd/47/1/41019441.pdf).
TIME (2009). National conference and workshop on technology and innovations in mathematics education, (http://www.math.iitb.ac.in/time2009/).

Modeling Modeling: Developing Habits of Mathematical Minds

John A. Pelesko, Jinfa Cai and Louis F. Rossi

1 Introduction

It is like gum. You chew gum and use it to freshen up your breath, but in the end, it's worthless and doesn't have any nutrition or vitamins. Math is used in school to determine your intelligence, but there is no need for it later. Unfortunately, this students comment is typical; all too often, students don't see the connection between mathematics and real-life (Cai and Merlino 2011). Even among those students who expect to become scientists, less than 75 % of those students believe that advanced mathematics or science courses are necessary for their future careers (Ma 2006). One of unfortunate reasons for developing such beliefs is a lack of experience in seeing the intimate connections between mathematics and the real world. Students are usually taught mathematical concepts and procedures in a vacuum. Little effort is made to connect mathematics with authentic real world applications and students gain little direct experience with how mathematics is used in the workplace (Bessot and Ridgway 2001; Freudenthal 1991).

Over the past several years, we have led a reform effort to develop undergraduate mathematics majors specific habits of mind associated with the effective practice of mathematics in applied and industrial settings. This effort has led to the creation of a new mathematical modeling capstone course possessing many novel features. How can we best develop students' habits of mind to see and interpret the

J. A. Pelesko (✉)
Department of Mathematical Sciences, University of Delaware, 501A Ewing Hall, Newark, DE 19716, USA
e-mail: pelesko@math.udel.edu

J. Cai
University of Delaware, 523 Ewing Hall, Newark, DE 19716, USA
e-mail: jcai@math.udel.edu

L. F. Rossi
University of Delaware, 524 Ewing Hall, Newark, DE 19716, USA
e-mail: rossi@math.udel.edu

A. Damlamian et al. (eds.), *Educational Interfaces between Mathematics and Industry*, New ICMI Study Series 16, DOI: 10.1007/978-3-319-02270-3_23,

world mathematically? This is the primary research question that has guided our reform effort. Through modeling the mathematical modeling processes in the course, we have used innovative curriculum materials and pedagogical strategies to develop seven habits of mind for students to make sense of the world mathematically. In this paper, we critically analyze and reflect on this reform effort and we provide concrete examples of best practices to connect mathematics and the real world. A unique feature of this contribution is its primary focus on developing students' habits of mind. Thus, the examples and ideas from our critical analysis of the reform effort can be easily adapted to other institutions and other courses.

2 Description of the Modeling Course

Our capstone course, Math 512: *Contemporary Applications of Mathematics* is the centerpiece of our undergraduate mathematics curriculum at the University of Delaware. It provides a hands-on learning experience in contemporary applications of mathematics, and involves work with investigators from industry, national laboratories, and other departments. Mathematical topics may include ordinary and partial differential equations, systems of differential equations, transform, asymptotic, and numerical methods. Recently, it has become very popular for majors outside of mathematics because of the connection between mathematics and real world. More importantly, it has been the target of our efforts to reform how we instill in students what we believe to be habits of mind that all effective mathematicians share. Our course was strongly influenced by the annual mathematical problems in industry (MPI) workshop, and mathematical contest in modeling (MCM) sponsored by the Consortium for Mathematics and its Application (COMAP: www.comap.com).

MPI is a workshop where faculty and graduate students spend a week working on open problems contributed from industrial experts. Participants spend almost the entire meeting in breakout rooms doing new mathematics and critiquing one another in order to make progress by the end of the week. MCM is a similar experience for undergraduates except that it is a four-day contest. Student teams are given an open-ended modeling problem and are allowed to use any inanimate source. Their goal is to craft a comprehensive solution, which is submitted to contest judges. After completing the contest, most students comment that they accomplished more in those four days than they do in an entire semester.

Starting in 2001, we began reforming our capstone course to develop effective qualities in students that we had observed at MPI and MCM. In particular, we wanted to focus on small groups of students working on open problems arising from disciplines outside of the mathematical sciences. We designed our course around the notion that there were certain key activities associated with experiences like MPI and MCM, and that we would have to make it possible for students to engage in these activities in the context of a regular course. Since 2002, with the support of two internal educational grants at the University of Delaware, we have continuously revised and improved the course to better prepare students of all majors to see and interpret the world mathematically. Most significantly, we added a laboratory component to Math 512 using our modeling experiment and computation (MEC) Lab, which is a fully instrumented wet lab housed within the Department of Mathematical Sciences. Fundamentally, industrial mathematics involves connections between a process that must be quantified as some kind of mathematical object, mathematical techniques and methods, and increasingly computation and simulation. Our laboratory component adds a new dimension to this undertaking by connecting all these activities with direct observation and measurement.

3 Developing Habits of Mind

As we have developed and refined our capstone course, we have sought to identify the habits of mind that are characteristic of successful mathematical modelers (Gross et al. 1988). It is these habits we attempt to instill in our students. These habits are: Awareness; Attention; Resilience; Specificity; Curiosity; Vision and Communication. We first describe each of these habits in turn and explain their relationship to mathematical modeling. In the next section, we discuss the pedagogical principles used to develop these seven habits of mind.

Awareness—Mathematical modeling begins with a problem. This first step requires the successful modeler to be aware of the world around them, to possess the listening skills to enable them to learn about the problem from other scientists, to possess the reading skills to enable them to learn about the problem from the literature, and to possess the breadth of knowledge to enable them to understand how the problem at hand relates to the field at large. Clearly, within this broad category of awareness, several specific skills must be developed. In our courses, we aim most directly at developing the reading skills of students as this seems to be a large stumbling block on the path to their success.

Attention—Once a modeler has understood the context of a problem, it is time to focus on the phenomena before them. This requires attention. That is, they must learn to observe. This process can take many forms. It may involve gathering data, examining large sets of data, or more frequently, deciding what data to gather. In our courses, students are placed face-to-face with an experimental system and pushed to pay attention. What do they see? What do they not see? They are urged to drop preconceived notions of how a system behaves and to simply give their attention to the natural world.

Resilience—A successful modeler must be prepared to fail and to fail often. Assumptions often don't fit the data; there is often ambiguity about what to leave in and what to leave out, and often the natural world leads one in unexpected directions. This requires that the successful modeler develop a high level of comfort with ambiguity. This also requires that the student abandons the comfort of prepackaged problems and develop the resilience needed to tackle the process of formulating their own problems. In our courses, we provide the students with open-ended problems to which we do not know the answer. This allows us to model the habit of flexibility as we go through the investigative process side-by-side with our students.

Specificity—The bridge from observation to mathematics is built with the ability to ask concrete quantitative questions. The student observes the pendulum moving, the student observes that the motion is periodic, but now, they must learn to go beyond and ask "What is the period?" Or, "How does the period vary with the initial angle of deflection?" In our courses, this step is an important part of our "milestone" structure. After the problem is grasped, the context understood, the system observed, students are asked to formulate their own quantitative questions. While we may give examples and model this process in the context of another problem, students are asked to take ownership of their problem and formulate their own quantitative questions.

Curiosity—Linked with the ability to ask quantitative questions is the desire to do so. The successful modeler must possess the habit of curiosity. They must be driven to look at the world and ask Why? This is perhaps, the most difficult of the habits to instill in students. In our courses, we grow this habit by providing them with a range of inherently interesting problems and allowing them to follow their own path in choosing which to investigate.

Vision—The modeler that knows the context of a problem, observes, and asks quantitative questions, possesses the information needed to generalize their results and synthesize others results with their own. But, they must develop a broad vision that drives them to do so. In our courses, we develop this skill by pushing the students to constantly iterate. They know context, they have asked good questions, and they have answered them. Now, they are asked how their answers compare with those of others. What are the implications of this? How does this work shed light on broader questions in mathematics and in science?

Communication—Our first habit, awareness, ultimately focused on communication. But, at the start, the focus is on input. That is, the modeler listens, reads, and digests. However, science and mathematics does not take place in a vacuum. The successful modeler is able to communicate their results, no matter how tentative, with others. In our courses, we emphasize this habit through requiring group work, asking for informal and formal presentations in class on project progress, and requiring the final product of a written report to be of high quality.

4 Pedagogical Principles of the Reform Effort

Four principles have guided our reform effort from the beginning. Over time, we have made adjustments and corrections, but we have found that these basic principles, and the practices that we have implemented based on them, help instill the seven habits of mind in our students.

Maps matter—The heart of Math 512 is a collection of exciting open research questions originating from outside of the mathematical sciences. The course was designed to guide them through the modeling process. The problems vary from year to year, but our goal was kept the same, designing the course around effective practices for helping students develop habits of the mind, rather than teaching a rigid set of topics. Students are driven by a desire to learn, and they are keenly aware of how they will be graded. We carefully designed our course to align our expectations that students would aggressively pursue open-ended problems with a grading scheme that would reward our students' efforts. We realized that we had to design a grading scheme that formatively assesses the students' progress on their research project throughout the semester. Our response to this challenge is a set milestones placed at equal intervals in the semester. A visual road map guides students through the semester, even though they may work on different modeling projects. On the first day of classes, students see the milestone chart as a grading scale, with their milestone grade arranged in rows. They know they will need to complete their literature review for the first milestone and that it will be worth 80 % of their grade. They know that they will use the wiki to draft their report until the 4th milestone when they will start preparing a final copy in LaTeX. However, as the semester progresses, they see that milestones are markers on their way to a significant work product, and this work-in-progress is what is being graded. As instructors, we use the table to mark our own progress as we guide and lead students on their intellectual journey.

Process Emphasis—As instructors, we learned early on that we would have to teach process almost as much as we would have to teach content. Our students have all taken a semester-long composition course offered by the English Department, but we spend a significant amount of time teaching students to write collaboratively within their groups. In the writing process, students explore, understand, and learn how to pose significant questions to research, as well as investigate the ways to solve them. The use of the advanced technology facilitated the processes. This includes mastering the latest technology for facilitating such collaborations. Students must use the universal typesetting language for mathematics. Over time, we have used a variety of collaborative versioning systems including concurrent versioning system (CVS), subversion (SVN), and now students draft material using wikis enabled with embedded.

Hands-on Opportunities—Students must be able to observe the processes that lie at the heart of their problems, whether it is freezing drops of water or observing an organic self-assembled network, so we have a dedicated wet lab set aside for our course. It is possible for students to acquire data from the literature or the web, but these sources of information do not respond to student interrogation. As students refine their assumptions and modify their expectations, they design an experiment and observe its outcome. Students learn that every observation comes with a cost in time and energy, and they become very adept at creating definitive experiments to resolve key issues in their projects.

Formative Assessment—If we expect students to approach problems professionally, we have to provide feedback throughout the semester tapping into the learning that we observe as instructors as well as the learning that occurs within each group outside of class. Over the last two years, we have refined a peer assessment rubric wherein students evaluate one another (including themselves). Within each group, students take responsibility for different aspects of the project. After implementing this element to our course, we find that all students are more engaged in all project activities and project groups tend to find ways to match individual skills and abilities to required tasks.

5 Outcomes and Conclusions

Our reform effort has led to multiple positive outcomes for our students and our department. Our capstone course has helped to prepare many of our students to succeed in highly competitive graduate programs across the nation. Not only do we bring industrial problems into the classroom, but our students' work has attracted new industrial problems into the more professional MPI setting. Our course has been featured in the press (Pelesko 2008) and in numerous invited talks nationally and internationally where there is interest in applying our practices at other institutions.

An example of strong connections between Math 512 and industrial mathematics research activities comes from the study of air bearings. A bearing is any device that supports a load. Most students have had some exposure to ball bearings used on axles for bicycles and skate-boards. An air bearing is a load bearing device that relies upon a cushion of air. Most students have experienced an air bearing when playing air hockey. Modeling air bearings has many industrial applications including the motion of the read/write head of a disk drive and control of precision, industrial robots.

At the 2001 MPI workshop, an interesting air bearing problem was posed by Ferdi Hendriks, from IBM. For static air bearings, their engineering team had observed that scratching the bottom of the air bearing improved its performance in the sense that grooved surfaces could bear a greater load than smooth surfaces with the same input air flow. Optimization via computer simulation led to the same conclusion, and the optimal bearing surfaces exhibited a fractal branching structure similar to structures observed in natural systems such as the cardiopulmonary system in mammals, root systems in trees, and so forth. The key difference is that the air bearing surface was optimized by a computer algorithm whereas examples in the natural world are optimized by natural selection. The problem proved to be very challenging, but the workshop participants developed a model based on lubrication theory and studied axisymmetric profiles. The model predicted that minor surface variations could lead to large changes in bearing performance. However, branching structures are not axisymmetric, so the week ended without much more quantitative insight into why the observed groove patterns were optimal (Howell et al. 2001).

The problem had all the ingredients of a great capstone project. The problem was compelling because the observed optimal solution was not intuitively obvious and there are connections to the natural world. With an interesting problem, awareness and attention are not difficult to encourage. Most students had had experiences with air bearings without really giving them much thought. We were confident that our students had sufficient mathematical background to grasp lubrication theory relatively quickly in the first weeks of the course, and a laboratory model of an air bearing could be constructed within our model budget and facilities.

The air bearing problem was offered as one of three project topics during the first year of our reform effort in 2003. As instructors, we learned many lessons during this first offering and these experiences inspired many changes that have since because a permanent part of our capstone course.

Our experience transferring an industry workshop problem into an undergraduate classroom taught us an important lesson about preparation: The success or failure in creating these experiences for undergraduates requires instructors to lay a thorough foundation far beyond what is required for a typical classroom experience. Not only did the workshop presenter inspire an interesting problem, but the workshop report was a valuable document for the upcoming capstone class.

The workshop report was accessible to students and made it easier for us as instructors to package the problem for student consumption. However, this is not a sustainable approach for a course that will be offered once per year and will require new projects every time it is offered. Fortunately, we obtained an internal grant to hire students during the summer to explore prospective projects individually. The summer research experience lays the groundwork for the semester-long project and helps identify potential hazards in the projects. For example, a laboratory model of the project might require extra equipment or may simply be too complex to use in the course at all. We have learned to use these summer experiences to sift through the literature and identify key references, so that students can climb the learning curve as quickly as possible during the semester.

We note that the summer experience and the MPI workshop experience differ in important ways. The MPI workshop focuses entirely on mathematical modeling and runs on a very energetic 5-day schedule whereas the summer experience includes access to our lab and time to pursue a variety of avenues. The resources differ as well. The MPI workshop can draw upon the skills of a dozen or more faculty and graduate students. Our summer experience will typically involve a single understand working with a faculty member. Whether we use the industry workshop or a summer student, extra preparation is required if one is to bring industrial mathematics into the classroom.

Another key lesson learned during the first offering was that we needed formal ways to instill vision in our students. Project teams worked diligently gathering data on low hanging fruit, but this often leads to an aimless project. To encourage a collective vision in the teams, we use wikis where teams identify their achievements, near term and long term goals. Every week, they revise their goals based on their goals based on their most recent experiences. To complete the process, there are weekly presentations with a lively question-and-answer period where students can critique one another.

The milestone structure has been the core of our reform effort from the beginning, but we found that we had to add features to create a successful industrial modeling experience. In an industrial setting, teams tend to be interdisciplinary and roles are well understood from the composition of the team. In a classroom setting, effective interdisciplinary teamwork is a new experience for most students who, for better or worse, are focused on grades. Therefore, we implemented a peer assessment system where students were required to assume responsibility for different portions of their project. Their evolving project report remains the basis for the grades of all the team members, but scores are adjusted up and down based on peer assessments. Peer assessment involves numerical ratings of different students on different tasks. The numbers themselves are meaningful, but it is just as important that students write why they are giving a particular score. This form of reflection helps instill all of the habits of mind identified in this paper. On the rare occasion when a student does not provide adequate reasoning for a peer assessment, the student is asked to rework the assessment and respond to provide answers to leading questions about the team's activities.

In addition to driving the development of our reform effort through its first cycle, the air bearing problem resurrected itself in a surprising way. A few months after the semester ended, New Way Precision, a Pennsylvania company specializing in porous media air bearings, contacted us. All student project reports are posted on our course page, and they had come across our students' work. New Way Precision was looking for a mathematical understanding of their air bearings, and the fluid dynamics underlying porous media bearings are sufficiently different from solid air bearings. Future discussions led to New Way Precision presenting their problem at the 2004 MPI workshop. In this particular case, industrial mathematics has come full circle, from industry to the classroom and back into the industrial setting.

In conclusion, we presented initial ideas in our reform effort to develop seven habits of mind for students to make sense of the world mathematically, through modeling the mathematical modeling processes in an undergraduate course. A mathematical modeling course has the greatest potential for students to have the direct experience about the intimate connections between mathematics and the realworld. Pedagogically, it is important to model the modeling processes in classroom so that habits of mathematical minds can be developed. In this proposal, we briefly presented the pedagogical principles for the mathematical course, and how the seven habits of mind can be addressed pedagogically. Through the example of air bearing, we provided some details about our critical analysis of using specific modeling projects and provided concrete examples of best practices to connecting mathematics and the real world. Through these specific modeling projects and concrete examples of best modeling practices, we also discussed the lessons learned to instill the habits of mind.

References

Bessot, A., & Ridgway, J. (2001). *Education for mathematics in the workplace*. New York: Springer.

Cai, J. & Merlino, F. J. (2011). Metaphor: A powerful means for assessing students mathematical disposition. In D. J. Brahier & W. Speer (Eds.), *Motivation and disposition: Pathways to learning mathematics.* National Council of Teachers of Mathematics 2011 Yearbook. Reston: NCTM.

Freudenthal, H. (1991). *Revisiting mathematics education: China lecture*. Dordrecht: Kluwer.

Gross, G., Haenschke, B., & Tenzler, R. (1988). Characteristic behaviour orientations of young scientists: A comparison between young mathematicians/physicists and other natural scientists. *Higher Education, 17*, 391–396.

Howell, P., Kedzior, M., Kramer, P., Please, C., Rossi, L., Saintval, W., Salazar, D., & Witelski, T. (2001). *Shape optimization of pressurized air bearings*: *Proceedings of the Seventeenth Annual Workshop on Mathematical Problems in Industry.*

Ma, X. (2006). Cognitive and affective changes as determinants for taking advanced mathematics courses in high school. *American Journal of Education, 113*, 123149.

Pelesko, J. (2008, April). The MEC lab at the University of Delaware, DSWeb dynamical systems magazine.

The Evolution of Graduate Applied Math Courses in the Institute of Mathematics, University of the Philippines

Carlene P. C. Pilar-Arceo and Jose Maria L. Escaner IV

The University of the Philippines System is the Philippines' national university. It is composed of seven constituent units, among which the University of the Philippines Diliman (UPD) is the flagship campus. Among UPD's more than 20 colleges, the College of Science is the third largest in terms of faculty members and the second in terms of student enrolment. It houses eight degree-granting institutes of which the largest is the Institute of Mathematics (I-Math), with 80 faculty members, around 360 undergraduate students, and around 200 graduate students. Half of I-Math's graduate students are enrolled in the Applied Mathematics programs.

1 Academic Programs

The academic offerings of I-Math include one undergraduate program, BS Mathematics, and five graduate programs: MS Mathematics, MS Applied Mathematics (MSAM), MA Mathematics, Professional Masters in Actuarial Mathematics (PMAM), and Ph.D. Mathematics. This study will focus on the applied and professional programs.

C. P. C. Pilar-Arceo (✉) · J. M. L. Escaner IV
Institute of Mathematics, College of Science, University of the Philippines Diliman, Quezon City, Philippines
e-mail: cayen@math.upd.edu.ph

J. M. L. Escaner IV
e-mail: joma@math.upd.edu.ph

A. Damlamian et al. (eds.), *Educational Interfaces between Mathematics and Industry*, New ICMI Study Series 16, DOI: 10.1007/978-3-319-02270-3_24,

1.1 MS Applied Mathematics

The MSAM is designed for students who wish to obtain good training in applied mathematics. It prepares its graduates for either work in R&D or I-Math's Ph.D. program. The areas of concentration are

- Mathematics of Finance (MF)
- Mathematics in Life and Physical Sciences (MLPS)
- Optimization and Approximation (O&A)
- Numerical Analysis of Differential Equations (NADE).

The common core courses are Linear Algebra, Real Variable Function Theory I, and Numerical Analysis I, after which each area of concentration has a roster of specialization courses and a graduate seminar. After coursework, students either write a thesis or take the preliminary-qualifying pair of examinations. The preliminary examination is comprehensive and written, while the qualifying examination, taken upon passing the preliminary examination, is specialized and oral.

1.2 Professional Masters in Actuarial Mathematics

The PMAM is designed to equip students, in theory and in practice, with the essentials of the actuarial profession. It also prepares them for the professional actuarial examinations. The PMAM core courses are Linear Algebra and Numerical Analysis I, followed by 24 units of specialization, a graduate seminar, and the completion of a special project. The 24 units of specialization are made up of Actuarial Theory and Practice, Loss Models and Survival Models, Actuarial Science I, and Actuarial Science II, plus four related electives on statistics, economics and finance. It is the special project that distinguishes this professional program from of its precursor, and from the rest of I-Math's graduate programs. To graduate, preliminary and qualifying examinations must also be passed.

2 Historical Overview

From the 1980s, the MSAM Areas of Concentration were Computer Science (CompSci), Operations Research (OR), and Actuarial Science (ActSci). However, for several reasons to be discussed further on, by the year 2000, only ActSci had enrollees.

Why did MSAM CompSci "disappear"? CompSci was a joint offering of the College of Science, to which I-Math belongs, and the College of Engineering. The faculty complement on the side of the then Department of Mathematics slowly dwindled so that eventually, due to administrative concerns, the program became housed entirely in the Department of CompSci of the College of Engineering.

Why did MSAM OR "disappear"? This came about due to several reasons. First, OR majors, like the rest of the MSAM majors, were not deemed to be Ph.D.-ready, and thus seemed "second class" citizens in a Department that was "pro-pure" (not exactly by design, but by circumstance as well, as senior faculty in OR and CompSci were unable to sustain efforts to manage the programs). Second, junior faculty who finished their MSAM OR went to industry, thus restricting the continuity and dynamism of the program by creating a widening gap in the OR teaching pool. This gap was made worse by the absence of any faculty complement from industry.

Third, OR graduates reported a "lack of X-factor" in terms of employability. That is, an MSAM OR degree did not translate to an advantage in job-seeking; industry looked more toward Industrial Engineering graduates. Part of the reason, also, seems to be the lack of R&D presence/appreciation in Philippine industries. It is thus not surprising that there was no regular interaction between academe and the OR industry. Clearly, there was a lack of symbiosis between the academe and industry in the area of OR. Just as clearly, MSAM OR was headed, like MSAM CompSci, toward a "natural death".

After 5 years or so, in 2007, there emerged the MSAM with concentration in O&A. This became the umbrella course for OR, and would be on I-Math's Ph.D. track. This proved to be a "win–win" resolution, giving the O&A/OR area the theoretical depth that it was deemed to have lacked, while allowing its graduates to proceed to Ph.D. studies in O&A/OR and truly specialize in applications.

3 The Big Shift

A major contributor to the inception of MSAM in O&A is the partnership of I-Math with the EU-funded Asialink organization. With this group, led by Profs. Marc and Francine Diener of the University of Nice Sophia-Antipolis, and including several professors from various universities in France, Italy, and Spain, I-Math implemented the International Masters in Applied Mathematics and Information Science, or the IMAMIS. The IMAMIS was implemented over 3 years, 2005–2007, with focus on three inter-related and multidisciplinary areas—Numerical Analysis, Mathematical Finance, and Information Science. The post-IMAMIS revised MSAM, as mentioned earlier, would now have four areas of concentration. Aside from O&A, there are the MF, the MLPS, and NADE.

Another contributor to MSAM O&A is a collaboration with professors from Austria. In 2005 and 2006, a group of professors from the University of Graz, led by Prof. Franz Kappel, held modeling and numerics workshops at I-Math. Since then, advising and post-doctoral collaborations were made possible through Prof. Kappel and an Austrian academic exchange not only with the University of Graz but also with the University of Vienna. Several junior faculty did their dissertation work in these two universities, and they have since returned to I-Math, actively doing research and mentoring students. They are now involved in the review of I-Math's OR offerings.

To further illustrate the international support factor, our faculty member in charge of the review of OR course offerings is a beneficiary of a research visit at the Max Planck Institute for Biochemistry (MPIB) in Munich with Dr. Ricardo del Rosario (UPD; MPIB). At least two more similar visits for dissertation research had since been undertaken by I-math junior faculty at the Ludwig Maximilian University of Munich (LMU) with Prof. Eduardo Mendoza (UPD; LMU). These junior faculty are now researchers and mentors in the Life and Physical Sciences area of the revised MSAM program, and Dr. del Rosario and Prof. Mendoza continue to be active collaborator and mentor of I-Math faculty and graduate students.

The revision of the old non-Ph.D.-track MSAM program connotes I-Math's decision to no longer prioritize industry demands and employment, but to look to more theoretical depth and go with global trends in the applied mathematics areas, no matter how belatedly.

4 Another Big Shift

In the meantime, while the MSAM OR program was going through rough seas, the MSAM ActSci program continued to traverse very smooth waters. Indeed, it was and still is known as I-Math's "box office" program. Still, even in the absence of enrolment and faculty complement problems, it was given a new identity, partly an offshoot of IMAMIS deliberations: the Professional Masters in Actuarial Mathematics, or PMAM.

Unlike its former MSAM siblings which were put on the Ph.D. track, the Act Sci program was placed on a decidedly non-Ph.D. track. Unlike OR, it remained (successfully) industry-focused—a training ground for the Society of Actuaries examinations, industry employment, and actuarial practice. Moreover, a special project requirement was added, a distinction from the old MSAM requirements. To date, there have been projects on microfinance, portfolio management, and microinsurance.

As an explicit professional program, PMAM all the more enjoyed strong industry support. The faculty complement became stronger than ever, seeing an influx of young practitioners who were MSAM ActSci alumni as well as former instructors of I-Math (MSAM ActSci gave its graduates the X-factor that MSAM OR somehow could not). In any given semester, there are usually around six to seven participating lecturers from the actuarial industry, versus only one from I-Math's. Actuarial institutions have been donating professorial chairs to the Institute. Actuarial institutions also regularly send their staff to the Institute for enrolment in PMAM. In most actuarial departments, roughly 40 % (around 2 out of 5 personnel) are PMAM students/alumni or MSAM ActSci alumni, and they are among middle management personnel. Indeed for the MSAM ActSci, and now the PMAM, there has always been a clear and well-sustained symbiosis and mutual trust between the academe and industry. Another valuable point to make: PMAM is the only program of its kind in the Philippines. In fact, it is the only graduate program in ActSci/actuarial mathematics in the Philippines.

5 Current Results and Developments

The revised *MSAM* areas of concentration have been enjoying a steady stream of enrolment. MF, in particular, is enrolled in by faculty from neighboring universities planning to set up similar graduate programs. A recent publication by an MSAM MF adviser–advisee tandem of the Institute was named outstanding research paper by the National Academy of Science and Technology (2011 *NAST Awards for Outstanding Scientific Paper*: "*Smooth asymptotics for the price of a DIC barrier option*", *by Jose Maria L. Escaner IV with Oliver Ian C. Wee*).

MLPS has generated increasing interest in the Mathematical Biology series of courses. Young Ph.D. faculty who did their dissertation work through the Max Planck and LMU connections are now actively mentoring graduate students in areas such as population migration. The current crop of graduate students is doing their theses on topics such as dengue and dopamine metabolism. O&A had three successful thesis defenses in the First Semester of AY 2011–2012 by students of returnee Optimization Ph.D.s. It is now undergoing a review of its OR offerings.

PMAM Continues to enjoy high, if not higher, enrolment. In any given semester, there are 10–15 new enrollees; in a recent semester, there were as many as 25 new enrollees. Program handlers would like to improve the graduation rate, which is currently at 5–10 graduates per year (hopefully, there will be 13 for this school year). The background to this is that the PMAM majors are mostly industry personnel as well. Thus, this is more a case of a slow completion rate, and not that of low completion-high mortality. PMAM is currently undergoing a program revamp due to a Society of Actuaries syllabus revamp.

6 Future Directions

I-Math is committed to a periodic review of all its programs. Curricular updating will be done to keep up with global trends. Ideally, in I-Math's design of its programs, both undergraduate and graduate, there will be changes in the way of thinking of several pertinent corners: for the math students, in their appreciation of the use and relevance of mathematics; for the math teachers, in the manner that math is/should be taught with respect to global trends; for government, in their support of programs pertinent to the appreciation for and the corresponding teaching of math, and the possibility of starting a math appreciation campaign at the high school level (at the elementary level, even); for industry, in its linkages with the academe, especially with respect to information- and resource-sharing; for the international community, in its continued assistance and collaborative efforts; and, for the general public, in its positive perception of mathematics and recognizing that math is not just a subject for classroom experience, but a partner for everyday living.

References

I-Math Handbook of Graduate Programs.
I-Math enrolment records.
IMAMIS webpage.
Chancellor's Reports at the 186th and 187th UPD Executive Committee Meetings.
Interviews with Math Dept Chairs/I-Math Directors.

The Vertical Integration of Industrial Mathematics, the WPI Experience

Bogdan Vernescu

The need for modeling and simulating emerging technologies, financial products, biological, or social phenomena has dramatically increased in the recent years and thus has substantially expanded the need for mathematically trained professionals. At the same time, reports concerning 4-year college and university undergraduate mathematics programs (e.g. Lutzer et al. 2002) describe a drop in the number of mathematics bachelor's degrees in the US. Thus, the need to find new ways to attract students to mathematics is a very real challenge for educators. Several authors (Friedman and Littman 1994; MacCluer 2000; etc.) show how industrial problems can be used to introduce new topics in mathematics, while others motivate the need to prepare students for jobs outside academia (Benkowski 1994). The recent SIAM Report (SIAM MII 2012) gives a clear picture of the mathematics used in industry.

The industrial projects can provide at the same time an essential training for industrial careers, and can also have an additional impact and motivation that cannot be obtained in any standard course or academic experience. Many students are motivated by working with a company on an industrial problem: the problem is real and the company needs a solution; having a mathematical solution impact a corporate decision is both challenging and rewarding for the students.

At the foundation of its educational philosophy, Worcester Polytechnic Institute has had, from its very beginnings in 1865, the balance between theory and practice. The vision of WPI's Founders to emphasize the mutual reinforcement between theory and applications, reflected in the university's motto "Lehr und Kunst", is present in all campus activities. Starting from its successful experience of project-based curriculum, WPI has been at the forefront of the national movement to develop opportunities for mathematics students and faculty to gain experience with the applications of mathematics in real-world settings, in particular in industrial problems typical of those in which scientists and engineers would

B. Vernescu (✉)
Department of Mathematical Sciences, Worcester Polytechnic Institute, 100 Institute Road, Worcester, MA 01609, USA
e-mail: vernescu@wpi.edu

A. Damlamian et al. (eds.), *Educational Interfaces between Mathematics and Industry*, New ICMI Study Series 16, DOI: 10.1007/978-3-319-02270-3_25,

depend upon mathematics for solutions (Davis 1998, 1999). In 1997, the Center for Industrial Mathematics and Statistics (CIMS) was established, as the interface between the Mathematical Sciences Department and business, industry, and government; guided by an Advisory Board formed by industry executives, the faculty members affiliated with CIMS made a concerted effort to integrate industrial applications vertically at all levels in the mathematics curriculum (Berkey and Vernescu 2007). CIMS has provided a wide range of opportunities from industrial projects for mathematics seniors, to industrial projects and internships for Master's and Ph.D. Students, to the development of K-12 outreach programs in industrial mathematics.

1 Undergraduate Projects

Senior-year projects are part of the degree requirements of all WPI seniors. Most usually a team endeavor, these projects are a substantial piece of work, equivalent to three courses in terms of credits and spanning over 3/4–4/4 of the academic year. The projects provide a capstone experience in the student's chosen major that develops creativity, instills self-confidence and enhances the ability to communicate ideas and synthesize fundamental concepts. By completing the project, the students are expected to be able to formulate a problem, develop a solution, and implement it competently and professionally; be exposed to interaction with the outside world before starting their careers; be able to work in teams and communicate well orally and in writing. A cornerstone of the WPI education, the senior-year projects have been highly successful at involving undergraduates in significant research.

The first senior-year projects in industrial mathematics were developed in the early 1990s and were sponsored by PresMet, a manufacturer of pressed-metal parts for the automotive industry and Morgan Construction (currently part of Siemens), at that time the largest US manufacturer of steel rolling mills. Since then, faculty affiliated with the CIMS have developed over 70 senior-year projects, in collaboration with over 45 sponsors from industry, business, and government such as: Bose, Compaq, Deka Research, GE, IBM, Procter & Gamble, Travelers, United Technologies, Veeder-Root, etc. Together with the sponsoring company, the advisors provide the initial formulation of the problem, in such a way that the mathematical and/or computational modelings are the essential part of the project, at the same time maintaining its industrial relevance. The students start by understanding the problem which requires most often understanding the engineering language and learning possibly new mathematical theory. Students come to appreciate the difference between the textbook exercises and the industrial problems; in textbooks, problems are always well formulated and are always based on the material in the preceding chapters, while the industrial problems may require them to ask more questions for a complete formulation and always requires them to figure out what mathematics tools are needed. Students also understand that reformulating the problems is an essential part of the industrial experience; better understanding the problem can lead to a different formulation. Students realize that in the corporate world time is of essence: an answer in the timeframe given by the economic cycle is at times more valuable than the best answer given a few years later. In most of these projects, the need and benefits of teamwork is reinforced; students with different backgrounds bring different perspectives to the problems. And finally, the students are forced to improve their communication skills as they need to present the results in a language accessible to their corporate audience, most often a different language than the one needed for the presentations to their mathematics faculty advisors. The entire experience is different from any type of coursework or even other types of project work.

A few examples of past projects illustrate this experience of training students: *Mathematical Modeling in Metal Processing* (1996 and 1997), developed several methods to simulate the wear and optimal geometry of the Laying Pipe mechanism, used for coiling steel rods in rolling mills, using optimal control theory, and calculus of variations techniques; *Mathematical Models of Damage Spread in Networks* (2002), described mathematical models for how damage can spread through an organization, which was completed in collaboration with Lehman Brothers investment firm in the aftermath of the terrorist attacks in September 11, 2001; *Modeling of Torque for Screw Insertion Processes* (2006), analyzed and improved a mathematical model of the self-tapping screw insertion process so that it can be used in manufacturing processes at the BOSE Corporation.

2 Research Experience for Undergraduates

Since 1998, we have provided research opportunities for undergraduates from other universities through an NSF-sponsored Research Experience for Undergraduates (REU) in Industrial Mathematics and Statistics, the first of its kind, to our knowledge (Vernescu and Heinricher 2000; Heinricher and Weekes 2007). In this 8-week summer program, we have replicated for students from other universities the experience our own students have in their senior-year project. Teams of, usually four, students have a faculty advisor and an industrial advisor and work on an industrial mathematics project of interest for the sponsor. In addition to project work, we invite industrial mathematicians to share how mathematics is used in the real world. Mathematicians from companies such as Microsoft, John Hancock, Fidelity Investments, and United Technologies are invited to talk with the REU students, and we take the students to visit companies like Bose, GE Plastics, IBM T.J. Watson Research Center, The Mathworks, United Technologies Research Center, and DEKA Research and Development.

Since 1998, over 167 students from 119 universities across 37 states of the United States, Puerto Rico, England, and France have participated in the program. REU students have worked on over 47 different projects for over 23 different companies with 17 different faculty advisors. It is a measure of success that several companies, including Bose, John Hancock Insurance, DEKA Research and Development, Premier Insurance, State Street, and Veeder-Root have sponsored several projects. Here are some sample REU projects from past research summers (more are available on the CIMS web at www.wpi.edu/+CIMS): *Optimal Cession Strategies; Modeling Fluid Flow in a Positive Displacement Pump; Statistical Procedures for Failure-Mode Testing of Diagnostic Systems (2001 and 2002); Estimation and Optimization for Constructing Hedge Portfolios.*

3 Professional Science Master's

At the graduate level, with support from the Alfred P. Sloan Foundation, we have introduced in 2000 a Professional Science Master's degree in Financial Mathematics and one in Industrial Mathematics. Currently there are over 30 recognized PSM programs in mathematics in various areas(Carbonara and Vernescu 2013).

The relevant industry was involved from the very beginning in the design of these programs' curricula, to provide the necessary training that would make graduates successful in industry (see Tobias et al. 2001; Sims 2006). Specifically designed to provide the training for the workforce, these programs do not preclude students from continuing in a Ph.D. program. The programs preserve the core component of the traditional curriculum, and also require exposure to applications of mathematics in sciences and in engineering, experience in formulating and solving open-ended problems, and computational, communication, and teamwork skills. The Professional Science Master Programs require students to complete courses from other departments, introduce them to professional skills through a special seminar, and provide them with internship opportunities. In both programs, industrial experience is gained through an industrial project sponsored by local industry. Industrial summer internships are encouraged and facilitated by CIMS through its industrial partners.

The Financial Mathematics graduate program at WPI has been designed to lead students to the frontlines of the financial revolution of the new century. The program offers an efficient, practice-oriented track to prepare students for quantitative careers in the financial industry including banks, insurance companies, investment and securities firms; it features coursework on mathematics and finance, computational laboratories, industrial internships, and project work. The goal is to provide the knowledge, skills, and experience necessary for the quantitative positions in investment banks, securities houses, insurance companies, and money management firms. A strong mathematical background is built for developing mathematical models for risk in relationship to returns, trading strategies, structured contracts, and derivative securities. A strong collaboration has been built between our faculty and the financial services industry concentrated in the Boston-Hartford corridor. Our graduates have started their careers in jobs involving financial product development and pricing, risk measurement and control, and investment decision support or portfolio management.

The Industrial Mathematics Program is aimed at training students for professional careers in industrial environments. In addition to a strong mathematical background with depth in one area, the program emphasizes the breadth required by the industrial environments through elective coordinated modules of mathematics and engineering/science courses (e.g., physics, computer science, mechanical engineering, electrical and computer engineering, bioengineering), tailored to individual students' interests and needs. By also developing the students' communication and business skills, the program aims to creating successful professionals for the corporate world. Our graduates have started, or continued, their jobs in the manufacturing, software, and environmental industry.

4 Ph.D. Program

A multidisciplinary experience, rooted in the real-world is also part of the Ph.D. Program in Mathematical Sciences, which has as degree requirement, a 1–9 credit project outside the Department. Taking students out of their comfort zone and challenging them with real-world problems has a significant impact on their mathematics understanding and appreciation and at the same time can open new career options, so valuable in times of economic downturn. Ph.D. students worked on projects for the Kodak, Air Force Labs, IBM, Sandia and Lawrence Livermore National Laboratories, and United Technologies.

5 Workshops for Faculty

WPI was chosen to be the site for the first SIAM Mathematics in Industry Regional Workshop (Davis 1998), held in May 18–19, 1998. The purpose of the workshop was to help mathematics faculty study and implement some of changes recommend in the SIAM Report on Mathematics in Industry. The keynote speakers from IBM, Kodak, Lucent, and Lehman Brothers discussed the need for sophisticated mathematics modeling in industry, and participants presented successful university–industry project collaborations. This workshop was the first in a series of six regional workshops that brought together mathematicians from industry and academia to showcase their scientific and educational collaborations. It was followed by five other regional workshops at the University of Illinois in 1998 (Midwest), Claremont Colleges in 1999 (Western), North Carolina State in 1999 (Southeast), University of Washington in 2000 (Northwest), and University of Huston in 2001 (Southwest). These workshops have emphasized the need for better connecting the mathematics world with the industrial world and are now followed by the biannual SIAM Mathematics for Industry Conferences.

The Mathematics Problems in Industry Workshops (MPI) were hosted in 2003, 2005, 2008 and 2013 at WPI, as part of a joint effort between faculty members at

RPI, U. Delaware, NJIT, and WPI. The workshops attract leading academic mathematicians and graduate students who work for the week on problems posed by engineers and scientists from industry. In the past, these problems have included, but were not limited to engineering and product design, process design and control, environmental remediation, scheduling and optimization, and financial modeling. As outcomes, these workshops provide a better understanding of existing models and methods, access to advanced computing solutions, and open a dialog with mathematicians in academia and government labs that provides a new look and fresh ideas.

Some of the problems presented at WPI were: Safe Fuel/Air Slow Compression (Gilbarco/Veeder-Root, Simsbury, CT); Optimal Wear for a Laying Pipe (Morgan Construction, Worcester, MA); Need a Lift?: An Elevator Queuing Problem (UTRC, East Hartford, CT); Enhanced Leak Detection in Fuel Tanks (Gilbarco/Veeder-Root, Simsbury, CT); Stability of the Oil-Air Boundary in Fluid Dynamic Bearings of Hard Disk Drives (Hitachi GST, San Jose, CA); Lubrication Layer Perturbations in Chemical–Mechanical Polishing (Araca, Inc., Mesa, AZ). For a complete list see www.wpi.edu/+CIMS.

6 Middle and High School Outreach Program

Business and industrial applications of mathematics are clearly a valuable but often underutilized means of both teaching and motivating middle and high school students in mathematics. For too many students mathematics seems irrelevant, disconnected from their career interests. The main cause is that mathematics is often taught disconnected from applications and the real-world problems it is so well equipped to solve and thus mathematics becomes a barrier to overcome. Thus, it should come as no surprise that most students, particularly women and minority students, lose interest in mathematics. Research reveals (Fennema and Leder 1990; Moses and Cobb 2001) that young women and underrepresented minority children do not persist in mathematics because they appreciate neither its connection to careers nor its utility in helping to solve society's problems. One-time school visits by a working scientist or engineer able to explain how mathematics can be used is not always the best way to encourage students to continue their studies (Campbell et al. 2002). It has also been shown that student performance is greatly improved when teaching concepts and applying them to concrete problems as opposed to only drill and practice (Wenglinsky 2000). In addition, the supply of graduate students is seen as a problem that starts in K-12 education, as students lose interest in graduate school long before they ever have a chance to enter it (Ewing 2002). Therefore, professional development for teachers that relates to the content and materials used in the classroom has been shown to be more beneficial for teaching and learning than professional development on abstract concepts.

In June 2000, with support from the NSF through the SIAM, we organized our first Mathematics in Industry Institute (MII) for teachers; 36 teachers from New

England worked for 5 days on industrial projects. With support form the GE Foundation, this pilot work was expanded into the MIIs for Teachers, for four consecutive years: 2001–2005. The Institutes consisted of: workshops actively involving teachers in developing industrial projects suitable for use in the classroom; development of dissemination activities to bring these idea and resources to entire school districts; and incorporation of these projects into their curriculum. A follow-up workshop in the spring allowed teachers to share "best practices". With the completion of the 2005 MII, more than 240 teachers from 18 states and Canada have come to WPI to work on industrial mathematics projects. There are now 23 industrial mathematics projects for classroom use on the CIMS web site.

The feedback on the impact of the institutes showed that teachers have changed the way that they think about mathematics, talk about applications for mathematics, and discuss quantitative careers in the classroom. For example, we saw an increase from 25 to 58 % in the percentage of teachers who regularly use engineering and quantitative home works and examples. Approximately, 50 % of teachers reported an increase in the use of industrial math examples in the classroom. The broader impact of these activities was reflected in the development of new courses, programs, and student competitions in schools. In addition, versions of the projects were been included in the Benchmarks for High School Mathematics Education developed at the American Diploma Project (an affiliate of Achieve) and were used in the NSF-sponsored Mathematics and Sciences Partnership project connecting WPI, Boston University, University of Massachusetts Lowell, the Educational Development Center, and five Boston-area school districts.

7 Conclusions

The model developed at WPI is one of vertical integration of innovative industrial projects in the mathematics curriculum at all levels, from middle and high school, through undergraduate programs and up to graduate programs and faculty research. At each level, the industrial problems have a different impact, and the different components reinforce each other. For students, the educational impact cannot be replicated in the regular classroom setting; the industrial projects provide a different type of challenge and motivation and at the same time they provide the training needed for a more diverse set of career options. For faculty, these offer an important connection to industry, business, and government, and a challenge to use these in the educational program. For industry, business, and government these projects have a significant impact: they provide solutions and insight on important problems and they provide insight in and access to the educational system. By seamlessly integrating the industrial projects at all levels, an economy of scale has been created and the program can respond to both educational and scientific needs.

References

Benkowski, S. (1994). Preparing for a job outside academia. *Notices AMS, 41*, 917–919.

Berkey, D., & Vernescu, B. (2007). *A model for vertical integration of real-world problems in mathematics, Proceedings of the 2007 ASEE Annual Conference*.

Campbell, P. B., Jolly, E., Hoey, L., Perlman, L. K. (2002). Upping the Numbers: Using Research-Based Decision Making to Increase Diversity in the Quatitative Disciplines, GE Fund.

Carbonara, J., & Vernescu, B. (2013). *Creating Tomorrow's Mathematics Professionals, In Proceedings of the NPSMA Workshop*, Niagara Falls, NY, October 11-13, 2011, COMAP.

Chan, T.F. (2003). The mathematics doctorate: A time for change? *Notices AMS, 50*(8), 896–903.

Davis, P.W. (1998). WPI hosts first in series of SIAM's regional math in industry workshops, SIAM News, http://www.siam.org/siamnews/09-98/wpi.htm.

Davis, P.W. (1999). WPI builds on industrial ties to create REU program, SIAM News, http://www.siam.org/news/news.php?id=772.

Ewing, J. (2002). Graduate and postdoctoral mathematics education, American Mathematical Society, Providence, RI; http://www.ams.org/ewing/GradEducation.pdf.

Fennema, E., & Leder, G. (1990) (Eds.). *Mathematics and gender*. Teachers College Press.

Forman, S.L., & Steen, L.A. (1995). Mathematics for work and life. In I. M. Carl (Ed.), *Seventy-five years of progress: Prospects for school mathematics* (pp. 219–241). Reston, VA: NCTM.

Friedman, A., & Litman, W. (1994). *Industrial mathematics. A course in solving real-world problems*. Philadelphia: SIAM.

Heinricher, A. C., & Vernescu, B. (2002). How can mathematics help—mathematicians at work today. Pythagorean Review, Mu Alpha Theta on-line Journal.

Heinricher, A. C., & Weekes, S. L. (2007). *Industrial Mathematics and Statistics Research for Undergraduates at WPI. In J. Gallian (Ed.), Proceedings of the Conference on Promoting Undergraduate Research in Mathematics*, American Mathematical Society, 2007, 193–197.

Lutzer, D.J., Maxwell, J.W., & Bodi, S.B. (2002). Statistical abstract of undergraduate programs in mathematical sciences in the United States, Fall 2000 CBMS Survey.

MacCluer, C. (2000). *Industrial mathematics. Modeling in Industry, Science and Government*. New Jersey: Prentice Hall.

Moses, R., & Cobb, C. (2001). *Radical equations: Civil rights from Mississippi to the Algebra project*. Boston, MA: Beacon Press.

Sims, L. B. (2006). Professional Master's Education, A CGS Guide to Establishing Programs, Council of Graduate Schools, Washington, D.C.

The SIAM Report on Mathematics in Industry (2012). Society for Industrial and Applied Mathematics, Philadephia, Pennsylvania, http://www.siam.org/reports/mii/2012.

Tobias, S., MacCluer, C., & Svetic, R. (2001). The Growth of the Professional Master's in Mathematics, *Notices of the American Mathematical Society*, May 2001.

Vernescu, B., & Heinricher, A.C. (2000). *Research experience for undergraduates in industrial mathematics and statistics at WPI, Proceedings of Summer Undergraduate Research Programs, AMS Conference*.

Wenglinsky, H. (2000). How Teaching Matters: Bringing the Classroom Back Into Discussions of Teacher Quality, ETS, October 2000.

Part V
WG Education in Schools

Educational Interfaces Between Mathematics and Industry at School Level

Report on WG 4

Gabriele Kaiser, Henk van der Kooij and Geoff Wake

1 General Frame: Mathematical Modelling at School as a Frame for Educational Interfaces Between Mathematics and Industry at School Level

There exists an international consensus that the prolific connections between mathematics and industry need an active counterpart at school level. These connections should be of high importance within mathematics education at all levels from primary to tertiary education.

As described in other parts of this book, mathematics plays an important role in our daily lives: mathematics every day, in many places and on many occasions underpins what we do. However, the increasing importance of mathematics in understanding many problems of the world surrounding us has not necessarily led to an increased importance of mathematics within school curricula. One possible reason for this unsatisfactory situation lies in the well known and often discussed paradox that mathematics is progressively used in our modern society while it is slowly disappearing from societal perception and our usual practice.

In society many mathematical algorithms are not needed by individuals anymore due to the fact that technological tools or computer-controlled machines carry out this work; even within the banking and finance sector, employees do not need to do complicated calculations anymore. On the one hand, the introduction of the calculator at school level has significantly decreased the importance of

G. Kaiser (✉)
Faculty of Education, University of Hamburg, Hamburg, Germany
e-mail: Gabriele.Kaiser@uni-hamburg.de

H. van der Kooij
Freudenthal Institute, Universiteit Utrecht, 85.170, 3508 AD Utrecht, The Netherlands
e-mail: H.vanderKooij@uu.nl

G. Wake
University of Nottingham, Jubilee Campus, Wollaton Road, Nottingham NG8 1BB, UK
e-mail: Geoffrey.Wake@nottingham.ac.uk

A. Damlamian et al. (eds.), *Educational Interfaces between Mathematics and Industry*,
New ICMI Study Series 16, DOI: 10.1007/978-3-319-02270-3_26,

standard algorithms such as those for long-division or trigonometric calculations. On the other hand, the introduction of technology and calculators into schools, which is nowadays taking place all over the world, certainly offers the opportunity for students to deal with more complex and realistic examples with many of these now being accessible even at school level.

Analysing the relevance and importance of industrial mathematics in school worldwide, one can state that the situation is not satisfactory: real world examples being related to industry and the workplace have been introduced into ordinary mathematics teaching, but with significant geographical differences. The ICMI affiliated study group "International Community of Teachers of Mathematical Modelling and Applications" (ICTMA, cf. the recent proceedings edited by Kaiser et al. 2011, Stillman et al. 2013), which consists of applied mathematicians focusing on tertiary education and mathematics educators interested in teaching at school level, concludes the following: applications and real world examples play a significant role in mathematics education in European, South and North American countries, but have only minor importance in East Asian countries with important exceptions in some countries such as Singapore. Especially, the promotion of modelling competencies, i.e. the competencies to tackle real world problems using mathematics, is accepted as a central goal for mathematics education worldwide, especially because mathematics education aims to promote responsible citizenship. In many national curricula, such as in Germany or the U.S. (Common Core Standards), modelling competencies play a decisive role providing evidence that the importance of mathematical modelling is accepted at a broad international level. However, beyond this consensus about the relevance of modelling, how to integrate mathematical modelling into the teaching-and-learning-processes is still disputed.

The call to teach mathematics in an application-oriented way may be traced back to the famous symposium "Why teach mathematics so as to be useful" (Freudenthal 1968; Pollak 1968). Why and how to include applications and modelling in mathematics education has been the focus of many research studies since then. These have not led to a single unique vision of the relevance and role of applications and modelling in mathematics education. In contrast, the views developed remain quite diverse, reflecting those of educators with both pedagogical vision and those seeking an enhanced role for applied mathematics.

There have been several attempts to analyse the various theoretical approaches to teach mathematical modelling and applications and to clarify possible commonalities and differences. For example Kaiser-Meßmer (1986) in her analysis from the beginning of the recent international debate on modelling and applications (until the mid eighties of the last century) distinguishes two main strands, the so-called pragmatic perspective focusing on utilitarian or pragmatic goals, with the applied mathematician Pollak as main protagonist, and the scientific-humanistic perspective oriented more towards mathematics as a science and humanistic ideals of education, with the mathematician and mathematics educator Freudenthal as main protagonist. The different goals have consequences for the way to include mathematical modelling in the curriculum, namely either based on cyclic model building processes as requested by Pollak (1969) or as complex mathematising encompassing interplay between mathematics and the real world as described by Freudenthal (1973).

A few years later, in their extensive survey of the state of the art, Blum and Niss (1991) focussed on the arguments and goals for the inclusion of applications and modelling in curricula and discriminated five layers of arguments such as the formative argument related to the promotion of general competencies, critical competence argument, utility argument, picture of mathematics argument and the promotion of mathematics learning argument. They made a strong plea for the promotion of three goals, namely, that students should be able to perform modelling processes, acquire knowledge of existing models and critically analyse given examples of modelling processes.

Based on this position, they analyse various approaches on how to integrate applications and modelling in mathematics instruction at school and even university level. These include teaching mathematics and modelling in different courses or the "islands approach", where small islands dealing with modelling examples are interspersed into the general mathematics course. In the mathematics curriculum-integrated approach, the problems come first and mathematics to deal with them is sought and developed subsequently. The most advanced approach, the interdisciplinary integrated approach, operates with a full integration between mathematics and extra-mathematical activities. These various approaches can be differently implemented, either as short-term activities in one or two lessons during ordinary teaching time or as modelling projects, during which students tackle modelling problems over a longer period. Simpler modelling activities seem to be possible in short-term activities whereas authentic, more complex modelling problems need intensive teaching time and therefore seem to be suitable for more extensive projects. In the past decade, these kinds of modelling projects have become popular in several parts of the world, e.g. Germany and Australia.

In their classification of the historical and more recent debate on mathematical modelling in school, Kaiser and Sriraman (2006) point out that there exist several perspectives on mathematical modelling in the international discussion on mathematics education. They proposed a framework for the description of the various approaches, which classifies these conceptions according to the aims pursued with mathematical modelling, their epistemological background and their relation to the initial perspectives.

The following perspectives were described, which continue positions already emphasised at the beginning of the modelling debate:

- Applied or realistic modelling, fostering pragmatic-utilitarian goals and continuing traditions of applied mathematics;
- Epistemological or theoretical modelling, placing theory-oriented goals in the foreground and being of more pedagogically oriented traditions;
- Educational modelling as an integrated approach, emphasising pedagogical and subject-related goals, which combine aspects of both the applied and the epistemological/theoretical approaches.

More recently other approaches have been proposed, such as the model eliciting and contextual approaches, which emphasise problem-solving and psychological goals, the socio-critical and socio-cultural modelling fostering the goal of critical understanding of the surrounding world or as a kind of meta-perspective, the cognitive modelling approach foregrounding the analysis of students' modelling processes.

The understanding of mathematical modelling processes, the way the relationship between mathematics and the "rest of the world" (Pollak 1968) is described, play a decisive role within the modelling debate and can be seen as a key feature of modelling activities. Already from the beginning of the discussion, modelling processes have been used and described differently by the various modelling perspectives mentioned above. The perspectives emphasise either the solution of the original problem, e.g. by the applied modelling perspective, or the development of mathematical theory, e.g. by the epistemological approach. So, corresponding to the different perspectives on mathematical modelling, there exist various modelling cycles with their own specific emphasis. Nowadays, however, despite some discrepancies, one common and widespread understanding of modelling processes has been developed. In nearly all approaches, the idealised process of mathematical modelling is described as a cyclic process to solve real problems by using mathematics, illustrated as a cycle comprising different steps or phases. For example the modelling cycle developed by Blum (1996) and Kaiser (1995) and based on the work of Pollak, amongst others, emphasise the necessity to simplify the given real problem in order to build a real model of the situation prior to translating the real model into mathematics. Further important phases include the interpretation of the mathematical results and validating them. Other descriptions of the modelling cycle coming from applied mathematics (e.g. by Haines et al. 2000) emphasise the necessity of reporting the results of the process and including refinements of the model more explicitly.

A central goal of the debate on strengthening the relation between mathematics and the real world in school is the promotion of mathematical modelling competencies, i.e. the ability and the willingness to work out problems with mathematical means taken from the real world through mathematical modelling (cf. Maaß 2006 for a detailed definition of modelling competencies). A distinction is made between global modelling competencies and sub-competencies of mathematical modelling. Global modelling competencies refer to necessary abilities to perform the whole modelling process and to reflect on it. The sub-competencies of mathematical modelling refer to specific phases of the modelling cycle, they include the different competencies that are essential for performing the individual steps of the modelling cycle (Kaiser 2007).

2 Mathematics at the Workplace and Their Educational Consequences: Results of the Discussion at the Conference

Based on the above-described framework, the discussions of the working group focussed on several aspects. These aspects provide an important agenda for future developments.

The requirements of current and future workplaces make clear that an overall change of the current school curriculum is necessary in order to meet the requirements of the changed nature of work. The aims of mathematics education have to be reconsidered in order to meet the new requirements of labour and the workplace. A decisive element in the necessary changes is the different nature of workplace mathematics and school mathematics. Workplace mathematics can be described as being functional, contextualised and incorporated into work activities and competencies necessary for work. In contrast school mathematics can be characterised as being more abstract and separated from concrete working activities. In addition school mathematics is dealt with explicitly in contrast to workplace mathematics, which is often invisible and used as a tool or a system that underpins workplace activities. Fundamentally we have to bear in mind for the future debate that workplace mathematics is not school mathematics found in workplaces, it is distinctively different from school mathematics. Due to the many different contexts, in which workplace mathematics is used, the general nature of workplace mathematics is difficult to detect. To summarise, school and workplace may be regarded as communities of practice with two different aims, namely learning and production, although informal learning is also taking place in the workplace.

Central to the changes required is the need to recognise and privilege a range of newly defined competences that acknowledge the quickly changing technical rich environments in which twenty-first century workers operate. Based on ethnographic studies combined with interviews, mathematics education research has gained deep insight into the essential core of workplace mathematics. Such studies have developed ideas of situated cognition, situated abstraction, techno-mathematical literacy, general mathematical competences, mathematics containing competences and so on, all of which recognise that workers develop their mathematics in complex and often technology-rich settings. In addition, these studies emphasise that developing mathematics require workers to be mathematically competent in settings that are very different from schools. Mathematical activity in workplaces differs significantly from mathematical activity in school and is often obscured by context and usage of black-boxes, in which mathematics is hidden by technologies (Williams and Wake 2007). Furthermore, in workplaces much of the knowledge, including the mathematical knowledge, is tacit, so that workers and others participants do not always recognise it. However, in principle, some competences and strategies have generality across different workplace activities. These draw on mathematical content, but most importantly on other problem solving and modelling competencies and strategies. Education and industry together need to explore these ideas further and develop a clear understanding of the definition of competences, which make sense for a whole range of workplaces. In addition, these competencies need to have enough longevity by being general enough to take into account the rapidly changing practices. This leads to the proposal to further develop substantial research and work that will inform our communities of how best to describe these competences and how to define and implement educational courses that support learning. We note that this would provide a major and significant advancement and would support better communication between the two communities of education and industry.

Another central aspect, which needs to be discussed, concerns more general mathematical skills. Creative, critical mathematical thinking and problem-solving skills need to be promoted since, when used in a flexible way, they are essential to the effective understanding and handling of workplace mathematics. The traditional conception of mathematics as absolute and infallible is in marked contrast to the negotiability (when reasonable) and common sense approach that is valued in the workplace. Context in workplace mathematics is integral to activity and is intimately coupled with mathematical models of a workplace reality. These are by nature often complex, messy and introduce many constraints and idiosyncratic notations, artefacts and representations of mathematical thinking. To work effectively with such a kind of mathematics requires flexibility of the approach used and a clear understanding of how reality and mathematics are structurally connected. Skills in problem solving, creativity and critical inquiry in mathematics need to be valued and promoted at school level.

Existing and new fora need to be used to develop channels of communication that allow a two-way dialogue between mathematics education and the workplace. This dialogue shall promote a better understanding of workplace mathematics within the mathematics education community and issues relating to the development of work-related mathematical competences.

Much needs to be done if school mathematics is to better prepare people for workplace activities in often rapidly changing and technology rich environments. On the one hand, this requires that education communities (including at the levels of policy making, the education of teachers through to the classroom) better understand workplace mathematics. On the other hand, industry needs to understand the difficulties of education in capturing the essentials of workplace mathematical activity and adapting curricula in ways that can lead to prolific teaching and learning.

Measures to promote such a new understanding of the interface between workplace and school mathematics require implementation at several levels. Teacher education needs to be one focus. In order to enable teachers to include a new understanding of mathematics in workplaces into their teaching at school level opportunities should be offered in pre- and in-service teacher education for teachers to experience this for themselves. As a first step for strengthening the connections between mathematics education at school and industry, the inclusion of a variety of activities such as modelling days or weeks, project work, interdisciplinary projects and many small-scale activities seem to be appropriate. For real changes it is indispensable to find ways to value these new activities, to include them into assessment schemes; the so-called soft skills such as communication or team working skills especially need to be valued.

To summarise: an agenda for action is needed containing short- and long-term activities that strengthen the relation between industry and mathematics education at school.

3 Description of the Chapters of the Book Tackling These Aspects

Other chapters of the book present important aspects mentioned within this framework of the debate on the teaching and learning of mathematical modelling and the aspects discussed at the conference.

The chapter by Gravemeijer and Wake refer to theoretical aspects of the modelling debate. Gravemeijer takes a rather fundamental view and starts from the assumption that societal changes ask for adaptations of a foundational mathematics curriculum for all. He argues that the development of the information technology and globalisation of the job market has strong consequences for the employability of coming generations. He therefore pleads for a fundamental change of the goals of mathematics education and calls, for example, the inclusion of problem solving. Wake expands on these ideas, reflects on the nature of the activity in relation to mathematics in both schools and workplaces and, on the basis of these analyses, develops curricular consequences such as expansive learning and developmental transfer.

The chapter by Stillman and Ng continues the curricular debate, but with a stronger focus on school activities. The authors discuss different models of curricular proposals bringing authentic real world applications into secondary school curricula, either through inclusion of modelling examples in mathematics education across its whole range or as interdisciplinary project work.

In his chapter, Geiger explores the potential synergy between research in the areas of mathematical applications and modelling and new technology in mathematics education. This may offer the means to design school-based activities that provide a bridge between school mathematics and the world beyond.

The final chapter by Kaiser, Bracke, Göttlich and Kaland describes project-oriented activities that incorporate authentic complex modelling problems into mathematics teaching using the approach of modelling weeks. In the evaluation of these activities, their educational potentials are highlighted, not only for students but for the mathematics to be considered as well.

References

Blum, W. (1996). nwendungsbezüge im Mathematikunterricht—Trends und Perspektiven. In G. Kadunz, et al. (Eds.), *Trends und Perspektiven. Schriftenreihe Didaktik der Mathematik* (Vol. 23, pp. 15–38). Wien: Hölder-Pichler-Tempsky.

Blum, W., & Niss, M. (1991). Applied mathematical problem solving, modelling, applications, and links to other subjects—state, trends and issues in mathematics instruction. *Educational Studies in Mathematics, 22*, 37–68.

Freudenthal, H. (1968). Why to teach mathematics so as to be useful. *Educational Studies in Mathematics, 1*(1/2), 3–8.

Freudenthal, H. (1973). *Mathematics as an educational task*. Dordrecht: Reidel.

Haines, C., Crouch, R., & Davis, J. (2000). Mathematical modelling skills: A research instrument (Technical Report No. 55). University of Hertfordshire: Department of Mathematics.

Kaiser-Meßmer, G. (1986). Anwendungen im Mathematikunterricht. Vol. 1—Theoretische Konzeptionen. Bad Salzdetfurth: Franzbecker.

Kaiser, G. (1995). Realitätsbezüge im Mathematikunterricht—Ein Überblick über die aktuelle und historische Diskussion. In G. Graumann, et al. (Eds.), *Materialien für einen realitätsbezogenen Mathematikunterricht* (pp. 66–84). Bad Salzdetfurth: Franzbecker.

Kaiser, G. (2007). Modelling and modelling competencies in school. In C. Haines, P. Galbraith, W. Blum, & S. Khan (Eds.), *Mathematical Modelling (ICTMA 12): Education, Engineering and Economics* (pp. 110-119). Chichester: Horwood.

Kaiser, G., Blum, W., Borromeo Ferri, R., & Stillman, G. (Eds.), (2011). *Trends in teaching and learning of mathematical modelling*. Dordrecht: Springer.

Kaiser, G., & Sriraman, B. (2006). A global survey of international perspectives on modelling in mathematics education. *ZDM—The International Journal on Mathematics Education, 38*(3), 302–310.

Maaß, K. (2006). What are modelling competencies? *Zentralblatt für Didaktik der Mathematik, 38*(2), 113–142.

Pollak, H. O. (1968). On some of the problems of teaching applications of mathematics. *Educational Studies in Mathematics, 1*(1/2), 24–30.

Pollak, H. O. (1969). How can we teach applications of mathematics? *Educational Studies in Mathematics, 2*, 393–404.

Stillman, G., Kaiser, G., Blum, W. & Brown, J. (Eds.), (2013). *Teaching Mathematical Modelling: Connecting to Teaching and Research. Practices – The Impact of Globalisation*. Dordrecht: Springer.

Williams, J., & Wake, G. (2007). Black boxes in workplace mathematics. *Educational Studies in Mathematics, 64*(3), 317–343.

Mathematical Applications, Modelling and Technology as Windows into Industry Based Mathematical Practice

Vince Geiger

1 Introduction

The growing importance of mathematical skills to the workplace and industry and the increasing sophistication of mathematics related activity within the workplace (Hoyles et al. 2002) are placing greater demand on teachers, schools and designers of curricula to find approaches to teaching and learning that facilitate the transfer of school-based mathematical knowledge to the world outside of school. At the same time, it is generally recognised that technology in the form of ICTs, are transforming the fabric of society including the nature of learning and working at school and in the workplace (Zevenbergen 2004). This chapter explores the potential offered by synergy between mathematical applications and modelling, numeracy, and ICTs. This synergy offers approaches to the design of school-based activity that provide a bridge between school mathematics and the worlds beyond school.

2 Mathematics in the Workplace and Industry

Research into the use of mathematics in the workplace and industry has demonstrated an increasing demand for a mathematically capable workforce. In addition, the mathematical demand of what workers do within a wide range of industries is becoming increasingly complex. This situation was highlighted by Hoyles et al. (2002) in a study of a range of industries including: electronic engineering and optoelectronics; financial services; food processing; healthcare; packaging; pharmaceuticals and tourism. The authors of this report identify the following trends in data gathered from participant workers in the study:

V. Geiger (✉)
Australian Catholic University, Brisbane, Australia
e-mail: Vincent.Geiger@acu.edu.au

A. Damlamian et al. (eds.), *Educational Interfaces between Mathematics and Industry*, New ICMI Study Series 16, DOI: 10.1007/978-3-319-02270-3_27,

- team-based working is widespread;
- the need for mathematical skills is being progressively extended throughout the workforce as a result of the pressure of business goals and the introduction of IT;
- there is a growing need to communicate information effectively, based on mathematical data and inferences and involving colleagues, customers and external inspectors;
- there is a need for hybrid skills, e.g. combining technical and analytic knowledge with the ability to communicate analytical information (p. 12).

These trends have clear implications for both general education as well as vocational and technical training as they indicate that education/training in technical skills specific to a vocation or profession are not enough alone to operate effectively in the rapidly changing world of work and industry. At the same time, Straesser (2007) warns against narrow approaches to mathematics education and training. He views mathematics as a strategic tool that can be adapted for a range of contexts and settings. In particular, he signals a concern for the "black box" view of mathematics in the workplace where the underpinning features and functions of mathematics are subsumed into simple routinised practice. Straesser goes on to suggest that the type of mathematics that spans the gap between school mathematics and the workplace is "no longer part of mathematics or 'the rest of the world' alone, but are a new type of knowledge bridging the division between mathematics and the rest of the world" (p. 169).

Zevenbergen (2004) also notes the changing nature of the workplace and identifies the use of ICTs as a factor in the intergenerational gap between employers and experienced workers, and those being initiated into the workforce for the first time. The new generation of workers seems to have the capacities to make use of their personal mathematical knowledge and their confidence and capabilities with ICTs to solve on the job problems in more inventive ways than their more experienced co-workers.

These commentaries imply that the capacity to think adaptably, a disposition to continue to learn new approaches to solving problems as they arise, and the capacity to embrace the use of technological tools are as important as the type of mathematical knowledge traditionally taught in schools. These are elements of Straesser's "in between worlds". But what is this new type of knowledge and what would it look like if we were to see it in a mathematics classroom?

3 School Approaches to Learning and Teaching Mathematics in Context

While traditionally learning and teaching have been based on curricula that have placed little emphasis on the use of mathematics in the beyond school world (Damlamian and Straesser 2009), there are developing areas of research and practice which focus on the integration of the learning of mathematical knowledge and the utilisation of this knowledge in real world contexts. Approaches to situating the learning and doing of mathematics within authentic life relevant contexts appear to take two distinct but closely related forms: modelling and applications in mathematics (MAM) and numeracy. While these two approaches have developed independently, the focus of research in these areas are closely aligned.

Mathematical modelling—formulating a mathematical representation of a real world situation, using mathematics to derive results, interpreting the results in terms of the given situation and if necessary, and revising the model—is now a well established field of research and practice within school settings. While MAM is not a mainstream practice in schools (Stillman 2007), it offers opportunity for students to investigate issues in the social, political, environmental and economic worlds to which they will eventually contribute and to connect with mathematical practices that are utilised by industries built within these worlds.

Again, while not yet a well-established practice in all schools or a priority for all school systems, attention to numeracy, which is also known as numerical literacy in some international contexts, is increasingly seen as fundamental to developing students' capacities to use mathematics to function as informed and reflective citizens, to contribute to society through paid work, and in other aspects of community life (Steen 2001). This aspect of mathematics education has been recognised internationally through the OECD's Program for International Student Assessment (PISA). According to PISA's definition mathematical literacy is:

> An individual's capacity to identify and understand the role mathematics plays in the world, to make well-founded judgments and to use and engage with mathematics in ways that meet the needs of that individual's life as a constructive, concerned and reflective citizen (OECD 2004, p. 15).

As the focus of numeracy is one the use of mathematics within lived worlds, numeracy practices within schools should connect directly with the use of mathematics in the world of work and industrial contexts. This is recognised in the interpretation of numeracy that is widely accepted within Australian school systems, "To be numerate is to use mathematics effectively to meet the general demands of life at home, in paid work, and for participation in community and civic life" (Australian Association of Mathematics Teachers 1997, p. 15).

While these are broad interpretations of the concept of numeracy, they do not encompass the aspects of learning and doing that will allow the capacities they describe to be enacted in a real world context. In order to effectively practice mathematics in the world of work and in the community at large, individuals also require the disposition to see a problem or an issue from a numeracy perspective, to think critically and flexibly, and act confidently and persistently in order to progress a way forward.

4 ICTs and Learning and Teaching in Contexts

There is now substantive research which supports the use of ICTs in enhancing mathematical thinking and learning including: content domains such as number (Kieran and Guzma'n 2005); geometry (Laborde et al. 2006); algebra and calculus (Ferrara et al. 2006); and collaborative investigative practice (Beatty and Geiger 2010). This research, however, has mainly focused on how digital tools can enhance the learning and teaching of mathematics as it exists independently from the utilisation of this knowledge in the world outside of school. As Zevenbergen (2004) observes:

> While such innovations [ICTs] have been useful in enhancing understandings of school mathematics, less is known about the transfer of such knowledge, skills and dispositions to the world beyond schools. Given the high tech world that students will enter once they leave schools, there needs to be recognition of the new demands of these changed workplaces (p. 99).

In answer to this challenge there has been growing interest in the use of ICTs to enhance the learning and teaching of mathematics in life relevant situations. The following, for example, is drawn from the literature of mathematical modelling.

> Many technological devices are highly relevant for applications and modelling. They include calculators, computers, the Internet and computational or graphical software as well as all kinds of instruments for measuring, for performing experiments etc. These devices provide not only increased computational power, but broaden the range of possibilities for approaches to teaching, learning and assessment (Niss et al. 2007, p. 24).

Other researchers have sought to define more clearly specific ways in which the use of ICTs can enhance the study of mathematics in context. In a report on research into uses of ICTs in mathematical modelling in Brazilian schools Villarreal et al. (2010) concluded that ICTs were vital elements in the construction and validation of mathematical models for both secondary school and university students. In addition, ICTs allowed teachers to offer learning experiences that were beyond the scope of the official curriculum. Geiger et al. (2010) in a study of the use of Computer Algebra Systems (CAS) within mathematical modelling classes in two different Australian educational jurisdictions found that ICTs provided the opportunity to explore sophisticated authentic contexts that would not have been possible with pen and paper alone, and allowed for the scaffolding of less capable students in a way that enabled them to engage in challenging tasks.

In order to accommodate the changing nature of knowledge, work and technology, Goos (2007) developed a model for numeracy that incorporates attention to real life contexts, the deployment of mathematical knowledge, the use of physical and digital tools and consideration of students' dispositions towards the use of mathematics. The development of a critical orientation was also emphasised in relation to numeracy practice, for example, the capacity to evaluate quantitative, spatial or probabilistic information used to support claims made in the media or other contexts (Fig. 1). This approach, therefore, also incorporates many of the

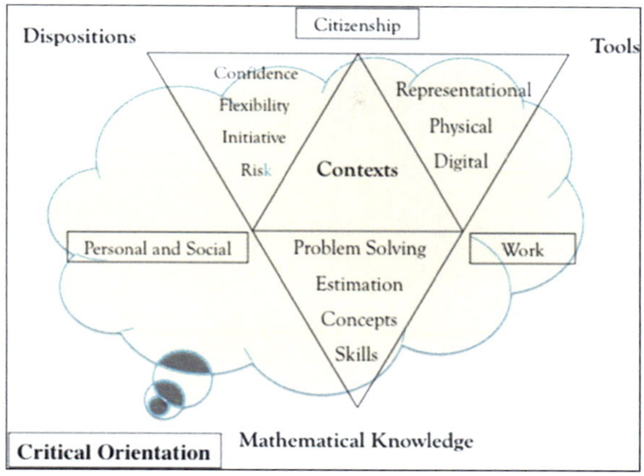

Fig. 1 A model for numeracy in the twenty first century (Goos 2007)

elements that are considered to be part of the process of mathematical modelling. This model offers a broader interpretation of the role of mathematics and ICTs in bridging the gap between school mathematics and the wider world. The use of this model to examine classroom practice will now be illustrated through an example drawn from a primary context.

5 An Example of Applications and Modelling From a Primary School Context

This example is drawn from a study of 20 middle school teachers (Years 6–9) working in pairs in 10 schools across South Australia during 2009 (see Goos et al. 2010). Participants were working on a state sponsored project that aimed to empower teachers to work with numeracy across curriculum areas. As part of the project, one teacher developed an activity within Physical Education (PE) where students investigated the level of their physical activity through the use of a pedometer that students wore during all waking hours over one week. The collected data, that is, the number of paces walked or run, were entered by students into a shared Excel spreadsheet every day. Students were asked to analyse their own data by using facilities within Excel, for example, the graphing tool, and then to compare their results with those of other students. As part of this analysis, students were asked to convert their total daily and total weekly paces into kilometres to gain a sense of how far they typically walked in the course of a day or a week. The task was also designed help to students realise that the distance they walked was not determined by the number of paces alone as an individual's pace length was also a factor. In order to make this conversion, students were required to design a process for determining the length of their own pace. After some discussion, which was guided by the teacher, students negotiated an approach which was acceptable to all

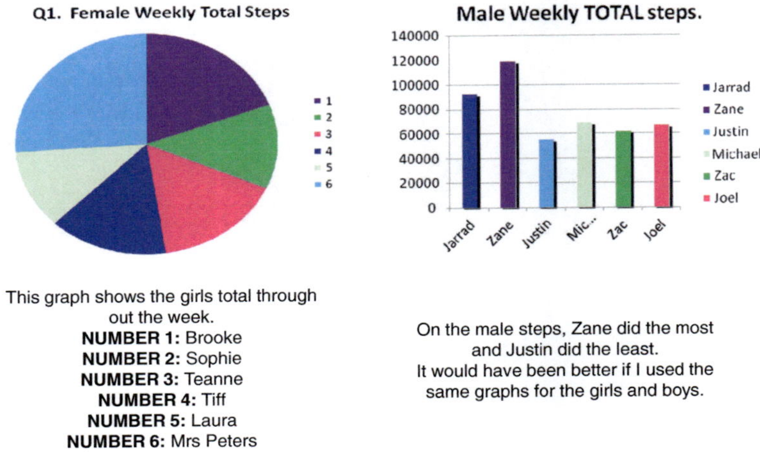

Fig. 2 A comparison of male's and female's weekly total steps

members of the class. This involved marking out a distance of 100 m along the footpath which bordered the school against which students counted the number of paces they each took to walk this distance. After demonstrating the procedures for obtaining the length of her pace, and converting paces in a day to kilometres from her own personal data, the teacher asked students to complete conversions of their own pace totals to kilometres. She also suggested that students compare their kilometric distances with each other and to discuss why they were different (see Fig. 2). The teacher finished the lesson by indicating the next session would include an investigation of the number of paces Usain Bolt takes during a 100 m sprint.

This activity utilised a range of mathematical knowledge including measurement (estimation and converting units), number (ratio) and chance and data (collection, organising and representing data). Students' learning was situated in the real life context of outdoor education that required them to convert personal information—paces walked in a week—into standard measures (kilometre). A range of tools were used through the lesson, physical tools such as tape measures to mark out 100 m and digital tools in the form of electronic calculators and the Excel spreadsheet used for recording students' data. The teacher encouraged positive dispositions towards the use of mathematics by challenging them to think of their own way of developing a conversion for their own pace length into kilometres and to think flexibly about the representation of data so that their personal details could be compared with that of others. This lesson also incorporated aspects of critical orientation as students made judgments about the reasonableness of results and were also asked to consider why the distances they had travelled differed in relation to both pace length and to students' different levels of activity over a week.

6 Concluding Comments

Designing for learning experiences that will promote the type of knowledge and skills. Straesser (2007) describes as bridging the divide between school mathematics and the rest of the world is a great challenge. Working in this "in between" world also requires more than mathematical knowledge and capacities with technology, but also the disposition to think flexibly and to have the confidence to persist when problems are not easily resolved are also vital elements in the bridge that leads from school mathematics to using mathematics in the world beyond school. This places greater demands on teachers as well as students to be able to change the direction of their thinking and explore a surprising result as the outcome of a problem. This chapter has offered a way of thinking about teaching and learning in Straesser's "in between world" through the numeracy model offered by Goos (2007). The model also has the potential to guide the design of teaching and learning activity in mathematics that bridges the gap between doing school mathematics and mathematics as a practice in the workplace and in industry.

References

Australian Association of Mathematics Teachers. (1997). *Numeracy = everyone's business. Report of the numeracy education strategy development conference*. Adelaide: AAMT.

Beatty, R., & Geiger, V. (2010). Technology, communication, and collaboration: Re-thinking communities of inquiry, learning and practice. In C. Hoyles & J.-B. Lagrange (Eds.), *Mathematics education and technology-rethinking the terrain* (pp. 251–284). New York: Springer.

Damlamian, A., & Straesser, R. (2009). ICMI Study 20: Educational interfaces between mathematics and industry. *ZDM, 41*(4), 525–533.

Ferrara, F., Pratt, D., & Robutta, O. (2006). The role and uses of technologies for the teaching of algebra and calculus. In A. Gutiérrez & P. Boero (Eds.), *Handbook of research on the psychology of mathematics education: Past, present and future* (pp. 237–273). Rotterdam: Sense Publishers.

Geiger, V., Faragher, R., & Goos, M. (2010). CAS-enabled technologies as 'Agents Provocateurs' in teaching and learning mathematical modelling in secondary school classrooms. *Mathematics Education Research Journal, 22*(2), 48–68.

Goos, M. (2007). Developing numeracy in the learning areas (middle years). Keynote address presented at the South Australian literacy and numeracy expo. Adelaide.

Goos, M., Geiger, V., & Dole, S. (2010). Auditing the numeracy demands of the middle years curriculum. In L. Sparrow, B. Kissane, & C. Hurst (Eds.), *Shaping the future of mathematics education (Proceedings of the 33rd annual conference of the mathematics education research group of Australasia)* (pp. 210–217). Fremantle: MERGA.

Hoyles, C., Wolf, A., Molyneux-Hodgson, S., & Kent, P. (2002). Mathematical skills in the workplace. Final report to the Science, Technology and Mathematics Council. Foreword and executive summary. London: Institute of Education, University of London; Science, Technology and Mathematics Council.

Kieran, C., & Guzma'n, J. (2005). Five steps to zero: Students developing elementary number theory concepts when using calculators. In W. J. Masalski & P. C. Elliott (Eds.), *Technology-supported mathematics learning environments* (pp. 35–50). Reston: National Council of Teachers of Mathematics.

Laborde, C., Kynigos, C., Hollebrands, K., & Straesser, R. (2006). Teaching and learning geometry with technology. In A. Gutiérrez & P. Boero (Eds.), *Handbook of research on the psychology of mathematics education: Past, present and future* (pp. 275–304). Rotterdam: Sense Publishers.

Niss, M., Blum, W., & Galbraith, P. (2007). Introduction. In W. Blum, P. Galbraith, H. Henn, & M. Niss (Eds.), *Modelling and applications in mathematics education: The 14th ICMI study* (pp. 3–32). New York: Springer.

OECD. (2004). *Learning for tomorrow's world: First results from PISA 2003*. Paris: OECD.

Steen, L. (2001). The case for quantitative literacy. In L. Steen (Ed.), *Mathematics and democracy: The case for quantitative literacy* (pp. 1–22). Princeton: National Council on Education and the Disciplines.

Stillman, G. (2007). Implementation case study: Sustaining curriculum change. In W. Blum, P. L. Galbraith, H.-W. Henn, M. Niss (Eds.), *Modelling and applications in mathematics education* (Vol. 10, pp. 497–502). Springer, US.

Straesser, R. (2007). Didactics of mathematics: More than mathematics and school! *ZDM, 39*(1), 165–171.

Villarreal, M., Esteley, C., & Mina, M. (2010). Modeling empowered by information and communication technologies. *ZDM-The International Journal of Mathematics Education, 42*(3), 405–419.

Zevenbergen, R. (2004). Technologizing numeracy: Intergenerational differences in working mathematically in new times. *Educational Studies in Mathematics, 56*(1), 97–117.

Mathematics Education and the Information Society

Koeno Gravemeijer

1 Introduction

"Where are the jobs?" asked BusinessWeek on the cover of its March 24, 2004 issue. At that time, the economy bounced back from a small recession, but although the economy recovered, employment did not rise. An explanation for this paradox was given by the economists Levy and Murnane (2006), who did an empirical study on how the job market in the US. developed between 1960 and 2000. Their analysis was that a crucial factor in the development of the job market is whether a task can be turned into a routine. Tasks that can be broken down into repeatable steps that hardly vary, will be consigned to computers, or will be handed over to lower paid workers outside the US.[1] They speak of a routine, when a task can be carried out by a machine, which follows formal rules that can be implemented in a computer program. This definition applies to a lot of production work, which in fact already has been taken over by machines. Mark, that the divide between routine and non-routine jobs, does not coincide with little or much education. Not all jobs that require a small amount of education can be computerized. Guiding a car through the traffic, for instance, cannot be computerized (yet). In contrast, many tasks such as well-educated accountants and computer programmers carry out can be entrusted to dedicated software.

The issue that we want to address in this chapter is, what implications such changes in the (future) job market have for a foundational mathematics curriculum. We will first look more closely on the changes in the job market under influence of globalization and informatization, and how they influence the requirements for future employability in a more general sense. Then we will investigate what the

[1] The extent to which outsourcing is a significant factor will vary in other countries; we do assume however that the transfer of jobs to computers will be a universal phenomenon.

K. Gravemeijer (✉)
Eindhoven School of Education, Eindhoven University,
P. O. Box 513 5600 MB Eindhoven, The Netherlands
e-mail: k.gravemeijer@tue.nl

A. Damlamian et al. (eds.), *Educational Interfaces between Mathematics and Industry*, New ICMI Study Series 16, DOI: 10.1007/978-3-319-02270-3_28,

implications are for the content of the mathematics curriculum by trying to get a handle on what mathematics is required in the modern workplace. Finally we will link these (preliminary) findings to ways in which information technology can help achieve those goals.

2 Jobs of the Future

To answer the question, what capabilities will be important for future employment, we have to look at the kind of jobs that will be offering good prospects. These jobs concern non-routine tasks, which are tasks that require flexibility, creativity, problem solving skills, and complex communication skills. Autor et al. (2003) refer to examples such as reacting to irregularities, improving a production process, or managing people. Empirical research in the US shows that employment involving cognitive and manual routine tasks dropped between 1960 and 2000, while employment involving analytical and inter-active non-routine tasks has grown in the same period. This change especially concerned industries that rapidly automatized their production. Parallel to the development in industry, similar changes occurred in other areas where a strong computerization took place. This change happened on all levels of education. Jobs with a high routine character are disappearing. The jobs of the future are the ones that ask for flexibility, creativity, lifelong learning, and social skills. The latter are jobs that require communication skills, or face-to-face interaction—such as selling cars or managing people. These changes do not only affect the decline or rise specific jobs; existing jobs are changing as well. Secretaries, and bank employees for instance have got more complex tasks since word processors and ATM's have taken over the more simple tasks.

It shows that the effects of computerization and globalization overlap and reinforce each other. Routine tasks can easily be outsourced, while information technology enables a quick and easy worldwide exchange of information. The latter also makes it possible to outsource business services, such as call-centers, or the work of accountants and computer programmers. Another effect of globalization is that it forces companies to work as efficient as possible. This requires companies to immediately implement computerization and outsourcing when it is economically profitable and strengthens the market position of the company. It also demands of the company to be on the lookout for opportunities to improve efficiency. As a consequence, working processes will have to be adapted continuously. This in turn, puts high demands on the workers, who have to have a certain level of general and mathematical literacy to be able to keep up.

3 Educational Change

Already in 1996, Kenelly pointed at the need for educational change:

> Business is a network of word processors and spreadsheets. Engineering and Industry are a maze of workstations and automated controls. Our students will have vastly different careers and we, the earlier generation, must radically change the way that education prepares a significant larger part of the population for information intensive professional lives (Kenelly 1996, 24).

Since then many reports and books have appeared that point to the need for educational change in response to the informatization and globalization of our society. They bear sweeping titles as, 'The Achievement Gap,' or, 'Rising Above the Gathering Storm.' The gist of these publications is that the current education in the US does not prepare students for the competition they are going to face in the global economy. First, future employees will have to compete with colleagues in other countries with similar skills who work for lower wages. Secondly, the skills that the current and future jobs require differ significantly from what even the best education offers. CEO's of large companies stress that they look for employees, 'who ask the right questions' (Wagner 2008). In line with the findings of Levy and Murnane (2006) problem solving and communication skills come to the fore as what is asked for, while schools focus on standard procedures and conventional skills.

If we look at these developments from the perspective of mathematics education, it shows that in the aforementioned documents much emphasis is put on general skills, while little attention is given to the content of the mathematics curriculum. One of the reasons may be that one of the effects of computerization is that mathematics becomes invisible. Mathematics is disappearing in black boxes, which makes it difficult to get a clear view on what mathematical knowledge is applied at the workplace. The mathematics is hidden in completely integrated systems, such as spreadsheets, electronic cash registers, and automated production lines. People who use these systems are expected to make decisions on the basis of the output of hidden mathematical calculations. Levy and Murnane (2006, 19) point to the significance of this development: "Because of computerization, the use of abstract models now permeates many jobs and has turned many people into mathematics consumers." They mention the manager of a clothing store who uses a quantitative model to predict the future dress demand, and a truck dispatcher who uses a mathematical algorithm to determine delivery routes, as examples. They point out that the computerized equipment often does the actual calculation is such cases. However, they go on to say, if the decision maker does not understand the underlying mathematics, he or she is very vulnerable to serious errors of judgment. If we follow this line of reasoning, we may discern two seemingly conflicting tendencies. On the one hand, we appear to need less and less mathematics, since various apparatus take over a growing number of mathematical tasks. On the other hand, we develop into 'mathematics consumers', who become increasingly dependant of the quantitative information and mathematical models, which we ought to understand.

This brings us to the central question of this chapter, what are the implications of the increasing informatization and globalization for mathematics education? Or more specifically, which goals and contents should we aim for if we want to prepare students for the information society? In this chapter, the focus is on employability. First, because the impact of computerization and globalization appears to be the biggest for the workplace and employability. Secondly, because we may assume that what one needs in the workplace will encompass what one needs for everyday personal life. In this respect, we will follow Levy and Murnane's (2006) notion of mathematics consumers that ought to understand the mathematical models they base their decisions on. We may add to their point of the risk of errors of judgment, the importance of innovation—in the light of global competition, and in response to societal problems such as global warming.

4 Techno-Mathematical Literacy's

When trying to relate the goals of mathematics education to the requirements of the workplace, an additional complication is that the use of mathematics at the work place is strikingly different from conventional mathematics (Hoyles and Noss 2003; Roth 2005). To describe this specific kind of mathematics, Hoyles and Noss (2003) coined the term 'techno-mathematical literacy's'—or TmL's for short. Those TmL's are defined as idiosyncratic forms of mathematics that are shaped by work place practices, tasks, and tools. Acting successfully at the workplace is dependant on a combination of mathematical knowledge and contextual knowledge. They identify competencies such as, "seeing the need to quantify, identifying and measuring key variables, representing and interpreting data" (Bakker et al. 2006, 355–356), and, reasoning about the models, "in terms of the key relationships between product "variables" (…) and their effect on "outputs" (…)" (Kent et al. 2007, 80).

Bakker et al. (2006) stress that the role of TmL's does not simply boil down to applying mathematical knowledge. In contrast to mathematical modeling—where contextual aspects are considered 'noise'—contextual knowledge is an essential element since it gives meaning to the decisions that are being made. Albeit, neither yields, that conventional mathematics is not needed or that common sense would be sufficient. However, students will have to be able to flexibly adapt their existing mathematical knowledge or adopt new knowledge. This asks for experience with a variety of non-canonical forms of mathematics. We may observe, however, that there is a tension between the benefits of canonical forms of mathematics for a longitudinal learning process, and the need for exploring varied and informal forms of mathematics to help students develop a kind of flexibility that may needed to develop TmL's. In a similar feign Steen (2001) observers that the work place asks for sophisticated use of elementary mathematics, while school mathematics focuses at elementary use of sophisticated mathematics. In addition, he observes a bias in favor of algebraic formulas as the preferred style of mathematics, instead of graphs, computers, and the like. In connection to this he advocates, among other things, more emphasis on data analysis and geometry.

5 Quantitative Models of Reality

Since computers and computerized appliances universally function as interfaces between users and the concrete reality, analyzing what this implies may also shed some light on what mathematics is needed in the information society. In relation to this, we may discern the following processes:

- reality is quantified to make it accessible for computers—as computers only work with numbers;
- these numbers are processed by the computer on basis of models that describe interdependencies between variables;
- the output, which often has some mathematical form, is interpreted.

Thus in order to understand how a computer deals with reality, one must,

- have some idea of what quantifying (or measuring) entails;
- understand at some level, what a variable is, how we can reason about interdependencies between variables, and how computer models represent reality;
- has to be able to interpret the output of computers.

An important aspect of a measuring is that there will always be some inaccuracy and uncertainty involved. In general students do not realize that there will always be some measurement error, or that a repeated measurement may result in a different outcome. Nor do they realize that variance is part of industrial production, even though they will be aware of the phenomenon of natural variance in nature. We would argue that it is important for students to develop this kind of understanding, for many of the numbers they will have to work with are the result of sampling. We may further follow Jones (1971) in his observation that measuring an object comprises assigning a value to a variable; what is measured is not the object but a property of that object. The significance of this observation comes to the fore when we look at co-variation. In such cases we do not compare the lengths and weights of Jim, Mary, Pete, for instance, but we study how 'length' varies with 'weight'. An important aspect in this context is that variables have to be understood dynamically, which is closely related with interpreting symbolic representations, such as graphs, dynamically. Fortunately information technology offers eminent educational tools to help students to come to grips with these ideas.

6 Information Technology and Instruction

A point which is, for instance, made by Kaput (Kaput and Schorr 2007), who argues that information technology allows for new ways to come to grips with what he calls the *mathematics of change*. Modern information technology, he argues, allows for dynamic representations. Computers can show the numerical results or graphical representations of measuring activities real time on a screen, but also offer representations of simulations that can be manipulated at will. These dynamic representations in turn allow for more qualitative ways of reasoning about change, which do not require algebraic calculations. This qualitative approach may become an alternative for conventional calculus for large groups of students. In this way we may be able to make reasoning about variables accessible for those students, for whom the algebraic calculations that calculus requires constitute an unsurpassable stumble block. For a select group of students, calculus will of course keep its value. But as far as basic education for all students is concerned, this more qualitative approach seems to offer unique possibilities.

Exploratory experiments in the Netherlands showed that mathematics education along those lines might already start in primary school (Galen and Gravemeijer 2008). This can be illustrated with the computer game 'Train Driver'. This game shows a dynamic picture of a rail track, on which a train is visualized as a moving red dot (see Fig. 1).

The students can speed up or slow down the train with the arrow buttons. The speed of the train is shown by the vertical bar below right. As the train moves, the computer produces a crude graph of the speed of the train by adding a copy of that small bar every second. Thus each bar in the graph signifies a separate measurement and there is a direct connection between the speedometer and the graph. This supports the students in reasoning about change in a very direct manner. Observations showed that many Grade 5 and 6 students could reason sensibly about acceleration by referring to the differences between subsequent bars, and about the relation between speed and distance covered by referring to an

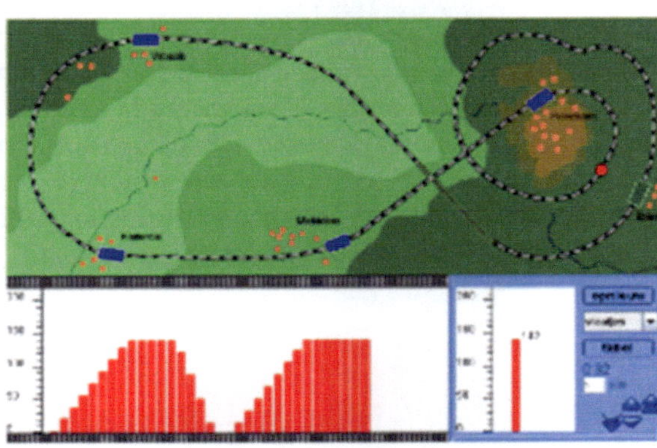

Fig. 1 Train Driver

addition the lengths of the bars. We realize this type of reasoning is still informal and qualitative, and that the graph is not a canonical speed-time graph, but what is important here is that the students could use this dynamic representation to reason about speed, time and distance in a manner that was grounded in the experiential reality created working with the computer game.

We see this as an example of the *mathematics of change* Kaput talks about (Kaput and Schorr 2007). It shows how the dynamic representation in which the graph emerges out of instantaneous pictures of the 'speed bar', supports the students' reasoning. We would argue that these experiences with the train simulation also help student to come to grips with the phenomena speed and acceleration. Dynamic representations can of course be used to explore a variety of situations. Design research by Gravemeijer and Cobb (2006), for instance, showed that, they could be used for a qualitative introduction in exploratory data analysis (see also Bakker and Gravemeijer 2004).

7 Conclusion

In this chapter, we investigated how a foundational mathematics curriculum would have to be adapted to the effects that informatization and globalization will have on employability. An important factor appears to be a distinction between routine and non-routine tasks, as the former will be consigned to computers, or handed over to lower paid workers. We further observed that employees exceedingly become mathematics consumers, and we argued they ought to understand the mathematics on which they base their decisions. We further touched upon the difference between mathematics in school and mathematics at the workplace that may be described as techno-mathematical literacy's. We finally concluded that in a foundational mathematics curriculum that prepares for the future, next to problem solving and communicating, mathematical contents, such as measurement, variability, and reasoning with (models of) relations between variables would be important. We closed by indicating how information technology might be employed for these very goals.

References

Autor, D., Levy, F., & Murnane, R. (2003). The skill content of recent technological change: An empirical exploration. *Quarterly Journal of Economics, 118*(4), 1279–1333.

Bakker, A., & Gravemeijer, K. P. E. (2004). Learning to reason about distribution. In D. Ben-Zvi & J. Garfield (Eds.), *The challenge of developing statistical literacy, reasoning, and thinking* (pp. 147–168). Dordrecht: Kluwer Academic Publishers.

Bakker, A., Hoyles, C., Kent, P. H., & Noss, R. (2006). Improving work processes by making the invisible visible. *Journal of Education and Work, 19*(4), 343–361.

Gravemeijer, K. P. E., & Cobb, P. (2006). Design research from a learning design perspective. In J. Akker, K. Gravemeijer, S. McKenney, & N. Nieveen (Eds.), *Educational design research* (pp. 45–85). London: Routledge, Taylor Francis Group.

Hoyles, C., & Noss, R. (2003). What can digital technologies take from and bring to research in mathematics education? In A. J. Bishop, M. A. Clements, C. Keitel, J. Kilpatrick, & F. Leung (Eds.), *Second international handbook of mathematics education* (Vol. 1, pp. 323–349). Dordrecht: Kluwer Academic.

Jones, L. V. (1971). The nature of measurement. In R. L. Thorndike (Ed.), *Educational measurement* (2nd ed.), (pp. 335–355). Washington: American Council on Education.

Kaput, J., & Schorr, R. (2007). Changing representational infrastructures changes most everything: The case of SimCalc, Algebra, and Calculus. In G. Blume & K. Heid (Eds.), *Research on technology in the learning and teaching of mathematics: Syntheses and perspectives* (pp. 211–253). Mahwah: Erlbaum.

Kenelly, J. (1996). Technology in mathematics instruction. In A. W. Roberts (Ed.), *Calculus: The dynamics of change* (pp. 24–27). Washington, DC: Mathematical Association of America.

Kent, P., Noss, R., Hoyles, C., & Bakker, A. (2007). Characterising the use of mathematical knowledge in boundary-crossing situations at work. *Mind, Culture and Activity, 14*(1–2), 64–82.

Levy, F., & Murnane, R. J. (2006). *How computerized work and globalization shape human skill demands.* Downloaded 2 Nov 2006, from http://web.mit.edu/flevy/www/.

Roth, W.-M. (2005). Mathematical inscriptions and the reflexive elaboration of understanding: An ethnography of graphing and numeracy in a fish hatchery. *Mathematical Thinking and Learning, 7*(2), 75–110.

Steen, L. A. (2001). Data, shapes, symbols: Achieving balance in school mathematics. In L. A. Steen (Ed.), *Mathematics and democracy: The case for quantitative literacy* (pp. 53–74). Princeton: National Council on Education and the Disciplines.

van Galen, F., & Gravemeijer, K. (2008). Experimenteren met Grafieken. *JSW, 93*, 16–20.

Wagner, T. (2008). *The global achievement gap*. New York: Basic Books.

Authentic Complex Modelling Problems in Mathematics Education

Gabriele Kaiser, Martin Bracke, Simone Göttlich and Christine Kaland

1 Theoretical Framework for Modelling in Mathematics Education

Within mathematics education it is consensus, that applications and modelling shall play an important role. However, how the implementation of these kinds of examples shall take place, which kinds of examples shall be used, how the modelling processes shall be implemented in school is highly debated. Influenced by the goals, which are connected to the teaching of applications and modelling and mathematics education in general, one can distinguish different perspectives of the modelling debate worldwide: amongst others, there are perspectives, which emphasise the use of authentic problems—named in a framework developed by Kaiser and Sriraman (2006) as realistic or applied modelling. Other positions emphasise more pedagogical goals such as the development of concepts or the structuring of learning processes. As an overall perspective a meta-perspective is discriminated, called cognitive modelling, which focuses on the cognitive processes taking place during modelling activities. Due to space limitations we refuse to go into more detail concerning the other approaches developed and refer to

G. Kaiser (✉)
Faculty of Education, University of Hamburg, Hamburg, Germany
e-mail: gabriele.kaiser@uni-hamburg.de

M. Bracke
Technical University of Kaiserslautern, Kaiserslautern, Germany
e-mail: bracke@mathematik.uni-kl.de

S. Göttlich
School of Business Informatics and Mathematics, University of Mannheim, 68131 Mannheim, Germany
e-mail: goettlich@uni-mannheim.de

C. Kaland
Department of Mathematics, University of Hamburg, Bundesstraße 55 20146 Hamburg, Germany
e-mail: Christine.Kaland@math.uni-hamburg.de

A. Damlamian et al. (eds.), *Educational Interfaces between Mathematics and Industry*, New ICMI Study Series 16, DOI: 10.1007/978-3-319-02270-3_29,

Kaiser and Sriraman (2006) and the extensive ICMI-study on applications and modelling (see Blum et al. 2007).

The approach described in the following belongs to the so-called realistic or applied modelling perspective. Based on extensive own empirical research (see for example Kaiser-Messmer 1986) we see the necessity of treating authentic, complex modelling problems interdisciplinary by nature, which promote the whole range of modelling competencies and broaden the radius of action of the students. The approach takes as essential starting point that the promotion of modelling competencies needs own experiences of the students carried out with authentic modelling problems. Similar proposals are developed by Haines and Crouch (2006) or already at the beginning of the modelling debate by Pollak (1969). Authentic problems are defined as problems that are only little simplified and are according to the definition of authentic problems by Niss (1992) recognised by people working in this field as being a problem, they might meet in their daily work. As reference framework of the approach, on which the realisation of modelling weeks are based, the realistic or applied modelling perspective can be stated. Central feature here is the usage of complex, authentic modelling problems in order to implement pragmatic-utilitarian educational goals like understanding of the real world or the promotion of modelling competencies.

Furthermore, our understanding of modelling can be characterised using the following well-know modelling cycle developed amongst others by Kaiser (1995, p. 68) and Blum (1996, p. 18): (Fig. 1)

The arrows symbolise sub-competencies of the overall modelling competency promoted by the modelling process: (a) idealise, structure, simplify (b) mathematise, translate into the language of mathematics, (c) mathematical work, operate and (d) interpret and validate. If the result is not satisfying, the steps will be iterated again—if necessary, the whole circle will be passed once again. Of course, this circle does not allow for a uniform definition of modelling competencies, but there is consensus that modelling competencies embrace both the abilities as well as the willingness to convert those into action.

The overall aim of the following chapter is the description of the practical implementation of modelling activities, i.e. to create a stimulating challenge so that pupils, students and teachers get motivated to start working on diverse interdisciplinary problems. With these activities we aim to communicate another view of mathematics—away from being boring and incomprehensible. We want to encourage young people to recognise mathematics as an interactive science by

Fig. 1 Modelling cycle

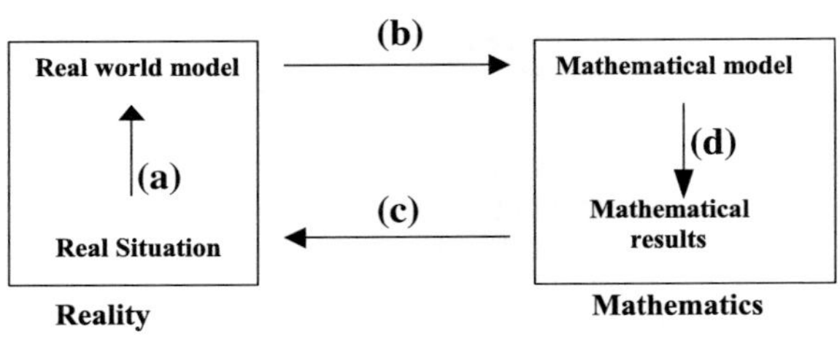

exploiting their mathematical knowledge. The problems dealt with in these modelling projects shall articulate the relevance of mathematics in daily life, environment and sciences and impart competencies to apply mathematics in daily life, environment and sciences. The conduction of modelling weeks or smaller modelling projects is considered to be a powerful and effective way to promote these examples.

2 Structure of the Modelling Week and the Modelling Project

The idea of modelling weeks has been developed at the University of Kaiserslautern by the working group of Helmut Neunzert, who are already running modelling weeks for more than a decade. The idea was adopted by other groups in Germany, e.g. at the Universities of Hamburg, Darmstadt or Munich. The structure of the modelling week or the smaller modelling projects follows a similar pattern at the Technical University of Kaiserslautern and the University of Hamburg, on which we will concentrate in this chapter.

Modelling weeks or shorter projects have been carried out at the University of Kaiserslautern since 1993 (modelling weeks, students' age 17–18) and 2000 (school projects, students' age 8–19), respectively. The only differences are duration and venue. The modelling weeks last for one week (from Sunday to Friday) and take place at youth hostels. On the contrary, short-time school projects last up to 3 days and are locally organised. The purpose of such a modelling event is to offer students (and teachers as well) an intensive workshop on real world problem solving. For instance, in case of modelling weeks, the school pupils and teachers (all coming from different schools) are divided into mixed teams of five pupils and two teachers. The pupils are allocated to teams on the basis of their areas of interest and mathematical expertise. The teachers are responsible for the management within the group, are a reliable partner regarding mathematical difficulties, and build a bridge to the academic staff members from the University responsible for the modelling week. The week starts with a brief presentation of several problems where the underlying industrial or scientific background is highlighted and motivated and every participant is allowed to select a problem of personal choice and is then assigned to a team tackling the problem. To identify and understand the real problem may take some time. Also the development of a mathematical model of the typically non-unique mathematical problem is not that easy. Usually, to get a solution, computer programmes (Excel or more sophisticated ones) must be applied. If the simulation results do not provide a satisfactory answer to the original problem, the model has to be slightly refined and the modelling iteration process starts another time. At the end of the week, the team prepares a final presentation about the results of their work, whereby each team member has to make a personal contribution. Last but not least, the pupils produce a written report that will be distributed among each other.

The project 'Mathematical Modelling in School' was established in 2000 by the Department of Mathematics in co-operation with Didactics of Mathematics at the Department of Education at the University of Hamburg. It was originally a university course project with future teachers for upper secondary level teaching and aimed to establish a conjunction between university and school. Within the project the future teachers supervise student groups from upper secondary level (aged 16–18 years) during their modelling activities. At the beginning, the modelling activities took place during ordinary mathematics lessons spanning over the first half of the school year. Due to organisational problems, the concept was changed in the direction of a modelling week in 2008. Up-to-date, two modelling weeks are carried out per year with about 400 students per year from upper secondary level from schools in Hamburg and its surrounding. Most of the schools send complete mathematics courses, i.e. weak as well as very talented students, only a few schools send their most interested students and select them from the adequate age cohort. The overall aim of the group at the University of Hamburg is to work with average students, not especially interested in mathematics, and change their view on mathematics and mathematics education. In contrast to the beginning, the modelling weeks are currently supervised by future teachers as field for gaining practical experiences and tutors from the Department of Mathematics, who work jointly together. The organisational forms are very similar to the way, how the modelling weeks are organised by the Technical University of Kaiserslautern, apart from the location, the modelling week in Hamburg takes place at the Department of Mathematics, an experience, which seem to be quite important for students already thinking about their university future study. In addition, so-called modelling days spanning over 3 days have been carried out, in which the whole age-cohort of year 9 students (age 14–15) of higher track schools (so-called Gymnasium) in Hamburg participate. In these activities future mathematics teachers participate as tutors and guide the groups within their activities. On the one hand these ongoing activities offer future teachers during their university study a practical field, in which they can gain practical school experience. On the other hand, the inclusion of whole schools shall promote these kinds of activities in the whole school and serve as mean for school development.

One important overall aim of the projects has been since its establishment, that the participating students will acquire competencies to enable them to carry out modelling examples independently, i.e. the ability to extract mathematical questions from the given problem fields and to develop autonomously the solutions of real world problems. It is not the purpose of these projects to provide a comprehensive overview about relevant fields of application of mathematics. The students shall develop the needed mathematical knowledge, which usually comes from different mathematical areas. This characteristic brings these kinds of problems into sharp contrast to many other modelling activities usual at school. Furthermore, it is hoped that students will be enabled to work purposefully on their own in open problem situations and will experience the feelings of uncertainty and insecurity which are characteristic for real applications of mathematics in everyday life and sciences. The students have to carry out the whole modelling cycle on their own, only supported by tutors or their teachers, sometimes they even have to carry out the modelling cycle several times, going back and forth quite often in so-called mini-cycles (for the structure of these kinds of modelling activities see Borromeo Ferri, 2011). An overarching goal is that students' experiences with mathematics and their mathematical world views or mathematical beliefs are broadened. This kind of approach can be described as holistic approach, using the terminology of Blomhøj and Jensen (2003), i.e. a whole mathematical modelling process is carried out covering all modelling competencies described above. Another central feature of these projects is the usage of complex authentic examples, i.e. many of the problems stem from industrial problems passed to the project organisers due to their close working relations to many industrial mathematicians. In total, all modelling problems—independent of the level of difficulty—provide a wide a range of examples—from industry to society and greatly benefit from the close cooperation to industry. The problems are only little simplified and often several ways, how to model the problem, are known. Quite often only a problematic situation is described and the students have to determine or develop a question that can be solved. The development and description of the problem to be tackled is the most important and most ambitious part of a modelling process, mostly neglected in ordinary mathematics lessons. Another feature of the problems is their openness which means that various problem definitions and solutions are possible in dependence of the norms of the modellers. The teaching-and-learning-process is characterised as autonomous, self-controlled learning, i.e. that the students decide upon their ways of tackling the problem and no fast intervention by the tutors or teachers takes place (or should take place).

Amongst others, the following modelling problems have been used, partly several times by the two groups; we will not go into details, but refer to already published descriptions of their implementation:

- Mathematics in private health insurance.
- Storage capacity of 2D pixel mosaics (Ableitinger et al. 2009a).
- Optimal capacity utilisation of airplanes (Ableitinger et al. 2009b).
- Playing darts (Bracke et al. 2013).
- Prediction of fishing quotas.
- Optimal position of rescue helicopters in South Tyrol (Hamacher et al. 2004; Kaiser 2005).
- Radio-therapy planning for cancer patients.
- Identification of fingerprints.
- Pricing for internet booking of flights (Kaiser and Schwarz 2006).
- Pricing of an Internet café (Kaiser and Schwarz 2006).
- Risk management (Kaiser and Schwarz 2006).
- Optimal mixture of chemicals in swimming pools (Kaiser et al. 2010).
- Optimal arrangement of automatic water irrigation systems (Bracke 2004; Kaiser and Schwarz 2010).

- Prediction of the spread of a sexually transmitted disease with ladybugs (Göttlich and Bracke 2009; Kaiser et al. 2011).
- Identification of Greek land turtles using firm characteristics of the tortoise shell (Bracke 2007; Göttlich 2007).
- Optimal planning of a wind park.
- Optimisation of roundabouts.
- Lighting of soccer fields.
- Optimal planning of bus lines and according bus stops (Kaiser and Stender 2013).
- New rules in ski jumping: the influence of a variable inrun.

3 Evaluation of the Modelling Week

At the University of Hamburg, regular evaluations of the modelling weeks have taken place, on which we will report in the following. Due to space constraints we will focus on the central question for our research, whether these kinds of complex authentic problems are feasible for students from upper secondary level, especially if they are average students, not particularly interested in mathematics (for details see Kaiser and Stender 2010).

For the evaluation, we used a questionnaire with four mainly open questions on the beliefs of students about mathematics teaching and three open questions on the appraisal of modelling examples tackled in the modelling week and five closed questions to be answered on a 5-point-Likert-scale. The questionnaire was filled in at the end of the modelling weeks, in March 2009 (a report on this evaluation with 289 students can be found in Kaiser and Schwarz 2010), March 2010, September 2010 and March 2011, the modelling week in September 2009 was not evaluated. Based on methods of the Grounded Theory (see Strauss and Corbin 1998), we have used in-vivo-codes for the open questionnaires, i.e. codes extracted out of the text written by the students as verbatim quotation and grouped them to related quotations under a theoretical perspective. In order to finally analyse the answers of the students in the open questionnaires, we transformed the grouped in vivo codes into theoretical codes. For quality assurance methods of consensual coding were relied on, which means that a coding team consists of two coders who conduct all steps described above together. In the following, we describe the appraisal of these modelling weeks restricted to a few aspects due to space limitations.

4 Learning Outcomes of the Modelling Week

We start with the appraisal of the modelling examples by the students. The first question dealt with the appreciation of the learning outcomes of the modelling week.

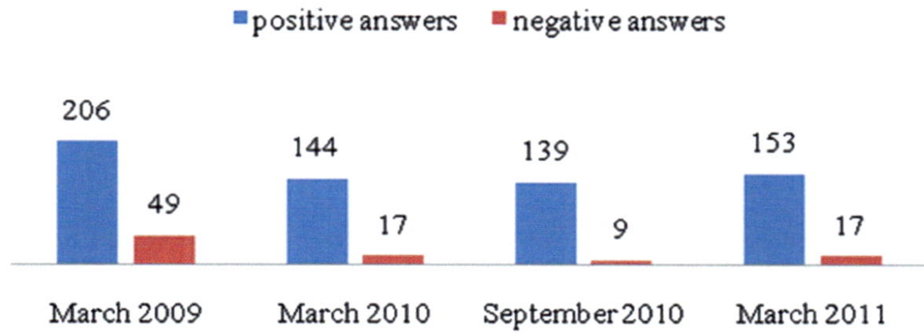

Fig. 2 Learning outcomes of the modelling weeks

The students answered on the question "From your point of view, what did you learn when dealing with the modelling example?" as follows: Between 80% and 93% of the students answered positively, with some changes between the different rounds of modelling weeks: (Fig. 2)

The following diagram displays the areas, in which the students reported learning success (several aspects could be named). Due to the change in the coding manual, the results of March 2009 are deleted (Fig. 3).

Particularly high learning successes have been achieved in the area of modelling competencies. With regard to modelling, many students stated that due to working on these modelling problems they understood how mathematics is applied and recognised its practical relevance, and relevance to everyday life. Furthermore, some students stated that, in fact, they were able to apply what they have learned in school. In the following, we exemplify the reasons by selected verbatim quotations of the students (in italics):

> …that mathematics not only can be applied in school, but also to specific examples in real life. (Female student, 19 years old)
>
> …that mathematics can also be applied to things in everyday life. (Male student, 17 years old)

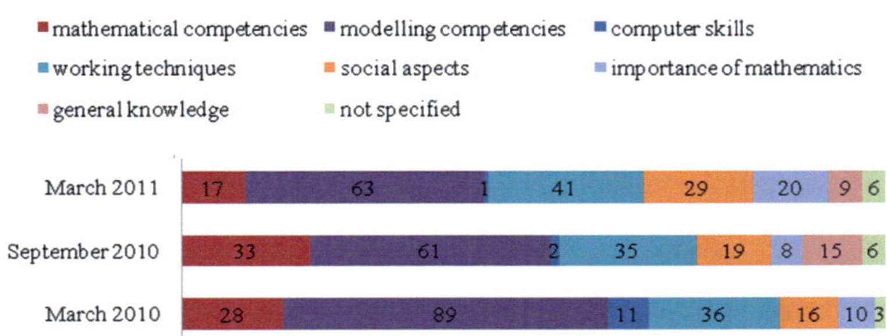

Fig. 3 Areas of the learning outcomes of the modelling week

Another highly important aspect relates to working techniques, many students reported that they learnt new strategies in problem solving and now understand how to approach problems which at first seem to be vague.

> ...that it is possible to approach a problem in various ways, based on different assumptions. (Female student, 17 years old)

In addition, many respondents mentioned that they have improved their ability to work independently, to have more perseverance, and that they now find it easier to structure and simplify.

> ...structured, co-ordinated and deliberate planning and working. (Male student, 19 years old)
> I learnt to compare different approaches in problem solving and to deal with a problem using quite varied methods and approaches. (Female student, 18 years old)
> ...to solve arising problems autonomously. (Female student, 16 years old)

Social aspects are emphasised as well: several students mentioned that during the week they not only got to know their fellow students and students of other schools but, most of all, learnt something about teamwork.

> ...to work well/better in groups and how important a good working atmosphere and teamwork is. (Female student, 18 years old)

It can be summarised, that most students described as their personal impression that they have achieved learning success during the modelling week. Furthermore, the range of aspects mentioned cover all goals connected with the inclusion of modelling problems.

5 Inclusion of Modelling Problems in Ordinary Mathematics Lessons

One central aspect is the question, whether the students want to have modelling examples included in their usual mathematics lessons or whether they want to keep these kinds of ambitious examples out of their usual lessons. This aspect is dealt with in the next question, where the students were asked the following question: "Should these examples be increasingly dealt with as part of regular maths classes or would you reject this?" The overwhelming positive answer on this question is displayed in Fig. 4:

Most frequently, the students mentioned as reason for the inclusion of these kinds of examples their relation to the real world. Many students emphasised that only with these kinds of examples they could develop a relation between mathematics and the real world:

> I think that these examples should DEFINITELY be dealt with in maths classes. Because of these examples one will only realise what mathematics is needed for. (Male student, 16 years old)

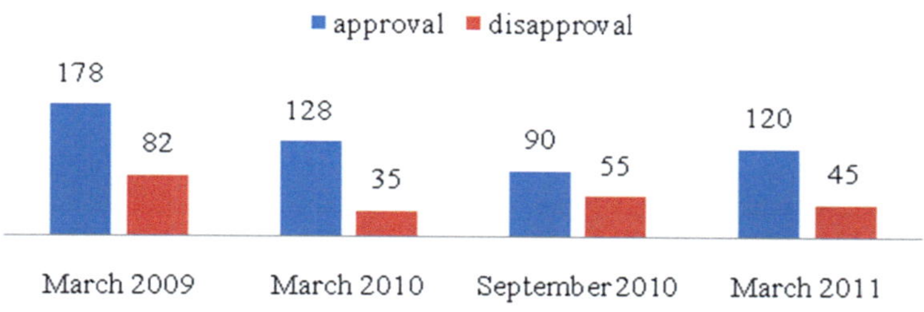

Fig. 4 Inclusion of modelling examples in ordinary lessons—positive answers

> I think so, because it illustrates the importance of mathematics for everyday life. (Male student, 16 years old)
>
> These examples should be dealt with in order to establish a stronger connection between the real life and what is taught at school. (Female student, 18 years old)

Motivation was the second most common reason for advocating the inclusion of these kinds of examples:

> I think it would be better to introduce less theory but more practical use in class to increase the students' motivation. (Female student, 18 years old)
>
> Yes, because this will make class more interesting. (Female student, 17 years old)

Many students mentioned the usefulness of these kinds of problems, being much higher than that of usual examples. Modelling examples were seen to promote complex mathematical thinking, to exercise learned topics and to allow insight into the work of mathematicians:

> Such problems SHOULD be dealt with in class, because they will improve the so-called 'competence in problem solving'. Furthermore, it will train the knowledge gained in previous grades. (Male student, 19 years old)
>
> I am in favour of dealing with these topics in class, because this will be good practice and they are good exercises. (Male student, 18 years old)
>
> Yes, because one automatically relates more closely to the topic and the relevant example. (Male student, 18 years old)

From our evaluation it is obvious that modelling weeks with complex authentic examples can be tackled successfully by ordinary students at upper secondary level. The students experienced high learning outcomes that reflect all the goals connected with modelling, ranging from psychological goals such as motivation to meta-aspects such as promoting working attitudes to pedagogical goals, namely enhancing the understanding of the world around us. The strong plea of the students for the inclusion of these kinds of examples in usual mathematics lessons support our position that it is appropriate to include these kinds of problems in ordinary mathematics lessons, clearly not everyday, but on a regular basis.

Acknowledgments We thank Sarah Mesrogli for the evaluation of the data.

References

Ableitinger, C., Göttlich, S., & Sickenberger, T. (2009a). Eine Modellierungsaufgabe zum Thema 'Optimale Auslastung von Flugzeugen'. In A. Brinkmann & R. Oldenburg (Eds.), *Materialien für einen realitätsbezogenen Mathematikunterricht, ISTRON* (Vol. 14, pp. 83–94). Hildesheim: Franzbecker.

Ableitinger, C., Göttlich, S., & Sickenberger, T. (2009b). Kann Papier sprechen? Eine Modellierungsaufgabe zum Thema 'Speicherkapazität von 2D Pixel-Mosaiken'. In A. Brinkmann & R. Oldenburg (Eds.), *Materialien für einen realitätsbezogenen Mathematikunterricht, ISTRON* (Vol. 14, pp. 113–123). Hildesheim: Franzbecker.

Blomhøj, M., & Jensen, T.H. (2003). Developing mathematical modeling competence: Conceptual clarification and educational planning. *Teaching Mathematics and its Applications, 22*(3), 123–139.

Blum, W. (1996). Anwendungsbezüge im Mathematikunterricht—Trends und Perspektiven. In G. Kadunz, et al. (Eds.), *Trends und Perspektiven. Schriftenreihe Didaktik der Mathematik* (Vol. 23, pp. 15–38). Wien: Hölder-Pichler-Tempsky.

Blum, W., Galbraith, P. L., Henn, H. -W., & Niss, M. (Eds.). (2007). *Modelling and Applications in Mathematics Education. The 14th ICMI Study*. New York: Springer.

Borromeo Ferri, R. (2011). *Wege zur Innenwelt des mathematischen Modellierens*. Wiesbaden: Vieweg + Teubner.

Bracke, M. (2004). *Optimale Gartenbewässerung—Mathematische Modellierung an den Schnittstellen zwischen Industrie, Schule und Universität. Mitteilungen der Mathematischen Gesellschaft Hamburg, 23*(1), 29–48.

Bracke, M. (2007). Why do turtles need a passport?—And what comes then? (Warum brauchen Schildkröten eigentlich einen Personalausweis?—Und was kommt danach?). In *Beiträge zum Mathematikunterricht* 2007 (pp. 328–331). Hildesheim: Franzbecker.

Bracke, M., Göttlich, S., & Götz, T. (2013). Modellierungsproblem Dart spielen. In R. Borromeo Ferri, G. Greefrath, & G. Kaiser (Eds.), *Mathematisches Modellieren für Schule und Hochschule* (pp. 147–162) Wiesbaden: Springer Spektrum.

Göttlich, S. (2007). Mathematical modelling at lower secondary level: passport for turtles. (Mathematische Modellierung in der Mittelstufe: Personalausweis für Schildkröten.). In *Beiträge zum Mathematikunterricht 2007* (pp. 324–327). Hildesheim: Franzbecker.

Göttlich, S., & Bracke, M. (2009). Eine Modellierungsaufgabe zum Thema: 'Munterer Partnertausch beim Marienkäfer'. In *Beiträge zum Mathematikunterricht 2009* (pp. 93–96). Münster: WTM-Verlag Stein.

Hamacher, H. W., Korn, E., Korn, R., & Schwarze, S. (2004). *Mathe & Ökonomie: Neue Ideen für den praxisnahen Unterricht*. Wiesbaden: Universum Verlag.

Haines, C. R., & Crouch, R. M. (2006). Getting to grips with real world contexts: Developing research in mathematical modelling. In M. Bosch (Ed.), *Proceedings of the Fourth European Society for Research in Mathematics Education CERME4* (1634–1644). IQS FUNDIEMI Business Institute.

Kaiser, G. (1995). Realitätsbezüge im Mathematikunterricht Ein Überblick über die aktuelle und historische Diskussion. In G. Graumann, et al. (Eds.), *Materialien für einen realitätsbezogenen Mathematikunterricht* (pp. 66–84). Bad Salzdetfurth: Franzbecker.

Kaiser, G. (2005). Mathematical Modelling in School—Examples and Experiences. In H.-W. Henn & G. Kaiser (Eds.), *Mathematikunterricht im Spannungsfeld von Evolution und Evaluation. Festschrift für Werner Blum* (pp. 99–108). Hildesheim: Franzbecker.

Kaiser, G., Lederich, C., & Rau, V. (2010). Theoretical Approaches and Examples for Modelling in Mathematics Education. In B. Kaur & J. Dindyal (Eds.), *Mathematical Applications and Modelling. Yearbook 2010, Association of Mathematics Educators* (pp. 219–246). Singapore: World Scientific Publishing.

Kaiser, G., & Schwarz, B. (2006). Mathematical Modelling as Bridge between School and University. *ZDM—The International Journal on Mathematics Education, 38*(2), 196–208.

Kaiser, G., & Schwarz, B. (2010). Authentic modelling problems in mathematics education—examples and experiences. *Journal für Mathematik-Didaktik, 31*(1), 51–76.

Kaiser, G., Schwarz, B., & Buchholtz, N. (2011). Authentic Modelling Problems in Mathematics Education. In G. Kaiser, W. Blum, R. Borromeo Ferri, & G. Stillman (Eds.), *Trends in Teaching and Learning of Mathematical Modelling: ICTMA 14* (pp. 585–594). New York: Springer.

Kaiser, G., & Sriraman, B. (2006). A global survey of international perspectives on modelling in mathematics education. *ZDM—The International Journal on Mathematics Education, 38*(3), 302–310.

Kaiser, G., & Stender, P. (2013). Complex Modelling Problems in Co-operative, Self-Directed Learning Environments. In G.A. Stillmann et al. (Eds.), *Teaching Mathematical Modelling: Connecting to Research and Practice, International Perspectives on the Teaching and Learning of Mathematical Modelling* (pp. 45–60). Dordrecht: Springer.

Kaiser-Messmer, G. (1986). *Anwendungen im Mathematikunterricht.* Bad Salzdetfurth: Franzbecker.

Niss, M. (1992). *Applications and modelling in school mathematics—directions for future development.* Roskilde: IMFUFA Roskilde Universitetscenter.

Pollak, H. (1969). How can we teach applications of mathematics. *Educational Studies in Mathematics, 2*(1), 393–404.

Strauss, A., & Corbin, J. (1998). *Basics of Qualitative Research.* Newbury Park: Sage.

Embedding Authentic Real World Tasks into Secondary Mathematics Curricula

Gloria Stillman and K. E. D. Ng

1 Introduction

According to Damlamian and Sträßer (2009, p. 525), "at all educational levels, students typically have been taught the tools of mathematics with little or no mention of authentic real world applications, and with little or no contact with what is done in the workplace." There are, however, several models of curriculum embedding of real world situations that could be contemplated (Niss et al. 2007). First, mathematical modeling of real world situations and use of real world applications can act as an underpinning of (a) all the mainstream mathematics courses (e.g., the Queensland Years 11–12 mathematics curriculum in Australia and the Ontario secondary mathematics curriculum in Canada) (b) special courses for high or low achieving students with the mainstream mathematics course still being the standard abstract mathematics. Secondly, real world situations become additions to the curriculum (a) within mathematics courses as school-based assessment in the form of extended tasks or (b) as projects crossing subject boundaries (so called interdisciplinary or multidisciplinary projects). In several educational jurisdictions such as Singapore (Kaur and Dindyal 2010), Germany (Maaß and Mischo 2011), and Denmark (Andresen and Lindenskov 2009), there has been a shift to expect that teachers will include real world tasks in mathematics and in interdisciplinary and multidisciplinary activities which involve mathematics.

Since the 1960s, there has been a continual ebb and flow in secondary school curricula with real world applications and modeling gaining in prominence and

G. Stillman (✉)
Education VIC, Faculty of Education, Australian Catholic University (Ballarat),
1200 Mair Street, Ballarat, Australia
e-mail: gloria.stillman@acu.edu.au

K. E. D.Ng
MME AG, National Institute of Education, NTU, 1 Nanyang Walk,
Singapore 637616, Singapore
e-mail: dawn.ng@nie.edu.sg

A. Damlamian et al. (eds.), *Educational Interfaces between Mathematics and Industry*, New ICMI Study Series 16, DOI: 10.1007/978-3-319-02270-3_30,

then retreating in some systems (e.g., Victoria see Stillman 2007) at the same time as the opposite is happening in other educational systems (e.g., Queensland see Stillman and Galbraith 2009). As a backdrop to this, there are many educational systems where no change has happened for many years where mathematics is indeed taught as "a dead science and a finished product" (Damlamian and Sträßer 2009, p. 530) but there are also, but much less frequently, instances where innovative curricula have been promoting realistic applications and modeling as an integral part of the curriculum for many years and teachers are expected to do this and actually do. In this chapter, we address the issue of curriculum change where the purpose is to include real world applications and modeling (a) in the mainstream mathematical course or (b) through interdisciplinary activities with mathematics as the anchor subject. The state of play with respect to industrial applications in these settings will also be explored.

2 Background

During the 1960s and early 1970s marked changes occurred in the social fabric of many countries all over the Western world. Youth and students in many countries revolted against what they saw as the unnecessarily restrictive shackles of the values of those in authority. A long-lasting ideological legacy of this time has been a desire among young people to be convinced of the efficacy of any activities they are asked to engage in instead of being willing participants following directions because these are given by someone in authority (Niss 1987, p. 491). At the secondary and tertiary levels of education students began to actively question the relevance of the content and form of the mathematics they were studying; "and right from the beginning relevance was interpreted by students, teachers and educationalists as *applicability*" (Niss 1987, p. 491). The nature of this applicability varied greatly from general societal applicability to applicability in students' everyday lives or expected future roles but where this was meaningful to students.

At the same time, there were other influences such as employer dissatisfaction with graduates of university mathematics departments (see McLone 1973) and educational reform movements. In *Everybody Counts: A Report to the Nation on the Future of Mathematics Education*, for example, the National Research Council (1989) in the U.S. flagged that in order to change mathematics education dramatically before 2000 it would have to go through several transitions including a shift from an "emphasis on tools for future courses to greater emphasis on topics that are relevant to students' present and future needs" (p. 83). Modeling was identified as one of the "distinctive modes of thought" (p. 31) mathematics offers and mathematics was said to play a special role in education because of "its universal applicability" (p. 31). However, model-building was identified as an additional topic area rather than a framework for teaching the application of mathematics.

3 Systemic Focuses on Modeling and Applications in Mathematics Curricula

From these influences arose genuine attempts in educational systems in several countries to include real world applications of mathematics and mathematical modeling more centrally within secondary mathematics curricula in the late 1980s and early 1990s. Some innovative curricula enjoyed a brief spectacular success and then were lost for various reasons (see Stillman 2007, for an example). Others have persisted to this day given new impetus as digital technologies became to be seen as a means of enabling modeling (Stillman and Galbraith 2009; Suurtamm and Roulet 2007).

In this chapter, one substantial initiative covering over 20 years is overviewed—mathematical modeling in Queensland, Australia, senior secondary mathematics courses. Applications and mathematical modeling were first introduced in 1990–1991 into senior mathematics curricula in a limited number of schools; however, by 1995 all secondary schools across the state were using such syllabuses. The development of the initiative from the perspective of implementing classroom teachers and key teachers responsible for its introduction and evolution has been the subject of a longitudinal research study, *Curriculum Change in Secondary Mathematics* (CCiSM) which is ongoing.

The objectives of the 1989 Trial/Pilot Syllabuses included the identification of the assumptions and variables of a mathematical model, formulation of a model, derivation of results from a model, and interpretation of these in terms of the given situation. Mathematics C, the most challenging Mathematics subject, also required the modification and validation of the model. While syllabus refinements have taken place since and some changes of emphasis have occurred, the essence of the mathematical modeling component has remained essentially stable for over two decades although different emphases in wording and in placement of modeling and application aspects in organising categories for the general objectives has brought to the fore different aspects in implementation over this time frame.

The Queensland system of assessment at the senior secondary level is entirely school based which means that the production of assessment tasks and the awarding of levels of achievement is in the hands of teachers in individual schools, with panels at district and state levels performing critical reviewing roles to assure comparability of outcomes across schools and regions. In keeping with the school-based nature of the Queensland context, individual schools and teachers design individual work programs (including assessment tasks) under syllabus requirements.

The tasks vary from applications of mathematics to open modeling tasks incorporating real situations such as glucose infusions, dust storms as part of erosion events, and emptying a flooded lagoon. Applications require students to carry out specified mathematical calculations in a real world context without any necessity to assess whether the proposed models are meaningful in this context. In these application examples essential activities, central to modeling, are absent. The purpose of applications is purely illustrative of uses of mathematics.

Modeling tasks are much more open, with the making of assumptions, choice of mathematics, and interpretation in context being aspects which students carry out. The contexts used vary greatly depending on the experiences of the teachers or their network of sources, which could be limited by geographical isolation of schools (Stillman and Galbraith 2011).

When applications and modeling were first introduced into the Queensland system, "a major shift in thinking" for both teachers and students (key teacher 1, CCiSM interview November, 2005) was necessary. "Rather than have, I suppose, students just practising endlessly, they wanted them to use it and see where it could be used and I think they felt that might engage students more (classroom teacher 1, CCiSM interview, November, 2005). The focus on modeling was generally not evident in implementation in the initial years although applications were very much present particularly in assessment (Fuller 2001; Stillman 1998), which now included a substantial weighting toward successful solution of application or modeling tasks in examination and alternative assessment contexts. This was attributed by many of those interviewed for the CCiSM project to an initial lack of understanding by teachers of the difference between these or a reluctance to take up modeling as it was considered too time consuming or too removed from their current practice. However, for others modeling had to be included:

> There were certainly people who were doing plenty of modeling. If you were in [Teacher named]'s school you would have done modeling. You had no choice. (Classroom teacher 7, CCiSM interview, October, 2007)

Initially, many teachers had to be convinced of the efficacy of the change from purely abstract mathematics with a major focus on concepts, skills, and processes. However, teachers quickly came to realize there were many benefits and the major ones, as seen from one teacher's perspective, were:

> The opportunity of showing kids a type of mathematics that they could see as relevant. The opportunity to allow kids to work and show what they could do in mathematics under different conditions I think some people really embraced. The schools who did that we saw kids doing some remarkable things. In contrast with the perspective which just says 'Awh, it should just be for kids who struggle with Maths', some of the work that we saw from high ability kids revealed some really insightful and rich understandings of what mathematics is about and what they could do with it was really, really impressive. (Key teacher 2, CCiSM interview, November 2005)

Over the years, modeling has slowly gained a place in classroom practice where genuine modeling is conducted to various degrees (Stillman and Galbraith 2009). This evolution has been strongly supported by the assessment monitoring panel system and insistence on use of alternative assessment. Technology has also been another driver (Stillman and Brown 2012) rapidly enabling progress in what is achievable using real world contexts in the classroom:

> I don't think anybody really had to be convinced that we should change our syllabus from pure algebra, pure calculus, pure geometry … to applications rather than pure mathematics. Also the technology became available which made it easier to do this. [Graphing calculators] were valuable tools in any modeling exercise…. With the advent of the TI80

> you pushed one button … that removed the grinding mathematics which was the aim of a lot of people and allowed people to discuss applications without being bogged down in the arithmetic of doing it. (Classroom teacher 7, CCiSM interview, October 2007)

Thus, teachers in the senior secondary area in Queensland do use applications and modeling of real situations in their teaching and assessment. The scope of the real situations used in these is limited to the experiences of the teaching staff at a school and their network of colleagues (Stillman and Galbraith 2011). It would appear to be important that if these teachers' awareness of industrial applications and modeling situations is to be raised that case studies from industry be made available to teachers at an accessible level as has happened in the past and that teachers and teacher educators be involved in programs where they experience industrial applications first hand. In addition special enrichment programs for secondary students (e.g., AB Paterson Modeling Challenge in Queensland) need to be fostered with industrial mathematicians contributing to these.

4 An Interdisciplinary Project Approach to Making Connections with the Real World

A second approach to involving the tackling of tasks with real world elements in the mathematics classroom is including interdisciplinary activities. In 2001, the Singapore Government accepted the recommendation of the Committee on University Admission that project work be included as part of university admission to take effect from 2005. The purpose of such project work is to "inculcate and measure qualities including curiosity, creativity and enterprise. Projects also nurture critical skills for the information age and cultivate habits of self-directed inquiry" (Ministry of Education, 1999). Interdisciplinary project work was then introduced as an educational initiative in primary, secondary, and pre-university institutions in 2000 in order to prepare students for contemporary workplace demands (Quek et al. 2006). The promotion of flexible and adaptive application of mathematical knowledge is thus an objective of interdisciplinary projects involving mathematics. A second objective is to ensure explicit links between different subject knowledge are made so students learn to "appreciate the inter-connectedness of disciplines and see the relevance of classroom learning to their current or future interests" (Chan 2001, p. 1).

Ng (2009) researched the implementation of a design-based interdisciplinary project she designed involving mathematics, science, and geography. Sixteen classes of students ($N = 617$) from grades 7 and 8 (aged 13–14) in two educational streams (high and average) across three Singapore government secondary schools were involved in this study. Ten student groups from these classes were the focus of intensive observation and data collection. The project was conducted through weekly meeting sessions facilitated by teachers in normal curriculum time over a 15-week period. The project aimed at enhancing students' environmental consciousness and exploration of environmental conservation in Singaporean life. It required student-groups to decide on the type, purpose, location, and facilities of a building of their own design including environmentally friendly features before making physical scale models of their buildings from recycled materials, based on scale drawings. The context of designing and building scale models of buildings was considered to be within the life experiences of students. Mathematical tasks in the project included: (a) *decision making* about the various aspects of the building (i.e., size, dimensions, location, purpose, environmentally friendly features, and design), (b) *cost of furnishing* and fitting out a selected area in the building, and (c) hand-drawn *scale drawings* of the actual building.

Some focus group students did not perceive school subjects to be useful during the project. Certain students viewed the project task as non-mathematical and did not consider using mathematics. Subsequently, they reported during interviews that mathematics was not useful for the project. These students did not participate when mathematical knowledge was being applied by others in their group or did not draw on their knowledge and skills for the project in a connected manner. Thus, the expected mathematical knowledge and skills were not entirely applied. Several student-group members did not consciously and consistently monitor the accuracy and reasonableness of their application of mathematical knowledge and processes. Certain students had difficulty identifying connections between mathematical concepts such as choice of scale and estimations of actual areas during application of mathematical knowledge for real world purposes.

Largely, coverage of mathematics by each group during the tasks depended on sensitivity to task features, engagement with the task, task scaffolding, and a shared repertoire of mathematical concepts, skills, and domain knowledge such as environmentally friendly features of building design. It was assumed that students working in groups would complement each other on the aspects mentioned and promote higher quality mathematical outcomes but this did not eventuate. This does not happen automatically but depends on close monitoring by facilitating teachers. At times, students struggled to engage mathematically with the task. Yet, too regimented scaffolding might not do justice to the open-ended nature of such tasks for creative problem solving and inter-connected meaningful learning. It is thus an open question as to how teachers "balance" scaffolding during such tasks and retaining the mathematical rigour of tasks in their eyes.

In addition, there was limited activation of real-world knowledge by the focus groups despite the presentation of the project within a real world setting. Only three high ability groups applied real world knowledge during all three mathematical tasks. On the positive side, there was at least one member within each focus group who attempted to remind other group members of real-world constraints during mathematical decision-making. In addition, all but one student group used some form of inter-subject connections during the project in contrast to the findings of Chua (2004) in a primary school setting.

It appears that the weighting of project work for tertiary entrance or its remoteness from grades 7 and 8 is not sufficient incentive for many students and teachers to view interdisciplinary projects as an integral part of teaching, learning and assessment, and of equal importance to school based end-of-term and end-of-semester tests, which are used within schools to determine progress, selections, or placement (Quek and Fan 2009). Perhaps in the future when students have more experience of interdisciplinary project work and more recent changes (i.e., introduction of applications and modelling as component processes in the framework of the school mathematics curriculum) have had time to be established, there is a chance for a breaking away by students and teachers from a focus on being exam smart and showing little interest in applications. With the new emphasis on application and modeling advocated at the pre-university level and the introduction of technology into primary and secondary school, it is likely "students will be able to solve real-world problems with messy data and to undertake mathematical modelling" (Wong et al. 2009, p. 36) in the future. As in other implementations that have successfully incorporated an applications and mathematical modeling approach to teaching and learning mathematics, there needs to be an "extensive period of exposure and gestation during which teachers" develop and share new ideas and are allowed to build the momentum required to experiment and change practice (Suurtamm and Roulet, 2007, p. 495). This is now occurring (see Ang 2013 for an example). For teachers to be able to incorporate modeling of industrial problems to project work, however, support is needed to broaden their awareness of industrial practices and mathematics related to the production of goods and services of relevance to the curriculum they are expected to teach.

5 Concluding Remarks

The examples presented provide some insights into the question of how "authentic real world applications" (Damlamian and Sträßer 2009, p. 525) might be embedded into secondary mathematics curricula. This is a complex question which has meant such applications have been ignored in many systems, given lip service by teachers when they were meant to be embedded in curricula, or come into the curriculum but now disappeared. For curriculum change in this area to involve industrial mathematics examples it appears that both support from industrial mathematicians and the allowance of time for change to take hold in the mindset of teachers and students are necessary before the potential expressed in curriculum documents can be fulfilled. However, it is important to realize the prominent roles both technology and assessment play in facilitating this process.

References

Andresen, M., & Lindenskov, L. (2009). New roles for mathematics in multi-disciplinary, upper secondary school projects. *ZDM—The International Journal on Mathematics Education, 41*(1–2), 213–222.

Ang, K. C. (2013). Reallife modelling within a traditional curriculum: Lessons from a Singapore experience. In G.A. Stillman, G. Kaiser, W. Blum & J. P. Brown (Eds.), *Teaching mathematical modelling: Connecting to research and practice* (pp. 131–140). Dordrecht: Springer.

Chan, J.K. (2001). *A curriculum for the knowledge age: The Singapore approach*. Paper presented at the 8th annual Curriculum Corporation Conf., Sydney, Australia.

Chua, J.J. (2004). *Differential learning outcomes of project work between streams and gender in a primary school*. Unpublished masters thesis, National Institute of Education, Nanyang Technological University, Singapore.

Damlamian, A., & Sträßer, R. (2009). ICMI Study 20: Educational interfaces between mathematics and industry. *ZDM—The International Journal on Mathematics Education, 41*(4), 525–533.

Fuller, M. (2001). The graphics calculator and mathematical modelling: Creating an integrated learning environment. In J.F. Matos, W. Blum, S.K. Houston, & S.P. Carreira (Eds.), *Modelling and mathematics education: ICTMA 9 Applications in science and technology* (pp. 143–150). Chichester: Horwood.

Kaur, B., & Dindyal, J. (2010). A prelude to mathematical applications and modelling in Singapore schools. In B. Kaur & J. Dindyal (Eds.), *Mathematical applications and modelling* (pp. 3–18). Singapore: World Scientific.

Maaß, K., & Mischo, C. (2011). Implementing modelling into day-to-day teaching practice—The project STRATUM and its framework. *Journal for Didactics of Mathematics, 32*(1), 103–131.

McLone, R.R. (1973). *The training of mathematicians*. London: Social Science Research Council.

Ministry of Education. (1999). *Project work guidelines*. Singapore.

National Research Council. (1989). *Everybody counts: A report to the nation on the future of mathematics education*. Washington: National Academy Press.

Niss, M. (1987). Applications and modelling in the mathematics curriculum–state and trends. *International Journal of Mathematics education, Science and technology, 18*(4), 487–505.

Niss, M., Blum, W., & Galbraith, P. (2007). Introduction. In W. Blum, P. Galraith, H-W. Henn, & M. Niss (Eds.), *Modelling and applications in mathematics education: The 14th ICMI study* (pp. 3–32). New York: Springer.

Ng, K.E.D. (2009). *Thinking, small group interactions, and interdisciplinary project work*. Unpublished doctoral dissertation, The University of Melbourne, Australia.

Quek, C., Divaharan, S., Liu, W., Peer, J., Williams, M., Wong, A., et al. (2006). *Engaging in project work*. Singapore: McGraw Hill.

Quek, K.S., & Fan, L.H. (2009). Rethinking and researching mathematics assessment in Singapore: The quest for a new paradigm. In K.Y. Wong, P.Y. Lee, B. Kaur, P.Y. Foong, & S.F. Ng (Eds.), *Mathematics education: The Singapore journey* (pp. 431–433). Singapore: World Scientific.

Stillman, G.A. (1998). The emperor's new clothes? Teaching and assessment of mathematical applications at the senior secondary level. In P. Galbraith, W. Blum, G. Booker, & I. Huntley (Eds.), *Mathematical modelling: Teaching and assessment in a technology rich world* (pp. 243–253). Chichester: Horwood.

Stillman, G. (2007). Implementation case study: Sustaining curriculum change. In W. Blum, P. Galraith, H-W. Henn, & M. Niss (Eds.), *Modelling and applications in mathematics education: The 14th ICMI study* (pp. 497–502). New York: Springer.

Stillman, G., & Galbraith, P. (2009). Softly, softly: Curriculum change in applications and modelling in the senior secondary curriculum in Queensland. In R. Hunter, B. Bricknell & T. Burgess (Eds.), *Crossing divides*. Proceedings 32nd Conference of the Mathematics Education Research Group of Australasia, Wellington (Vol. 2, pp. 515–522). Adelaide: MERGA.

Stillman, G., & Galbraith, P. (2011). Evolution of applications and modelling in a senior secondary curriculum. In G. Kaiser, W. Blum, R. Borromeo Ferri, & G. Stillman (Eds.), *Trends in teaching and learning of mathematical modelling* (pp. 689–699). New York: Springer.

Stillman, G. & Brown, J. (2012). Get down and get dirty in the mathematics: Technology and mathematical modelling in senior secondary. In J. Clarke, B. Kissane, J. Mousley, T. Spencer

& S.Thornton (Eds.), *Mathematics: Traditions and {new} practices*. Proceedings of the AAMT-MERGA conference, Alice Springs. Adelaide: AAMT/MERGA.

Suurtamm, C., & Roulet, G. (2007). Modelling in Ontario: Success in moving along the continuum. In W. Blum, P. Galbraith, H-W. Henn, & M. Niss (Eds.), *Modelling and applications in mathematics education: The 14th ICMI study* (pp. 491–496). New York: Springer.

Wong, K.Y., & Lee, N.H. (2009). Singapore education and mathematics curriculum. In K.Y. Wong, P.Y. Lee, B. Kaur, P.Y. Foong, & S.F. Ng (2009). *Mathematics education: The Singapore journey* (pp. 13–47). Singapore: World Scientific.

Drawing on Understanding of Workplace Practice to Inform Design of General Mathematics Curricula

Geoff Wake

In this article, I explore how we might develop general mathematics education curricula that reflect understanding of the nature of activity in relation to mathematics in both schools and workplaces. In doing so, I like many other researchers in the field of workplace mathematics, adopt a sociocultural theoretical perspective. Ideas of expansive learning and developmental transfer appear to offer potential for vocational and possibly well-focused prevocational education. For more general mathematics education, it is the nature of mathematical activity in horizontal and vertical senses that perhaps provide a way forward. Further, I suggest that learning communities need to not only consider the content of the curriculum but also to reconsider the didactical contract in their mathematical activity.

1 Introduction: The Transfer Problem

This chapter reflects on research that explored the boundaries between workplace practice and school/college mathematics (Wake and Williams 2001). In the development of about a dozen case studies, each of which focused on the practice of a particular worker, in a range of different settings we engineered 'breakdown' moments (Pozzi et al. 1998) by asking workers to explain their activity to researchers and students with their teachers. This generated understanding not only of the activity of the worker and its relation to mathematics but also provided insight into the use of academic mathematics and the affordances and constraints of current curricula. More recently, my involvement in research into proposed curriculum changes in mathematics in England, and into transitions into

G. Wake (✉)
School of Education, University of Nottingham, Jubilee Campus, Wollaton Road, Nottingham NG8 1BB, UK
e-mail: Geoffrey.Wake@nottingham.ac.uk

A. Damlamian et al. (eds.), *Educational Interfaces between Mathematics and Industry*, New ICMI Study Series 16, DOI: 10.1007/978-3-319-02270-3_31,

'mathematically demanding' courses in university has provided additional insight into the central issue of transfer: that is, the use, or application, of mathematics in a range of different settings.

In the last two decades or so many researchers have contemplated the overarching question of how 'transfer' might be conceptualised and better supported. Extreme positions in the debate might be characterised as those of, on the one hand, proponents of situated cognition (that is, that knowledge is developed in social settings by individuals in interaction with others and is dependent on the cultures, traditions and values of the community: see, for example, Lave (1988) and Lave and Wenger (1991)), whilst on the other hand, proponents of the classical psychology and information processing perspective (with knowledge being abstract generalisable and applicable in a range of different situations (e.g. Anderson et al. 1996). Our own research in many ways appeared to support the former contention as we found that students appeared ill-equipped to use their knowledge of school mathematics to understand workplace practices. We found their mathematical competence to be very much situated in a school culture that values technical competence with well-defined procedures for solving problems in familiar, and often mathematical, settings. This particular genre of mathematics appeared not well suited to support transferability, or transformation, of mathematics into unfamiliar settings. To illustrate this consider the following brief vignette which I have written about in more detail elsewhere.

As part of one case study we investigated the work of Alan, a railway signal engineer, who explained some of his day-to-day workplace activity to a group of students on a prevocational engineering course. As part of his work, Alan checked calculations of colleagues of where to place signal boards to give advance warning to train drivers that, at a signal ahead, they may be required to bring their train to a halt. The example training calculations that show how to find the average gradient over a number of sections of track proved particularly problematic for the students to understand. Alan explained that such averaging is required so that account can be taken of the gradient of the track when positioning signal boards because uphill gradients assist, and downhill gradients oppose, the braking of the train. Crucially the students were unable to bring together their understanding of the concepts of gradient and of averages to make sense of the process illustrated. This brief extract from the transcript of the discussions between researcher and one student from a group of three illustrates this.

Researcher	Yes… So can you just explain what's going on in there [indicating a table which systemised using different gradients to find the total rise/fall over sections of track of different lengths]
Student	used different gradients for each slope and he's averaged it out…
R	yes can you sort of explain the detail …
S	you started adding them together—adding the gradients together and divide by two
R	Perhaps if we describe what each column is doing

Here, the student appears to associate finding an average with the school mathematics procedure of "adding the values together and dividing by the number of values" rather than finding the total fall of track and dividing by its total length. The ensuing discussion was lengthy requiring the researcher to explain the basic concept of gradient, by drawing a number of diagrams.

Throughout our case studies we found students similarly ill-equipped to understand how workers were using mathematics in relation to their day-to-day work. In this article, therefore, I revisit the findings from the project that generated this vignette and reflect on these in light of current understanding of the nature of workplace activity involving use of mathematics with the aim of considering how this might inform future development of mathematics curricula in general education. At a time when many countries seek to encourage more of our young people to be motivated towards further study in science and technology-based subjects this provides an important challenge in strategic design (Burkhardt 2009).

2 The Nature of Mathematics in Workplaces

In summary we noted the following important features relating to the practices of workers as they went about those of their day-to-day activities that involved use of mathematics (in some way):

- Knowledge is often crystallised (e.g. Hutchins 1995) in artefacts, including tools and signs, often as a result of reification by workplace communities (Wenger 1998).
- Use of mathematics is often 'black-boxed' (Williams and Wake 2007) and engagement with mathematics often only occurs at 'breakdown' moments.
- The fusion (Meira 1998) of mathematical signs (in the sense of Pierce) with the reality they represent reduces cognitive effort.

Further, we identified the following important issues relating to mathematics content and competences that might usefully inform future development of both (pre-)vocational and general mathematics curricula:

- School/college mathematics is just one genre of mathematics and should be recognised as such with attention being drawn to the diversity of ways in which mathematics might appear elsewhere. This suggests that it is important to focus clearly on key mathematical concepts and principles and for students to experience how these can be applied in a variety of different situations using a range of different notations, inscriptions and so on.
- Mathematics is used in a rich variety of contexts both in workplaces and more generally in communicating information in all walks of life; these contexts are often complex and detailed, although often simplified to allow mathematical analysis. Mathematics curricula should allow time and space for students to

experience using their developing mathematical knowledge, skills and understanding in increasingly complex situations.

- Students appear armed with competencies in relation to mathematics that sees them particularly inadequately prepared to engage in using mathematics in workplaces. Particularly important in this regard is their lack of skill in making sense of the "mathematics of others". This is something that many workers have to do, given that they often take over parts of the work process that have previously been established. We note that our research pointed to a number of strategies useful in this regard (Wake 2007) and these should be highlighted in curriculum specification.
- Workers are often so immersed in their practice that the mathematics becomes 'fused' with the workplace reality it models. Underpinning assumptions are not made explicit but workers fully understand how a change in these will affect outcomes in terms of workplace processes. Curricula should provide students with experiences of working with mathematics in complex situations that mirror such scenarios with particular attention being paid to interpretation, variation and adaptation of models.
- Our research identified seven general mathematical competences (for example, interpreting large data sets, costing a project) (Wake and Williams 2001) each of which we saw in use across a number of different workplaces. Fundamental to these is the expectation that technology is integral as a tool when mathematics is being applied. This is mirrored in other recent research that identifies and organises mathematics around techno-mathematical literacies (Hoyles et al. 2007). Curricula should recognise and emphasise such competences.

At the time of this research, and in much of my research in schools/colleges since, it seems the case that mathematical content, often due to curriculum specification, is seen as compartmentalised around major content themes such as number, algebra, geometry, statistics and probability with statements of requirements and resulting curriculum implementation often being atomised. Resulting pedagogies are often transmissionist (Pampaka et al. 2011) and concerned with the development of instrumental, rather than relational, understanding (Skemp 1976). Consequently consideration should be given to how connections can be made across mathematical content areas cognisant of how concepts are often blended in workplaces and other areas of application.

3 Theoretical Perspectives

To understand the different practices we observed, and to explore the relationship between workers, mathematics and their socially constituted workplace practices, our analysis of case studies from a sociocultural perspective drew in particular on Cultural Historical Activity Theory (CHAT). Before attempting to synthesise the above to reach a strategic overview that might inform future curriculum development I, therefore, reflect further on considerations and understanding of 'transfer' from a CHAT perspective.

The well-known schema of Fig. 1 (Engeström 1987) draws our attention to the complexity of the interacting factors that mediate the actions of an individual within the activity of a system such as a workplace or school community. Building on Vygotsky's (1978) conceptualisation of artefact mediated actions of the individual in the upper triangle, in second generation Activity Theory, Leont'ev (1981) and followers expanded the unit of analysis to take into account the activity of the collective, as represented by the lower part of the triangular structure. This draws attention to how an individual's actions are socially constructed and mediated by rules (both implicit and explicit and which are historically evolved) and the division of labour between members of the community. Further, in third generation activity theory, Engestrom (2001) and others consider the interaction of two or more activity systems leading to notions of boundary objects (Star 1989) (artefacts that have 'currency' in each system) and boundary crossing (Engeström et al. 1995) by individuals who move between systems.

It is in the 'boundary space' between workplace and college that Hoyles et al. (2010) identify the potential for workplace training or perhaps, 'education for the workplace'. In such situations there is the potential for expansive learning and developmental transfer as participants in activity systems question and develop their practices in each and with each system consequently learning from the other. However, in general curriculum provision for mathematics that might better prepare students to be able to transform, mathematics into a *range* of different practices in which they might engage in the future, the development of such a boundary space does not at first sight appear to provide a solution. It is this unknown future participation that poses the problem: how can we second-guess the nature of the knowledge, skills and understanding that might be needed?

I suggest that the way forward is to consider the generality of the actions of individuals that emerged in the boundary space activity that our research, like that of others, developed. In particular, I draw attention to how, as an outsider introduced to a novel practice, one is required to strive to make sense of the mathematical activity as it has been historically and culturally constituted. This requires skills that allow one to de-construct these existing practices; in other words, to decouple the mathematics and the reality it models, prior to being able to reconstruct and if required build on these practices. It is also important to recognise the mathematics itself may comprise of a blend of mathematical concepts across domains as in the vignette reported here. This suggests that ideas of horizontal and

Fig. 1 Activity system schema

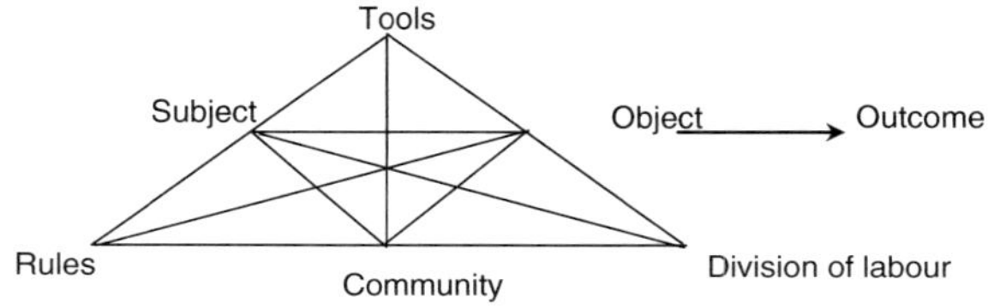

vertical mathematisation, in the sense of Freudenthal and colleagues (e.g. Treffers 1987), are important, with 'models of' situations being generative of 'models for' further mathematical development. Crucially here mathematical models and mathematical modelling competencies here have an important role to play.

This is perhaps best exemplified by referring back to the illustrative vignette introduced above. Here we have a workplace activity relying on the blending of two important mathematical concepts (each potentially difficult in their own right): those of gradient and average. The students in the 'boundary space' afforded by the research activity struggled to unearth their understanding of these concepts beyond a most elementary procedural understanding. A more productive attempt to make sense of the mathematics would have understood that what had been developed was in effect a *model for* an average gradient over the whole length of track. It is in the development of this model that we detect vertical mathematisation as models *for* gradients of contributing sections of track are deconstructed to provide values for total rise/fall and total length of track. In terms of mediated actions we might view this as the learner having to switch from mathematics as being initially a tool for understanding the workplace artefact of gradient for a section of track and then becoming the object of study itself as the student/researcher uses it to reconstruct average gradient over a number of sections. In either of these interpretations of student action it is essential that the student is able to switch back and forth between the 'model for' and 'model of' the situation with ease and with understanding of one reinforcing understanding of the other.

This suggests that we require a curriculum formulation that

1. develops understanding of mathematical concepts *in addition to* procedural fluency with techniques
2. prioritises competencies in mathematical modelling and applications
3. requires students to engage with making sense of, and developing existing mathematical models (of others) of non-trivial situations.

In general, this suggests that mathematical modelling needs added emphasis in future curricula, and that in addition to developing a meta-cognitive understanding of modelling, it is also important that students engage with, and grapple with coming to understand, the models of others. It is important that students develop the enquiry skills that those in our research project lacked as it is such critical inquiry that is quintessential to the curriculum development proposed.

This draws attention to the nature of the activity of the learning community and how this needs to be reformulated: in essence a renegotiation of the didactic contract (Brousseau 1997). It was noticeable how in our workplace research and more recent research in classrooms, workshops and lectures in schools, colleges and universities in their implementation of the curriculum teachers adopt mainly transmissionist teaching practices (Pampaka et al. 2011). This suggests that teachers generally believe that competence in transfer is achieved through learners firstly acquiring technical and procedural competence prior to application (which in many cases can be an afterthought if it is suggested at all). The proposal here that the curriculum requires, at least in part, substantial connection with a non-mathematical reality, and that exploring this becomes the focus of collective activity building from individual actions, suggests a very different approach and consequently adoption of different roles by both teachers and learners affecting the division of labour in learning communities as it is generally currently constituted.

4 Conclusion: The Challenge for Curriculum Design

In summary my proposal is for future general mathematics curricula to introduce new practices for students that prioritise making sense of, and developing further, the mathematical models of others. This provides a major challenge in strategic curriculum design and I draw attention to three major factors that such design needs to take into account.

1. Specification. Mathematics curricula are often specified by stating mathematical content that should be learned. In developing a curriculum as suggested here it is important to emphasise expected outcomes in terms of new skills and competencies that learners require in understanding, and being able to develop, mathematical models. There is currently limited understanding of how learners might best de-couple mathematics and reality in ways that allow them insight into each. This an under-researched area, but one that might be informed by the substantial body of research into use of mathematics in workplaces.
2. Support. Because of the novel nature of the proposed curriculum there is a need to identify a range of rich resources to support the required activity: again there is perhaps the potential to draw on workplace research case studies to develop some of these.
3. Pedagogy. It is important that teachers do not consider how they might reduce the expected enquiry activity to a set of new heuristics. It is likely that they may need to re-conceptualise their role to become a supporter of joint enquiry rather than transmitter of mathematical expertise. This suggests a major shift in the typical didactic contract with mathematics learning being considered a workshop-situated and inquiry-based activity.

The approach proposed here tackles issues of transfer directly by placing the study of how mathematics models reality at the core of the curriculum. Uncoupling mathematical concepts from the situation they model and making sense of each and their interrelatedness becomes central to the mathematical activity of the student. Consequently students will be expected to explore how mathematics can be transformed to meet the needs of a range of diverse situations, with at times a focus on the coupling, and at other times a focus on the development of the mathematics itself. This suggests the need for continued research and development in this area of major importance to future worker expertise and adaptability.

References

Anderson, J. R., Reder, L. M., & Simon, H. A. (1996). Situated learning and education. *Educational Researcher, 25*(4), 5–11.

Brousseau, G. (1997). *Theory of didactical situations in mathematics*. Dordrecht: Kluwer.

Burkhardt, H. (2009). On strategic design. *Educational Designer, 1*(3). Retrieved from http://www.educationaldesigner.org/ed/volume1/issue3/article9.

Engeström, Y. (1987). *Learning by expanding: An activity-theoretical approach to developmental research*. Helsinki, Finland: Orienta-Konsultit.

Engeström, Y., Engeström, R., & Kärkkäinen, M. (1995). Polycontextuality and boundary crossing in expert cognition: Learning and problem solving in complex work activities. *Learning and Instruction, 5*, 319–336.

Engestrom, Y. (2001). Expansive Learning at Work: toward an activity theoretical reconceptualisation. *Journal of Education and Work, 14*(1), 133–156.

Hoyles, C., Noss, R., Bakker, A., & Kent, P. (2007). *TLRP research briefing 27—techno-mathematical literacies in the workplace: A critical skills gap*. London: Teaching and Learning Research Programme.

Hoyles, C., Noss, R., Kent, P., & Bakker, A. (2010). *Improving mathematics at work: The need for techo-mathematical literacies*. London: Routledge.

Hutchins, E. (1995). *Cognition in the wild*. Cambridge, Mass: MIT Press.

Lave, J. (1988). *Cognition in practice: Mind, mathematics and culture in everyday life*. Cambridge, UK: Cambridge University Press.

Lave, J., & Wenger, E. (1991). *Situated learning: Legitimate peripheral participation*. Cambridge: Cambridge University Press.

Leont'ev, A. N. (1981). *Problems of the development of the mind*. Moscow: Progress Publishers.

Meira, L. (1998). Making sense of instructional devices: The emergence of transparency in mathematical activity. *Journal for Research in Mathematics Education, 29*, 121–142.

Pampaka, M., Williams, J., Hutcheson, G. D., Davis, P., & Wake, G. (2011). The association between mathematics pedagogy and learners' dispositions for university study. *British Educational Research Journal, 38*(3), 473–496.

Pozzi, S., Noss, R., & Hoyles, C. (1998). Tools in practice, mathematics in use. *Educational Studies in Mathematics, 36*, 105–122.

Skemp, R. R. (1976). Relational and instrumental understanding. *Mathematics Teaching, 77*, 20–26.

Star, S. L. (1989). *Regions of the mind: Brain research and the quest for scientific certainty. Stanford*. CA: Stanford University Press.

Treffers, A. (1987). *Three dimensions: A model of goal and theory description in mathematics instruction—the Wiskobas project*. Dordrecht: Reidel.

Vygotsky, L. S. (1978). *Mind in society: The development of higher psychological processes*. (M. Cole et al., Trans.). Cambridge, Massachusetts: Harvard University Press.

Wake, G. D. (2007). Considering workplace activity from a mathematical modelling perspective. In W. Blum, P. L. Galbraith, H.-W. Henn, & M. Niss (Eds.), *Modelling and applications in mathematics education. The 14th ICMI study* (pp. 395–402). New York: Springer.

Wake, G. D., & Williams, J. S. (2001). *Using College mathematics in understanding workplace practice. Summative report of research project funded by the Leverhulme Trust*. Manchester: Manchester University.

Wenger, E. (1998). *Communities of practice: Learning, meaning, and identity*. New York: Cambridge University Press.

Williams, J. S., & Wake, G. D. (2007). Black boxes in workplace mathematics. *Educational Studies in Mathematics, 64*(3), 317–343.

Part VI
WG Mathematics-Industry Communication

Communication and Collaboration

Solomon Garfunkel, Rolf Jeltsch and Nilima Nigam

1 Introduction

This chapter reports on the outcome of the Working Group 3 (WG3), which treated the topics mentioned in Section on *"Communication and Collaboration"* of the EIMI Discussion Document. Some further input came from the Macau Workshop,[1] held November 3–4, 2011.

The EIMI Discussion Document formulated four questions: Q.1 *How to identify which societal and/or industrial problems should be worked on*? Q.2 *How to better communicate within multidisciplinary working groups*? Q.3 *How to communicate the underlying mathematics to the problem owners and/or general public*? Q.4 *How to achieve greater quantitative literacy among school leavers, workers, and the general population*?

Co-chairs: Solomon Garfunkel [USA], Rolf Jeltsch (Rolf Jeltsch was supported by a grant from the FCT Fundação para a Ciência e a Tecnologia, Portugal, and the HC Andersen Academy of the University of Southern Denmark.) [Switzerland], Nilima Nigam [Canada]
Working Group Members: Adérito Araújo, Jorge Buesco, Jaime Carvalho e Silva and José Francisco Rodrigues [Portugal], Hanne Christensen and Briynhild Totland [Norway], Thomas Grandine and Peter Turner [USA], Keono Gravemeijer [The Netherlands], Djordje Kadijevich [Serbia], John Keogh and Terry Maguire [Ireland], Ajiit Kumar [India], Nilima Nigam [Canada], Wolfgang Schlöglmann [Austria], Rudolf Sträßer [Germany], Geoff Wake [United Kingdom] and Tine Wedege [Sweden].

[1] See also the presentation done at the in conference in Macau on the webpage: http://www.sam.math.ethz.ch/jeltsch/

S. Garfunkel (✉)
COMAP, Bedford, USA
e-mail: sol@comap.com

R. Jeltsch
Eidgenössische Technische Hochschule, Zürich, Switzerland

N. Nigam
Department of Mathematics and Statistics, McGill University, Montreal, Canada

A. Damlamian et al. (eds.), *Educational Interfaces between Mathematics and Industry*, New ICMI Study Series 16, DOI: 10.1007/978-3-319-02270-3_32,

During the Study Conference, the WG added: Q.5 *How to better understand/enhance communication between the educational system (schools) and industry*? Q.6 *How to better understand/enhance communication between (academic) mathematics and industry*? Q.7 *What about issues related to the inner functioning of communication between the different groups*?

In the workplace, mathematics is seldom an individual activity. Mathematical work, mostly on modeling and problem solving, is almost always a group activity and frequently the groups involved are made up of individuals with diverse expertise and expectations. Communication (listening, writing, speaking, and the use of communication technologies) between and among such groups is crucial. In order to answer the questions we specify what we mean by those various groups.

Industry: Following the Discussion Document, all sizes—from small and mediums size businesses up to large industrial companies—are included. Additionally, the term ‘industry’ applies not only to the private sector but also to government organizations, international organizations, and non-governmental organizations. The objectives of such industry is the production of goods or services. This includes the financial sector, banks, and insurance companies. On some occasion, we even include the general public, because it sometimes behaves in the same way as an industry in the narrow sense. The communicators can be persons or teams within a multidisciplinary working group, but also a person at the workplace with their colleagues, the supervisor, or employee. However, various companies may have to communicate concerning mathematics or education of mathematics. A group of industry may communicate to the government or the public about mathematics education, the need of mathematicians and engineers.

Educational System: Here, we have the classical system starting with preschool, primary education, and secondary education. These are sometimes called “K-12” education, since at the end, a pupil has been educated for 12/13 years. Finally, education may include post-secondary or University education in which we include professional and vocational training. Even the education of Ph.D. students is considered as education. Clearly, further education of a worker on the workplace within industry should also be considered. However, we did not include it in the WG’s discussion, as it was covered by other WGs. Consequently, question Q4 was not discussed. Clearly, the level of mathematics involved in the educational system varies dramatically from the primary school to the university level. The communicators here are Educators in Mathematics at all levels of the educational system but also groups such as, for example, professional organizations (unions) of high school teachers in mathematics.

Mathematics: We include in this definition all levels of mathematics, from the elementary to very high degree of abstraction. In this context ‘mathematics’ ranges from the very applied simulation of complex systems using powerful computers to the mental mathematical construction where an immediate application may not be immediately foreseen and may actually only appear to be used in a 100 years after the invention (or even never!). The communicators of mathematics can be individuals but also groups, e.g., Mathematics Departments of Universities and professional and learned societies. However, in most cases, we restrict ourselves to the mathematicians from Academia and abbreviate this group by Math in our diagrams. For our diagrams, the professional organizations of high school teachers in mathematics are assumed to belong to the Educational system.

All three systems described above consist of various groups of people which can be very heterogeneous. Discussing the communication and collaboration between all these different groups goes beyond the scope of this study. Hence, we shall restrict ourselves to interactions where industry is involved. We will discuss the communication within multidisciplinary working groups with a particular viewpoint on groups in industry. The most important observation is that communication and collaboration is a two-way action. Listening is as important as speaking. For the purpose of the chapter and since communication between various groups occurs in myriad formats, we do not differentiate between different media:

- Oral communication between individuals, focused on a particular problem (direct interaction, via phone, video meeting, etc.)
- Publications focused on a particular problem (newspapers, TV, movies, theater plays, websites, blogs, discussion forums)
- Interactions in which a particular problem of shared interest is not yet identified. This includes visits (e.g., high school mathematician visits with a math class, an industry, or research institutes).

All of the above represent interaction at the interface between mathematics and some other group; all represent both opportunities and challenges for communication. We shall need to distinguish between interactions which occur within a group, and those which occur between groups.

Consequently, we organize the questions above into two broad categories:

- Communication and Collaboration *between* groups (1-Industry or the Public with Mathematics and 2-Industry or the Public with Education) and
- Communication and Collaboration *within* groups (such as discussions within industry).

Section 2 considers the Communication and Collaboration in heterogeneous bodies, Sect. 3 deals with Communication and Collaboration between Mathematics on the one hand and Industry or the Public on the other, and Sect. 4 is about Communication and Collaboration between Education on the one hand and Industry or the Public on the other. Each Sects. 2, 3, and 4 is divided into observations and recommendations.

2 Communication and Collaboration in Heterogeneous Bodies

We start this section by identifying barriers to communication within heterogeneous groups. They are nearly universal, in the sense that these are perceived as obstacles to communication in multidisciplinary groups regardless of a specific industry/country/culture. The major difficulties identified as barriers are: lack of mutual respect; lack of shared goals; lack of shared vocabulary and culture; lack of realization that hybrid problems need hybrid solutions.

Lack of mutual respect: It is extremely important in multidisciplinary or heterogeneous situations that members of different subgroups respect each other. Feelings of superiority of subgroups are detrimental to the functioning of a heterogeneous group. Even if this is an example of a very human tendency toward tribalism: one identifies with one's 'tribe' (e.g., mathematician, engineer, manager) more strongly that as a member of a team assembled to solve a problem. It may also be that the members of one subgroup have not previously worked with members of the other subgroup, adding to mutual apprehension and misunderstanding. These tendencies lead to a lack of mutual respect, and need to be overcome. The discussion will not be focused on a solution but rather on establishing a hierarchy or relative superiority. Groups may have hidden agendas and want to use the communication and collaboration to push their agenda, e.g., get better funding or support. They may want even to solve the problem alone.

The lack of a shared vocabulary can be misused to express the lack of mutual respect. The lack of mutual respect can also be observed by the unwillingness to listen to the other group. It could also be that one group feels inferior to the other one. This can create the problem that the 'lower' group tries to use the language of the 'upper' one.

Lack of shared goals: Sometimes, a heterogeneous working group does not have an obvious shared goal. It is extremely important that, from the very beginning, it be made very clear by the leadership that the different groups actually do have a shared goal. In some cases this is obvious, e.g., when the USA brought the first man to the moon. In these cases, it is very important that all individuals involved know that their best work is needed to achieve the goal. One needs engineers, physicists, mathematicians, mechanics, medical doctors, psychologists to achieve the goal common to the whole group. It is important that each individual member knows that the goal cannot be achieved without his/her skill and experience but also the skill and experience of the others. Of course different groups may have different goals. As an example, there is the case when mathematicians had been asked to optimize the garbage collection in a city. The problem was that the trucks always arrived at the garbage depot at the same time, producing a traffic jam. Mathematicians proposed a schedule to avoid this jam, but did not take into account that the garbage collectors wanted to get to the depot for a break and to meet each other. This shows that the institution running the garbage collection did not share the same goals as the actual collectors.

Lack of shared vocabulary and culture: It is very clear that due to their different education, subgroups within a multidisciplinary setting can have very different vocabularies and a different culture regarding how to do things. For example, engineers have more of a culture of combining knowledge from different areas to invent some new device or solve a problem, while more science oriented academics want first to understand fundamentals before using the information. Even within a given group, individuals are different. A difference between industry and academia is the different time scales in which they work and their differing aspirations. Industry is interested in getting results as quickly as possible. Industry is very often satisfied if the solution is a reasonable approximation of the solution of the problem at hand while academics wants to provide an exact solution. For example, mathematicians often want to prove that there exists a unique solution as a first step toward solving a problem.

Another difference is the culture of personnel management. In industry, personnel change much faster than in academia. Professors usually stay for long periods, even decades, at the same university while in industry, persons may move to another team or leave the company or even whole parts of the company are sold to another company. It has happened that a contract is signed and 2 month later the leading researcher of the company leaves the company; 1 year later the whole section of the company is sold to another company. In academia, the 'worker' is often a Master's or doctoral student working on a thesis. Hence, her/his period of working on the topic is limited from half a year to about 4 years. Moreover, the objective of a student is completely different; it is to complete the thesis successfully.

Due to their background or the environment in which they live, two communities often have no shared language. For example, many engineers have a solid mathematical education. Nevertheless, a certain algorithm may have a different name in the engineering science and in numerical analysis. In the banking industry, many employees have a banking career and thus do not know the mathematical tools behind certain banking products. On the other hand mathematicians may not know the financial expressions. Similarly, biologists and medical doctors usually do not have the mathematical education to understand the mathematics behind a measuring technique, not to speak of developing new tools together with mathematicians. Conversely, a mathematician usually lacks the experience of a medical doctor to distinguish a cancerous cell from a healthy one.

While many examples above have been chosen for the interaction between industry and mathematics, even within industry there are wide differences. It often happens that sales persons in a software company sell a product which the software engineers still do not know how to produce. The result is often that the delivery of the software is delayed for years or worse, is canceled because of excessive costs. Clearly, such sales people have a different culture than the software engineers.

Hybrid problems need hybrid solutions: Many problems are concerned with very complicated systems. It is obvious that solutions require different skills from different groups. Complex systems, such as modern airplanes, oil-drilling platforms, even simple cars, need the contributions of diverse groups.

We offer some suggestions and recommendations on how to improve communication within a team comprised of people with different mathematical expertise and backgrounds:

Breaking the ice—Create the 'Geoduck' Synergy: The Geoduck is a mollusc native to the Pacific Northwest. Why do we have this here in the title? In a company in the Pacific Northwest of the USA, two groups had to solve a problem together. Each group wanted to solve the problem with its own techniques and was reluctant to cooperate with the other group. The problem already started with finding an appropriate name for the project. The atmosphere changed when somebody said that it should be called the Geoduck project. Since the name was close to neither group, this created a common thinking. There are many other techniques to make individual members begin to identify with a new developing team rather than their individual background. This is a process of team management. There are many references to how to build, form, and lead a team (see for example Bruce W. Tuckamn, http://www.infed.org/thinkers/tuckman.htm).

Focus on shared goals and outcomes: Once the team has been able to move beyond the self-interest of individual members, it is important to actively create cooperation by defining a common goal.

The advantage of being able to listen: A must condition for forming a team is that each member and each group learns to listen to other team members. Just acoustically listening is not enough. Part of listening is also learning a common language. This is required not only of the actors who listen but also of the speakers. They have to learn to use a language which may be academically on a lower level than they would use in their 'standard' community of practice. For example, when a mathematician has to present a project in mathematics to a group of researchers working in other fields, more common expressions have to be used. If an engineer or a psychologist needs to collaborate with a mathematician, he or she should not try to use mathematical terms. In most cases, it is better just to describe the phenomena in terms accessible to an outsider, in line with an article in a newspaper with a general audience. It is also helpful for the partner to repeat what he or she understood using other terms, like "what you just said looks to me as being similar to…," "let me repeat what you said, so you will know what I understood." Also, the listener should always be allowed to ask questions if he or she does not understand a term, a technical expression, an abbreviation. The functionality of a process is much more important than the technical name.

Accepting that not all problems are complicated: This recommendation basically is directed to academic groups or group members. Many academic persons have the idea that a solution is better if it is more sophisticated. It is important to note that the goal is to solve the problem, not to use one's own techniques and intellectual capacity. The goal of some mathematicians may be to show their cleverness in using mathematics and in creating mathematics, whereas industry needs an optimal solution for their problem, independently of whether complicated or only simple mathematics is needed. For the mathematician, it is very important to manage to explain the result to the persons who will implement the result.

3 Communication and Collaboration Between Mathematics and Industry

3.1 How to Better Understand/Enhance Communication Between (Academic) Mathematics and Industry?

We start with some observations:

Industrial mathematicians downplay the specific application while discussing with academic mathematicians. Often the mathematician in industry will want to downplay the application because he/she feels that the academic mathematician is not really interested in the application. However, it is extremely important that the academic mathematician really understands the application and its requirements in order to be able to judge how the problem should be modeled.

Industrial mathematicians try to describe the problem in an academic language. From academic education, the industrial mathematician tends to think that the academic language is "better" or more precise. The industrial mathematician may want to show to non-mathematicians in the team that she/he is still able to use academic expressions. This would make him/her indispensable to the team as being the only one capable of communicating with the academic mathematician. She/he may also think that the academic mathematician prefers the academic language and that using the academic language gives the impression of a more challenging academic problem at hand. Nevertheless, the problem may not need an academic, sophisticated formulation. It may also be the case that the needed mathematics did not exist at the time when the industrial mathematician was educated. A very nice and old example is the Kõnigsberg Bridge Problem. With the graph formulation invented by Euler, it becomes a trivial problem and can be explained to almost anyone. This also shows that the problem should be presented in a colloquial language and the mathematician should have the freedom to do the abstraction, filtering out unimportant information, and preserving the important structure.

Academic mathematicians want to solve "academically" hard problems. This observation is often triggered by the education. "Academically" hard problems are much more celebrated within academia and by the general public. The proof of a long-standing conjecture is much more appreciated than that given using standardized techniques. From this, students acquire a value system which is likely to be life long. This is true for mathematicians in industry as well as for those in academia. Of course, there are situations where a certain academic level has to be maintained. In particular, if the problem eventually leads to a Master or Ph.D. thesis in mathematics, it is clear that the objectives of both groups (industry and academia) are different from the start. However, both should then accept that their objectives are different. This should be spelled out from the very beginning of the communication process.

Academic mathematicians try to modify the problem so that in the end it may no longer be the problem to be solved. There are several reasons why this happens. One reason is that with appropriate modifications the problem can become easier from a mathematical point of view. One way to make problems easier to solve is to simplify them. However, there is the danger of oversimplification. Another way of making a problem more tractable can be by generalizing it. In many theories it is easier to analyze a problem in a high dimensional space. However, problems of industry actually occur in the three-dimensional space we live in. For example, industry is interested in fluid flow in three space dimensions. Hence, they are not interested in results of pure mathematicians on solutions of Navier–Stokes equations in four space dimensions. Generalizing the problem, on the other hand, may lose important details.

The fact that the math is simple does not mean that the problem is not serious. In this sentence it is clear that the word "serious" is meant from the industry point of view while the mathematical community means "simple." The title expresses in a sense the same ideas as in the previous section.

There are few opportunities for industrial mathematicians to interact with 'pure' mathematicians and students. One reason for this situation is certainly the cultural difference between industry and academia and education, as already stressed. There is the difference in what is a "serious problem" in the language of industry compared to a "hard problem" in pure mathematics. Each culture has its "important" problems. The cultural differences are due to a different value system. Industry wants to solve problems to create products cheaper, better, and faster. A mathematician may just be driven by curiosity, by the enjoyment of thinking about or imagining deep problems, connecting different theories. Unfortunately, few academic mathematicians are willing to spend time for interaction with industry. Concerning the interaction between industry and the educational system, the objectives are again very different. Education in mathematics for non-mathematicians in high schools or in universities is done mostly by teachers and university professors whose education has focused more on pure mathematics. Hence, the tendency to neglect the usefulness of mathematics and to convey the value system of pure mathematics to the students. Often, teachers and professors of mathematics have no interest in getting involved in applications of mathematics.

The time factor is crucial for all three groups. Each group is generally so busy within their own occupation that it is difficult to find time to interact with the other group. This is stressed even more by the fact that one may not appreciate the other groups' objectives. One needs an extra effort to communicate with the other communities. In addition superiors or colleagues often do not consistently support this activity. In academia, a mathematician having contact with industry may be confronted with the comment: "You are an engineer."

With respect to the Recommendations, we start with some for the interaction with the industry. Further down, we offer some the recommendations concerning Education.

Industrial mathematicians should bring the things to the academic mathematicians that are unique to industry (including stories, anecdotes…). The industrialists should bring their specific story to the academic mathematician. For example, in Formula 1 racing, "We have to adjust the racing car to the specific track to be able to do the fastest lap and we have only a few days to do this." Hence, a driver on the first day has to run 20–30 laps exactly the same way. Sensors will record a few crucial measurements during these laps. This information is sent to the home base where a testing ground has to be adapted by solving an inverse problem to reproduce this particular track as precisely as possible. One can then experiment with different parts on this testing ground. One can even speed up this process if certain parts of the car are represented analytically, e.g., with the help of computer model. Such a description will motivate the academic mathematician to think about the whole process and she/he may come up with additional ideas.

Academic mathematicians should be in listening mode. By this we refer to the interactive listening described in the previous section in the recommendations on working in a team.

Industry should keep constancy of purpose. This recommendation looks easy to satisfy but in fact it is difficult to achieve. If industry changes its purpose and both, the original and the modified problem are simple from a mathematical point of view, this does not generally create difficulties. However, if the first and the second task need a longer period of research by the mathematician, then this may cause problems. In particular, this is the case when a 3-year contract should lead to a Ph.D. thesis and the industry changes the topic substantially after 2 years. As described before, such difficulties have to be discussed prior to making the contract

Problem solving and mathematical modeling. The recommendations aim at exposing the students to problems originating from outside of mathematics, leading them to the mathematical formulation and in the end to the mathematical solution. Finally, the result should be interpreted for the underlying original problem. These recommendations differ basically in the amount of time a student is engaged and advised by an instructor. Does she/he work almost completely independently? These activities can also differ in the size of the problem and the time spent to solve and interpret the result. The activities can be a modeling course where a participant solves a small problem each week, or an industrial study week where one spends a week in a group to solve problems from industry in a period of 2 or 3 days or a term paper, bachelor or master thesis for which the interaction between the problem owner and a student could last several weeks. One can let students work in groups, pairs, or alone. A student could equally well spend some time in industry in between terms, e.g., in the form of internships.

Course focus on concrete applications, e.g., case studies. The idea is that a course consists of case studies. A mathematician in industry, an engineer, or a scientist explains a problem and shows where and how mathematics is relevant for analyzing, maybe solving it. This may be taught jointly with a mathematician. Such a course is more to show the breadth of the applications of mathematics rather than to have the students perform the application themselves.

Career advice. Students of mathematics on the tertiary level may need career advice. This could be done by special events followed up by individual coaching if requested by students. For female students in particular, it would be excellent to also present female colleagues as role models.

3.2 How to Communicate the Underlying Mathematics to the Problem Owners and/or General Public?

We start with two *observations* then offer two *recommendations*.

The general public recognizes the importance of mathematics. The general public recognizes that mathematics is important but may often feel that it is not personally relevant. There are several reasons for this. One is that mathematics is often identified more with basic arithmetic rather than with concepts and logical reasoning. The more sophisticated use of mathematics is hidden to the public. Examples are digital pictures, movies and high definition TV, security codes, search tools on a laptop, search engines, weather forecast, and financial instruments, just to name a few. The general public can also have a wrong impression of the importance of mathematics. We give an example: in a public discussion in Switzerland a participant said that the problem with pension plans is due to the mathematicians. A German mathematician then answered that mathematicians had advised the German government some 30 years earlier that due to the age structure there would be a problem with pensions plans in the future. However, it was the government that did not act in time.

The problem owner may not recognize that there is a mathematical dimension in the problem. Already at the begin of this section, it was indicated that problems should be formulated in colloquial terms by the problem owner and it is the mathematician's task to figure out which mathematical structure can describe the problem or to invent new mathematical notions.

Raising public awareness. The first recommendations are basically programs to raise public awareness. Probably, the biggest effort was started during the World Mathematical Year 2000. Many societies now have special programs to raise public awareness, e.g., the German Mathematical Society. The European Mathematical Society recently created a committee to raise public awareness (with a popular mathematical web page http://www.mathematics-in-europe.eu). In Berlin, the Universities created a competition with the Math Christmas Calendar (see http://www.mathe-spass.de/advent/index.htm). In Canada, MITACS set up a webpage "math en jeu," (see https://www.mathamaze.ca/index.php/en). In the "Philosophers' Café," people meet to discuss topics rather than listen to a lecture. Such events take place in the whole Vancouver metropolitan area. The topics include also mathematics. The Canadian Mathematical Society is making a Math in the Mall Display, see http://cms.math.ca/Education/MallMath/. The Science Centre in Singapore organizes lectures under the heading "Science in the Café." Such lectures are not in typical lecture rooms but in a much more relaxed atmosphere and attract an interested audience from people coming from high schools to retired people. At a break with snacks and a bar, the speaker mingles with people, after which a discussion takes place. For its 150 anniversary, ETH Zurich set up small tents for up to 30 persons, at very busy locations in the middle of the city. People just walk in and listen to short talks in science, engineering, and mathematics. The idea is to go to the public rather than to ask the public to come to the university. Similarly, one can go to schools on every level and make pupils aware of what mathematics is and where it can be used. Clearly there are many more possibilities, e.g., movies, plays on mathematics or mathematicians, books on the topic, and the like. All of the above examples contribute to making people aware of mathematics, where it is hidden, how it can be useful. It is also important to convey the notion that new mathematics keeps on being created and that there are many unsolved problems.

Explain the underlying mathematics at the work place. Mathematicians should go out to the work place and look at the processes. They should listen to the expertise of the worker. The workers feel much more comfortable to present what they are doing at the work place rather than in a foreign, e.g., academic environment. In this way, mathematicians may discover hidden mathematics and may be able to reveal it to the workers.

3.3 How to Identify Social/Industrial Problems to be Worked On?

We start with two *observations*, followed by a series of *recommendations*.

Mathematicians from academia try to explain to industry that it needs mathematics. There have been many initiatives which try to explain to industry that it needs mathematics, such as MACSInet (see http://www.macsinet.org/index.html), OECD Global Science Forum "Mathematics in Industry" (see http://www.oecd.org/), European Science Foundation, Forward look on Mathematics and Industry (see http://www.esf.org/index.php?id=6264). Such projects are almost exclusively initiated by academic mathematicians even if they do have some participants from industry. Sometimes a book is produced or a report. However, the long-term effect is difficult to measure. For example, MACSInet was created about a dozen years ago and it looks that the last event recorded on the web page is 8 years old. No more information can be obtained from the MACSInet web page…

Mathematicians from academia and from the educational system ask people/society to identify problems to be worked on in education. Sometimes this is done by questionnaires. The problem is that this interaction is only in one direction.

There were some *recommendations* on how to find the problems in industrial mathematics:

- Do not ask people/society to identify mathematical problems, which should be worked on (in education);

- Try to find out their problems, then dig out the inherent mathematics;
- Digging out cannot be done by questionnaires alone; approaches like workplace visits, case studies, ethnographic studies, internships, Study Groups should be used;
- Digging out can be very difficult, tedious, time consuming;
- Find a common language;
- It can happen that the user does not even understand what problem she/he really has;
- The digging out may need several loops.

Find a true interaction between the problem owners and the mathematician in the educational system and in academia, respectively. The main part of the recommendations is that mathematicians from the educational system and from academia should really get in close contact with the problems owner, be it industry, the workers, the society. By digging out we mean an interactive dialog between the problem owners and the mathematicians as we have already described in the previous section. The main part is to get to know the problem owners problem in their language, find a common language.

Finally, a few *recommendations* were suggested for deciding which problems to consider. From now, we assume that "we" understand the problem and the inherent mathematics.

- The decision on what should be worked on depends on the mathematical level and composition of the population involved. For example, school, academic teaching, vocational training, researchers.
- The level of sophistication will be different in teaching mathematics in upper secondary class from university teaching. Additionally, the truly useful solution to a problem may not be at a mathematically advanced level. The focus should be on providing solutions that are useful to the end user.
- Due to the pervasive use of technology, including web-based applications, data simulation, and processing software, it may be impossible to dig out all mathematics in use. These "Black Boxes" are needed to keep the focus on the problem being solved, instead of the techniques being employed.
- A democratic and highly industrialized society needs a lot of black boxes, but also competent and critical citizens who are potentially capable to open up black boxes if they wish or feel the need to do so.

4 Communication and Collaboration Between Education and Industry or the General Public

Note that parts of this subject are similar to parts of Sect. 3, but Mathematics is replaced by Education. To avoid duplication, here we try to answer the Question: *How to better understand/enhance communication between the educational system (schools) and industry*? In principle this question has been answered in the previous section too. In addition, we have to recognize that workplace mathematics and school mathematics are different. Hence, one has to work together to better understand these differences. The development of a common language and the way to communicate has been mentioned earlier.

An important distinction between problem solving in educational and industrial settings is in the overall goals. For the former, it is knowledge transfer (education) or creation (research), whereas in the latter the important outcome is the successful resolution of the problem at hand. This creates a tension in situations where people from these different communities work together. In principle it is important for students to see mathematics in action from an early stage, and this can be achieved by exposing students to the use of mathematics in industry. In practice, this is a situation fraught with complexity. How is such an interaction to occur in a meaningful manner? How are the (occasionally competing) imperatives of pedagogy and industrial practice to be reconciled?

Internship and co-op programs, and vocational training opportunities, are already implemented in many systems. These provide an existing infrastructure upon which to build mathematics-in-industry curricula. Some efforts in this direction already occur, notably in the EU and the UK. Some universities in the US also have such programs. Harvey Mudd College in the US invites industrial experts to present open problems to their students, and the students then apply all their existing knowledge—not merely mathematical—to solve the problem. This is a successful program, and the companies often recruit the students. The most successful interactions of this type often occur in universities such as Stanford and MIT, with deeply entrenched cultures of innovation and industrial outreach. However, since there are vast disparities in educational systems from country to country, providing targeted recommendations is challenging. Analogous to discussions elsewhere in the book, we recommend the creation of 'mathematics in industry case-studies' for curriculum, as a resource for educators to use with students.

5 Presentation of Additional Papers

Within this Working Group, three contributions were presented and were selected to illustrate different aspects and experiences on the Mathematics-Industry Communication:

Engineering, Mathematics Communication, and Education: reflections on a personal experience, by Jorge Buescu, shows how popularization essays in a periodic magazine with an Engineering audience, in addition to helping bridge the two communities, can have an "overspill effect" when crossing the interface to the outside world and may have a real impact on the educational community;

In A View on Mathematical Discourse in Research and Development, Vasco Alexander Schmidt argues that analyzing the discourse in industrial mathematics is key to understanding how mathematics innovates, where obstacles occur, and how innovation can be organized more systematically in the future. He considers that ethnographic research with focus on language uses may reveal best practices and help develop consulting and training offerings for mathematicians who work in R&D units or study mathematics.

Using Popular Science in a Mathematical Modeling Course, by B.S. Tilley, summarizes a student-driven approach to a mathematical modeling course. The approach consists of students choosing their topic based on an article from a popular science medium, spending time researching the background information of the article, constructing a mathematical model, solving it, and either writing a report or giving a brief presentation before a critical audience. Through this experience, students learn the underlying techniques of formulating relevant mathematical models.

Engineering, Mathematics Communication, and Education: Reflections on a Personal Experience

Jorge Buescu

1 Introduction: Communication of Mathematics, a Personal Journey

In 1995, I was invited to write a regular column on the general subject of the popularization of Science for the professional journal of the Portuguese Engineering Society, *Ingenium*.[1] Being a mathematician, these essays gradually morphed into short pieces, sometimes with a slight journalistic flavor, on Mathematics proper. Conceiving these essays was always a rewarding challenge. As opposed to strictly journalistic pieces, they had the enormous advantage of addressing a mathematically educated audience interested in learning about subjects with meaningful mathematical content, albeit in an informal setting. The questions addressed had a wide scope, ranging from Wiles' proof of the Fermat theorem to exotic n-spheres, from check digit schemes to the mathematics behind Sudoku. These would be much more difficult to reach if addressing a more general audience. The emphasis is always to convey some meaningful mathematical content through "real-world" problems, which would appeal to an audience consisting mostly of engineers.

At some stage the interest on these publications overflowed the intended audience. Clearly, there was an interest in seeing Mathematics in action in real-world problems, in contexts where we would least expect to find it—be it our ID cards, our Euro banknotes, the Sudoku puzzles in the paper, the graph of the Internet, coding theory, and CD scratches. I started having lots of requests for authorization for classroom use by high school and college teachers. Many readers suggested enthusiastically that these essays should be published in book form.

[1] Ingenium, Boletim da Ordem dos Engenheiros (Bulletin of the Portuguese Engineering Society). Available online at http://www.ordemengenheiros.pt.

J. Buescu (✉)
Departamento de Matemática, Universidade de Lisboa, Lisboa, Portugal
e-mail: jbuescu@math.ist.utl.pt

A. Damlamian et al. (eds.), *Educational Interfaces between Mathematics and Industry*, New ICMI Study Series 16, DOI: 10.1007/978-3-319-02270-3_33,

In 2001, I published my first book thus derived, *O mistério do BI e outras histórias* (Buescu 2001). The title comes precisely from the piece on the check digit of the ID card, about which I still receive questions today. It was a best-seller by Portuguese standards, standing today at its 11th edition and over 10,000 copies sold. It was followed in 2003 by *Da falsificação de Euros aos pequenos mundos* (Buescu 2003), in 2007 by *O fim do Mundo está próximo?* (Buescu 2007) and in 2011 by *Casamentos e outros desencontros* (Buescu 2011). The number of original papers in *Ingenium* is by now well over one hundred.

These books had great impact in the educational community, since they conveyed, in clear but rigorous language, examples of use of Mathematics to solve real-world problems. Examples of this impact in schools are many and, in some sense, unusual. I would briefly mention the following:

1. Utilization of some of the problems described (ID card, Monty Hall problem, Euro banknotes...) in College-made exhibitions of Mathematics (e.g., IPL Leiria 2003).
2. Dozens of yearly and standing invitations to speak on Mathematics in high schools, technical schools, Colleges, and Universities all over the country. Between March and May alone I receive about a dozen speaking invitations in high schools.
3. Proliferation of Java applets in the Web with the algorithms I describe, with reference.[2]
4. Many high schools had students elaborate school works on some of the topics covered, and more than one chose to profile me as a mathematician.

2 First Case Study: The Portuguese ID Card and Check Digit Schemes

Sometime in the early 1990s, the responsible Portuguese governmental agency started to include an "extra" digit following the national ID card number, in an isolated box. Obviously, this was a check digit, in most likelihood a checksum digit. However, care was not taken to explain to the public what the new digit stood for. As a consequence, by the late 1990s the wildest urban myths about this recently introduced extra digit were floating around. The most popular one was that it would represent the number of people with the same name as the card bearer. This urban myth was extremely widespread at the time (and in fact still survives). But there were others, ranging from the number of outstanding traffic tickets (a reporter, interviewing me in 2000, said "and in my case, it is correct!")

---

[2] Projecto Atractor. Available online at http://www.atractor.pt/mat/alg_controlo/index.htm.

to the number of family members having been jailed during the pre-1974 dictatorship.

In the meantime, a fellow mathematician at the University of Coimbra, professor Jorge Picado, had gone to work on disproving this myth. He collected the ID numbers and corresponding check digits of a few dozen persons and, under the assumption that the check sum algorithm was similar to the ISBN algorithm, programmed his computer to reverse engineer the problem and discover the checking algorithm. Some surprises were in store. First of all, only 10 digits were used as checksums: 0–9. It is well-known [see e.g., Gallian (1991)] that the additive algorithms require a prime number of check digits; thus, for instance, the ISBN algorithm uses the set of 11 digits {0, 1, … 9, X}, since 11 is the smallest prime number greater or equal than our usual numbering base 10. Now Picado's data included only a set of 10 check digits; there was no 11th digit. This might lead us to think that non-additive algorithms, such as non-commutative algebraic codings based on dihedral groups, like the Verhoeff scheme adopted by the Bundesbank (Gallian 1991; Picado 2001) might be in use.

Secondly, a very strange thing happened. When Jorge Picado ran the algorithm detection code, the program did not converge, whatever the supplementary heuristic hypotheses applied. This seemed to indicate either an inconsistency in the data or a faulty error-detecting algorithm (e.g., one which would not injectivity of the check digit).

In fact, the data revealed an obvious anomaly: the relative frequency of the digit 0 was double the frequency of the other digits. This suggested that something could be wrong with the check digit 0. In fact, simply eliminating all the 0s from the data and running the algorithm-detecting code ensured immediate convergence and allowed Picado to discover the identification system used by the Portuguese agency.

This discovery was a double-edged sword. On the one hand, it was shown that the identification scheme is (almost) precisely the ISBN scheme. On the other hand, since there is no 11th check, the scheme cannot work efficiently. Indeed, what was observed is that the Portuguese ID card uses a version of the ISBN identification scheme *with a mathematical bug*: the non-existent 11th digit for checksums is replaced by a second (false) zero. Thus, the identification system is not injective, and ½ the occurrences of the digit 0 as a checksum are false.

This is a rather embarrassing situation, since it implies that indeed the check digit cannot be relied upon for error-detecting purposes, therefore defeating the purpose of its own introduction and rendering it useless! In actual practice, no official agency using the ID card number ever bothers asking for the check digit—not even the agency, which issues the cards themselves.

Things get even more curious. Although the whole ID card was recently changed, with the issue of a Citizen card incorporating the most modern biometrical and physical technologies, the ID numbers did not change, neither did the check digits. Thus, the ID card bug propagated to the new identification scheme. Moreover, the exact same identification scheme is used in the Fiscal ID number—and thus the exact same mathematical bug occurs.

In May 2000, I published a column in *Ingenium* about this question, *O mistério do Bilhete de Identidade* (The mystery of the ID card) which instantly became my most widely read article. In a short span of time I got reactions of awe, disbelief, and the article was widely reproduced in the media and the Internet, where it still can be found today. As mentioned above, in 2001 I published a book, mostly a collection of some the columns I had publishes in *Ingenium* (Buescu 2001). Its publication had a large impact and made the material available to a very different audience, the general public. The title of the book, the problem itself, and the question of the bug in the identification system led to a large impact in the educational community: everybody has an ID card and number, and it is easy to explain and implement the mathematics and the algorithm behind the ISBN scheme.

The ramifications and implications of this work are quite surprising. Many high schools and Colleges had students doing school works about identification numbers and the ID card. Many students, either in Colleges or Universities, developed computer applets to compute the ID card number (as well as the Fiscal ID number). One College (ESTL, at Leiria) developed an exhibition of Mathematics for high-school students based partly on material from my book. Both me and Jorge Picado were contacted very frequently to give talks on these (and other subjects in schools). In fact, even to this day I am asked to give talks in high schools specifically on this subject.

And, last but not least, both me and Jorge Picado were contacted by the financial branch of a large multinational corporation because of our "expertise about the identification problem in Portuguese ID cards" (we did not, however, become private consultants)! This shows quite clearly how communicating Mathematics to an Engineering audience can have an "overspill effect" which quickly crosses the interface to the outside world and has a real impact on the educational community. Indeed, it is quite likely that without the original article in *Ingenium* Picado's discovery might still be largely unknown to the world outside mathematicians, with a loss both for Mathematics Education and the general public (the urban myths about "the extra digit" still exist to this day!).

3 Second Case Study: The Euro Banknotes

As a second case study, it is relevant to mention the Euro banknotes. With the introduction of the Euro in 2002, the European Central Bank faced the obvious problem of facing counterfeiting of banknotes in the biggest market in the world. The most advanced technology was used in the banknotes, employing special paper end silver bands, lasers, and holograms; at least two of the anti-counterfeiting measures were kept secret by the ECB. One could expect the identification scheme used for Euro banknotes would be equally sophisticated, truly twenty-first century, as it was being equally kept secret.

After the circulation of the Euro banknotes on 1st January 2002, I undertook a somewhat quixotic personal project: to discover the mathematical algorithm of identification of, which passed thorough me, and recruited some friends as well. In the process of programming and entering the data, I noticed the structure of the numbers: they were L-DDDDDDDDDDD, where L stands for a letter of the alphabet from J to Z and D stands for a digit from 0 to 9, so the number is a alphanumeric string made up of one letter and 11 digits. However, a strange thing happened: the last digit is never allowed to be 0. This alone indicated a mathematically special role for the last digit as a check digit.

At that point I published an article in *Ingenium* explaining the problem, giving preliminary results, and inviting readers to submit their recorded data. Response was enthusiastic: with this collective effort, I gathered thousands of data. In the process of programming and entering the data, I realized experimentally the following.

If a fixed numerical value is attributed to the letter L at the start of the identification number, then the (last) control digit is simply determined by imposing that the sum of all the 12 numerical values of the digits is congruent with 0 (mod 9). Or, in plainer terms: the identification scheme of Euro banknotes is simply the thousand-year old process of casting out 9s!

This is all the more surprising since it is an extremely inefficient error-detecting algorithm, as we all know from elementary school (success rate <90 %). So the banknotes with the most advanced physical systems against counterfeiting had also, embarrassingly, the most mathematically unsophisticated algorithms for error-detection! The reader can check this for him/herself by pulling out a Euro banknote. All that is required is to know the correspondence between the leading letter L and its numerical value: J 2 UK; K 3 Sweden; L 4 Finland; M 5 Portugal; N 6 Austria; O 7—none; P 8 Netherlands; Q 9—none; R 1 Luxembourg; S 2 Italy; T 3 Ireland; U 4 France; V 5 Spain; X 6 Denmark; W 7 Germany; Y 8 Greece; Z 9 Belgium.

The numerical values of the letters are simply 1–9 in ascending order [again with no attribution of 0, since this is arithmetic (mod 9)]. Moreover, a letter has no deeper meaning than simply identifying the country of origin of the banknotes. In fact, even though the UK, Denmark, and Sweden are not (yet?) part of the Euro zone, they already have allotted places by this process, should they wish to join!

This collective quest with the readership of *Ingenium* was a great success. We even managed to identify what may be called a second "mini-code" on the flip side of the notes, in very small print, consisting of L-DDD-LL in very small print. This was an even bigger mathematical disappointment: it is a mere serial number from the typography where the banknote is issued and has no mathematical content at all. All this information is gathered in book form in (Buescu 2003) and, to the author's knowledge, nowhere else.

This makes for wonderful material for communicating mathematics in popular lectures. Everybody (in the Euro zone) handles banknotes, even small children. So it is possible to tell a story with meaningful mathematical content, by adapting the theoretical aspects of coding theory and identification systems to the audience, but always keeping in mind a very real example—the banknotes people handle.

In this way a very unlikely interface between Mathematics Education and Engineering was created. The audience of mathematically educated Engineers provided the data. A mathematician discovered the solution. This solution is ideally suited for communication and popularization of Mathematics. A very unlikely connection indeed!

4 Some Conclusions from an Unexpectedly Useful Interface Between Engineering and Mathematics

Admittedly, my experience as a Mathematics communicator is not at all typical (if such a thing exists at all). Starting from writing about Mathematics for the professional society of Engineers, it quickly spread into the educational community and thus established an unexpected interface between Industry (in the broadest sense of the term) and Education. Although this has been a very rewarding journey, it could hardly serve as a model. However, there are some useful lessons I think can be drawn from this personal experience.

A first point which professional mathematicians are very prone to miss: the educational community is in fact eager for meaningful examples of real, everyday-world applications of Mathematics satisfying both the following conditions:

1. The problems they describe should be easily understandable, namely by high-school students preferably through some everyday-life appeal;
2. They should have real mathematical content.

These kind of problems cannot in general be supplied by high-school teachers. The Mathematics they learned are generally either superficial, outdated, or both. In fact a cursory browse through high-school textbooks shows a very limited set of more or less applications of Mathematics to the real world—Fibonacci sequences in pine cones and the golden ratio is probably a universal example.

It has become clear to me that problems with significant mathematical content above the more or less trivial must be supplied by mathematicians or scientists willing to invest the necessary time, work, and effort. Mathematics changes. Technology changes. Innovation occurs, and Mathematics a crucial part of it. The uses of Mathematics in Technology be it in cell phones, DVDs, or ATM machines, changes. It seems clear that the communication of emerging applications can only be made by those in the know—mathematicians, in particular—and there is no reason why we should expect high-school teacher should even be aware of these emerging subjects. In hindsight, it is obvious that it is the mathematical community, which should have the energy to make the outreach effort.

Both case studies above illustrate this situation quite clearly. They deal, each in their own way, with the same kind of mathematical problem—identification systems and check digits. This is a problem with which everyone can be assured be acquainted in one form or another: it arises in contexts from national ID cards (mandatory in Portugal) to credit cards, from Internet passwords to bar codes, from cell phones to DVDs, and increasingly so in the past 20 years. Moreover, it satisfies both criteria stated above (understandability and mathematical content) as characterizing problems, which should interest the educational community, so it seems ideally suited to establish such an interface.

However, it is not only unrealistic but altogether absurd to expect the typical high-school teacher to have any knowledge at all of the mathematics involved in identification systems, or even to be aware of such questions.

So the first lesson is that mathematicians, or at least some mathematicians, should be actively engaged in outreach activities.

A second interesting point, which acts somewhat as a counterpoint to the first one, is as follows. Although Mathematics Departments in general recognize that outreach is part of their mission, in practice this is not much more significant than lip service.

Somewhat paradoxically, generally the attitude of academic peers to the personal investment of time and effort into outreach activities and communication of Mathematics is not necessarily positive. There exist quite gratifying exceptions, but generically the average mathematician ranks as a first-rate activity scientific research and as a second priority teaching duties and academic chores. Outreach activities and communication of Mathematics to the general public generally are not considered "important" or, for the most fundamentalist people, even "serious" work, and in terms of the spectrum of scholarly activities its consideration lies somewhere between the "not serious" and the "waste of time."

It is not hard to imagine that the tension between (1) and (2) raises some very real and serious issues for the future of Mathematics Education. On the one hand, we want to show young students that today's Mathematics has real impact and meaning for today's world, and this can only be effectively done by professional mathematicians. On the other hand, there are no formal or informal reward mechanisms in the academic world to perform outreach activities—a situation that sometimes even perversely acts as a negative incentive, which can become at times quite frustrating.

There is clearly a tension in this interface that should at some stage, in the interest of all parties, be resolved. I think there is a larger issue involved that is not necessarily related specifically to Mathematics. The articulation between in fact, the whole of Science and society at large has probably never been a more pressing issue. Paradoxically, we have never lived in a society more dependent on advanced technology, with generalized reliance on energy, cars, computers, or cell phones; and at the same time never before has such a society been in such a state of generalized dependence on Science (and, in particular, Mathematics) for this technology and thus for its own functioning.

References

Buescu, J. (2003). *Da falsificação de Euros aos pequenos mundos*. Lisboa: Gradiva Publications.

Buescu, J. (2007). *O fim do Mundo está próximo?*. Lisboa: Gradiva Publicações.

Buescu, J. (2011). *Casamentos e outros desencontros*. Lisboa: Gradiva Publicações.

Gallian, J. A. (1991). The mathematics of identification numbers. *The College Mathematics Journal, 22*, 194–202.

Picado, J. (2001). A álgebra dos sistemas de identificação. *Boletim da SPM, 44*, 39–73.

Buescu, J. (2001). *O mistério do BI e outras histórias*. Lisboa: Gradiva Publications.

A View on Mathematical Discourse in Research and Development

Vasco Alexander Schmidt

1 Introduction

Awareness is rising that mathematics plays a crucial role for innovation in many industries, including logistics, finance, electronics, and the chemical and pharmaceutical industry. The growing demand of mathematical expertise in industry has led to a series of initiatives from mathematical communities in many countries. Aiming at a more systematic cooperation, new university programs for applied mathematics have been defined and research centers for industrial mathematics, special interest groups, and faculty positions focusing on industrial mathematics have been founded within the last decade.

These initiatives have had a positive effect, in some cases boosting the knowledge transfer into R&D departments. Nevertheless, a systematic exploitation of mathematical knowledge in industrial settings does not happen yet, at least not on a large scale. Reasons for this gap are seen in a different terminology and language use in mathematics and the application domain, deficits in the education of engineers, as well as in practical and organizational conditions (Grötschel et al. 2009:16). In addition, there are educational issues within mathematics that go beyond the sheer knowledge of mathematical theories and application domains. Many graduates pursuing a career in industry feel that they are not able to apply their mathematical knowledge in the industrial setting except their general ability of logical thinking. One reason might be the discourse that takes place in R&D units. There, mathematicians must carry through their ideas in a setting of conflicting views and different levels of mathematical knowledge, with constraints in time and budget and hierarchies to take into account.

This chapter argues that analyzing this discourse in industrial mathematics is key to understand how mathematics innovates, where obstacles occur, and how

V. A. Schmidt (✉)
Products and Innovation, SAP AG, Dietmar-Hopp-Allee 16,
69190 Walldorf, Baden, Germany
e-mail: vasco.alexander.schmidt@sap.com

A. Damlamian et al. (eds.), *Educational Interfaces between Mathematics and Industry*, New ICMI Study Series 16, DOI: 10.1007/978-3-319-02270-3_34,

innovation can be organized more systematically in the future. Analyzing the use of language—whether in oral or written form—may show how mathematicians use their mathematical knowledge in practice. It can even help to reveal a specific mathematical way of thinking. A new form of participatory research will be introduced. It is based on the role of a "useful linguist" who takes an active role in research and development teams as he helps with the creation and review of documents such as technical documentation and PR articles. In addition to interviews and observations, he uses the work on documents to make visible and document the discourse that is taking place in the research and development teams.

This Linguistic research may reveal best practices and help to manage a fruitful application of mathematics, including consulting and training offerings for mathematicians who work in R&D departments or study mathematics.

2 Role of Language

Focusing on language has been a promising approach for analyzing the nature of mathematics. Mathematics is a science that examines abstract objects and methods. It therefore relies on language when it comes to defining and communicating the objects under investigation and proving mathematical findings. This is why the reflection on mathematics must always take into account the language of mathematics and the language use of mathematicians.

There is a significant difference between both terms. The first one stands for an internal view on mathematics, whereas the second term allows for an external view on mathematics. Mathematics is often seen as an exclusive domain that is only accessible for those who can understand the formalisms that are used by mathematicians. This view is mirrored in numerous popular math books written by mathematicians. Best example is the classic book "What is mathematics?" (Courant and Robbins 1941). When reading the book title one could get the impression that the book is an essay about mathematics. But it is not. It is rather a textbook that invites readers to learn mathematics by doing it. The message is that learning the language of mathematics is a prerequisite for talking about mathematics.

This view has been challenged. Another classic mathematics book, "Experience Mathematics" (Davis and Hersh 1981), paved the way to external accounts on mathematics. It catches the essence of mathematics in an every day language without explaining mathematics in the traditional sense. It focuses on explaining how the mathematicians' practice looks like. This approach is the basis of the following argumentation that argues for a meta-research on mathematics using linguistic and ethnographic methods for analyzing discourse at the work place of mathematicians in industrial research and development. The idea is not to focus on the mathematical core of innovations, but to investigate how these innovations evolved. It takes a closer look on the conversations that precede and follow mathematical innovations. This view on industrial mathematics can be based on philosophical and sociological groundwork.

2.1 Philosophy of Mathematics

The early investigations on mathematics were conducted by mathematicians themselves. They portrayed the ideal use of mathematical language with emphasize on mathematical formalism. This was mirrored in the opinion that mathematics is a "hard science" that is error prone and produces findings of eternal truth. The body of mathematical knowledge is seen as cumulative, consensual, and historical invariant. This view was challenged in the second half of the twentieth century by philosophers that followed the ideas of social constructivism (and others). They stated that even mathematics is socially construed and therefore not free from human influence, for example, power, taste, and will. This thesis made possible a sociological and linguistic investigation of the practices of mathematicians (for example Ernest 1998, Heintz 2000). This research could not transform mathematics into a "soft science," not even from a theoretical point of view. But it opened the view to the basic characteristics of mathematics and how they are influenced by people and historic circumstances.

The essay "Proofs and Refutations" (Lakatos 1976) can be seen as a milestone. Lakatos constructed a fictive dialog of students with a teacher who moderates the conversation. The group talks about the Euler formula $E - K + F = 2$. They exchange ideas, claims, proofs of their statements, and counterexamples. The dialog is fictive, but it is not created from scratch. All contributions from the students are in fact historical statements from mathematicians who worked on the Euler formula. The statements are composed in the form of a conversation. The dialog shows that the invention of a mathematical proposition is not a linear process. It involves detours, errors, and controversies. The mathematical form and content of the proposition were developed under heavy influence of opinions, feelings, and taste of the involved persons. The main arguments are written in the natural language, not in the language of mathematics.

Analyzing mathematical discourse in industrial contexts, as proposed in this chapter, uses the idea of analyzing a mathematical discourse, but without creating it from historical sources. Similar conversations occur when mathematicians work in interdisciplinary teams with a common goal but team members who have different views on how to reach this goal. Analyzing discourse in industrial mathematics takes Lakatos' approach one step further since it is focused on the discussion in our time and analyzes the language use not within a discipline (as did Heintz 2000), but investigates the interface between mathematics and other disciplines. In doing so, assumptions become visible that underlie the common understanding of the application of mathematics.

2.2 Mathematics as Lingua Franca

One claim is that mathematical formulas are the lingua franca of science and technology—an opinion that is closely linked to the famous quote from Galileo saying that mathematics is the language of the book of nature. Two more sayings gained fame within the mathematical community. Eugene Wigner saw an "unreasonable effectiveness" of mathematics in science and technology (Wigner 1960), and the David Report concluded that high technology is always mathematical technology (David 1984).

Observing the daily work of interdisciplinary R&D teams lead to the impression that mathematics is indeed useful in industrial settings and can serve as lingua franca. But this is not the complete picture. When mathematicians talk at their work place, mathematical formulas are always embedded in natural language. Mathematicians use a mixture of formulas and words, which makes the natural language as crucial for industrial mathematics as formulas are. They use metaphors, examples, and stories to explain mathematical ideas to colleagues and to convince them that these ideas are the right ones. Natural language serves also as a means for searching the right abstraction of phenomena within the application domain. The goal of this mathematical discourse might be a mathematical formalism, but formulas are only reached through a discussion with extensive use of natural language.

2.3 Transfer of Mathematical Knowledge

The transfer of mathematical knowledge is often seen as a mechanical process, which covers the packaging of mathematical ideas and methods and their working into a technical product. This might include stimulation of mathematical research through the interdisciplinary work with engineers. Nevertheless, knowledge transfer is mainly seen in the opposite direction using preexisting mathematical knowledge in an application domain. A first view on mathematical discourse in an industrial setting shows that knowledge is not transferred in this sense. It rather changes while being applied, since it must be verbally constructed anew in discussions with engineers and managers of the application domain. Even more: Mathematicians must see to get their perspective and ideas applied. Engineers and managers are supplied with different knowledge and different views on the technical product in development. Technical products can be construed with less (or no) mathematics although more mathematics promises to make them better. To carry through the mathematical ideas is a central challenge for mathematicians working in R&D units. This challenge is taken up verbally in the interdisciplinary dialog. As fieldwork shows, this dialog includes rhetoric strategies for hiding mathematical content, showing its usefulness and proving its cost-efficiency. These characteristics of the mathematical discourse in R&D shed a light on the actual behavior of mathematicians, how they integrate themselves in interdisciplinary teams and which rules and strategies they use for positioning mathematics.

3 Methodology

Analyzing mathematical discourse requests methods that are stringent and provide general insights. They must go beyond examples that mathematicians tell from their individual experience and point of view. This is why we propose a participatory observation, including the work with documents that are created within development projects such as technical documentation and marketing collaterals. A linguist takes part in the project work as a "useful linguist"; aside from observations he prepares for example documentation and other writings and manages review cycles that allow mathematicians, engineers, and managers to articulate their views in written form (Schmidt 2009).

Fieldwork should include observation of the daily work of R&D groups in mathematical industries like finance or optimization in logistics and transportation, as well as the application of simulation and control theory across the industries. Collaboration is planned with university institutes as well as research centers designed for knowledge transfer and R&D departments of private companies.

3.1 Ground Work in Linguistics

Much work has been conducted in the area of analyzing public discourse on scientific results, showing that a funneling process takes place which shapes the presented knowledge (Liebert 2002) and which adds—by using natural language—specific views on this knowledge resulting in a semantic battle (Felder 2006).

Semantic battles usually take place in the public arena, when a group of individuals want to dominate the discourse on a topic, but others with an opposed view try the same. The semantic battle can concern topics that are per se controversial and belong to the sphere of politics, such as taxes or school education. Linguists have observed a specific language use in discourse about those topics. Each party tries to set their views dominant by using terms that support their views and criticize the view of others. Even if the used terms are neutral, they normally set one aspect dominant, which is used to influence the direction of the conversation. Research had been conducted for analyzing semantic battles in several domains, including public debates on biology, especially genetics (see Felder and Müller 2009).

When analyzing discourse in R&D units, the scope is of course different. Not public debates are analyzed, but discourse within an organization. One assumption is that the linguistic tools for observing public discourse can be applied to organizational communication. As already mentioned, this discourse contains also different views on a subject, and each team member in an interdisciplinary team brings in his specific knowledge, ideas, and goals which lead to semantic battles in a similar way.

When it comes to analyze documents with respect to their mathematical content, there is also linguistic work available. Text linguistics focuses on text structures, language use in texts, but there is also research conducted on mathematics and its popularization in different texts types (Schmidt 2003).

3.2 Ground Work in Sociology

The investigation of scientific knowledge and its creation has also a tradition in sociology. Groundwork for the proposed approach are studies that have challenged the opinion, natural sciences are sciences that are clean from human influence such as battles on power or pressure from outside the research teams (Knorr Cetina 1984). The proposed work applies studies that identify different scientific cultures across disciplines (Knorr Cetina 2002) and that define a research program for a sociology of knowledge (Keller 2005). It expands existing studies on (pure) mathematics which were conducted in this tradition (Heintz 2000).

The mentioned sociologists used the participatory observation to analyze the behavior of individuals and groups, organizational set-ups, and power structures. Knorr Cetina spent time at the CERN in Geneva regularly to talk to scientists, conduct interviews, watch them, and take notes. Heintz joint the Max Planck institute for mathematics in Bonn for several weeks and gained her insights also by watching and talking to the mathematicians there. In addition, both sociologists analyzed documents that had been written by the scientists and how they were reviewed.

The method of participatory observation is also at the heart of our approach. However, it will be adapted to the domain of industrial mathematics and to the purpose of analyzing innovations in this domain.

3.3 The Useful Linguist

The work on documents is an important part of the scientific work, since results must be published, and scientists must apply for grants. Knorr Cetina showed that all insights she got from participatory observation were mirrored in the joint work of the scientists on a scientific paper, including the text revisions, comments from reviewers and the kind of document cycling during the writing and review process.

In industrial research and development, documents have a similar importance as in natural sciences. Nevertheless, they are of another kind and variety. In software development, there are for example internal documents like specifications and design documents that are used to prepare the development of technical artifacts, such as algorithms or interfaces. In addition, there is project documentation including project charters, minutes of team meetings, and status reports. Other documents are prepared for the external audience. They include product documentation that is shipped with the technical product, for example installation guides or operating instructions. Companies prepare also marketing documents such as White Papers, Solution Briefs, or Leaflets about products.

Usually all those documents are written and reviewed by project members and other experts, normally leading to a number of revisions and several document versions. The revisions, especially in this variety of document types, make visible technical problems, discussions, and solution proposals as well as different views on how to position the later product in the market. That is why the work on texts serves as a tool for gaining a closer look on semantic battles that come with the application of mathematics.

The useful linguist joins research and development projects in order to draft and edit documents, and to organize the cycling of documents for reviews. He uses his role as technical writer to get to know the inner world of the project. As a project member he is at the core of the innovation and can observe how mathematics comes into play. He joins the project on a long-term basis so that he is able to dive deeply into the topics and to communicate at eyes level with the engineers and mathematicians. This helps to reveal what is happening in the project and to draw the right conclusions.

As a technical writer he does not belong to the inner group of colleagues in the project, since he is a co-worker with focus on language. Therefore, he has an internal, but distant view on the product development. He is not concerned with the product itself and also not with the mathematics in use. He focuses on the communication about the product, its features, and how the mathematicians were involved during development. As a linguist he can use the creation and review of document to control the document cycling and to enrich the participatory observation.

4 Lines of Investigation

Industrial mathematics is a diverse field. It takes place at university departments, mostly as project-based collaboration with companies. There are spin-offs that often productize one specific mathematical invention. Innovation in small and medium enterprises may come from local or regional collaboration with universities or public research institutes. Larger companies can afford an own research and development department, some companies even have units that focus on mathematical consulting. The different industries have their own culture and tradition, also from a mathematical point of view. For example, insurance companies build their business on statistics, others on operations research.

This diversity must be taken into account when conducting research about industrial mathematics. Since a full coverage of all possibilities of mathematical innovation is not possible, only exemplary studies are realistic. However, they need a central theme and guiding research questions.

The following questions may lead to a clearer picture of mathematical innovation in research and development:

- How do mathematicians argue for the use of mathematics? What barriers are conceived by the mathematicians that hinder mathematical innovation? How do they position their mathematical ideas in this context?
- How is mathematics sold? Do mathematicians use arguments from an economic point of view such as addressing costs and benefits of the use of mathematics? Which roles have patents?
- How do mathematicians find a mathematical model of the central objects of the application domain? What strategies are used for developing a common language? How are objects of the application domain changed or redefined to make them fit to the mathematical model? What issues influence the mathematical model? Are only aspects from the application domain relevant or also organizational issues like time constraints and the availability of budget?
- Which mathematical theories and tools are in use? Are they developed anew or reused, for example from a software library? Which level of proficiency do the project members have? Do the team members judge the level of sophistication of the used and proposed mathematical models?
- Which strategies are used to make the mathematical tool set visible or to hide the mathematical content? How is the mathematical content documented in the product documentation? What is explained and what is left out? Are mathematical artifacts visible on technical interfaces or user interfaces?
- Which role has proofs in industrial mathematics? Which standards from research mathematics are applied? Do mathematicians refer to truth, beauty, or similar concepts?
- Is there a mathematical way of thinking that goes behind the application of mathematical models and methods? How do mathematicians bring in their implicit knowledge and their experience with abstract mathematical structures?

5 Outcomes

When addressing the interface of mathematics and industry, the organizational development of industrial mathematics and education are without doubt the main issues.

The proposed analysis of mathematical discourse is meta-research that can support organizational concerns. It may contribute to both mentioned areas of activity and help to leverage the use of mathematics in industrial settings and to leverage communication skills in R&D teams. Linguistics and Sociology help to find best practices for the transfer of mathematical knowledge, which may lead to a better management of organizations for industrial mathematics and a better integration of mathematicians in R&D units.

Fieldwork may lead to the documentation of best practices; it can reveal success stories and can help to detail out shortcomings of today. In addition, it can be a means for specifying needs for mathematical research and education, addressing them to the mathematical research community. Outcome of the linguistic fieldwork can include the specification of technical tools, platforms for community building across mathematicians in academia and industry.

Last but not least, training can be developed that focuses on soft skills that are needed by mathematicians who work in R&D units. This can lead to a higher impact of mathematics in industry through people at their work place. Furthermore, industrial mathematics will be promoted as a whole, which helps to close gaps in the interface of mathematics and industry.

References

Courant, R., & Robbins, H. (1941). *What is Mathematics? An Elementary Approach to Ideas and Methods*. London: Oxford University Press.

David, E. E. (1984). *Renewing U. S. mathematics: critical resources for the future*. Washington.

Davis, P. J., & Hersh, R. (1981). *The Mathematical Experience*. Boston: Birkhäuser.

Ernest, P. (1998). *Social constructivism as a philosophy of mathematics*. New York: SUNY Press.

Felder, E. (2006). Semantische Kämpfe in Wissensdomänen. Eine Einführung in Benennungs-, Bedeutungs- und Sachverhaltsfixierungs-Konkurrenzen. In E. Felder (Ed.), *Semantische Kämpfe. Macht und Sprache in den Wissenschaften*. Berlin, New York: de Gruyter.

Felder, E. & Müller, M. (Ed.). (2009). *Wissen durch Sprache. Theorie, Praxis und Erkenntnisinteresse des Forschungsnetzwerks "Sprache und Wissen"*. Berlin, New York: de Gruyter.

Grötschel, M., Lucas, K., & Mehlmann, V. (Eds.). (2009). *Produktionsfaktor Mathematik. Wie Mathematik Technik und Wirtschaft bewegt*. Berlin, Heidelberg: Springer.

Heintz, B. (2000). *Die Innenwelt der Mathematik. Zur Kultur und Praxis einer beweisenden Disziplin*. Wien, New York: Springer.

Keller, R. (2005). *Wissenssoziologische Diskursanalyse. Grundlegung eines Forschungsprogramms*. Wiesbaden: VS-Verlag.

Knorr Cetina, K. (1984). *Die Fabrikation von Erkenntnis. Zur Anthropologie der Naturwissenschaft*. Frankfurt am Main: Suhrkamp.

Knorr Cetina, K. (2002). *Wissenskulturen. Ein Vergleich naturwissenschaftlicher Wissensformen*. Frankfurt am Main: Suhrkamp.

Lakatos, I. (1976). *Proofs and Refutations*. Cambridge: Cambridge University Press.

Liebert, W.-A. (2002). *Wissenstransformationen: Handlungssemantische Analysen von Wissenschafts- und Vermittlungstexten*. Berlin, New York: de Gruyter.

Schmidt, V. A. (2003). *Grade der Fachlichkeit in Textsorten zum Themenbereich Mathematik*. Berlin: Weidler.

Schmidt, V. A. (2009). *Vernunft und Nützlichkeit der Mathematik. Wissenskonstitution in der Industriemathematik als Gegenstand der angewandten Linguistik*. In E. Felder & M. Müller (Ed.).

Wigner, E. (1960). The unreasonable effectiveness of mathematics in the natural sciences. Communications on Pure and Applied Mathematics *13*, 1–14.

Using Popular Science in a Mathematical Modeling Course

Burt S. Tilley

1 Introduction

A student-driven approach to a mathematical modeling course is presented. Students choose their topic based on an article from a popular science medium, spend time researching the background information of the article, constructing a mathematical model, solving it, and either writing a report or giving a brief presentation before a critical audience. Through this experience, students learn the underlying techniques of formulating relevant mathematical models.

How do you know whether a story in the media is true? How do you evaluate quantitative evidence that is presented? How do you judge the conclusions drawn from this quantitative evidence? What skills does a person need in order to approach these tasks independently and confidently? The ability to mathematically model a problem, formulate a solution, and articulate this solution in the problem's original context effectively answers these questions.

The skill to translate a contextual problem into a quantitative framework is fundamental for citizens to perform their role in a technology-based democratic society. Quantitative courses in different disciplines focus on the mathematical methods that elucidate the underlying concept, such as in economics, engineering, or the sciences. However, the translation from context to mathematics has already taken place. Students see the mathematics involved as a tool for solving a problem in a particular topic and may not understand the connection of the mathematics to a significantly different context.

Ever since the first "study group" at Oxford University in 1968, applied mathematicians have been developing mathematical models to solve industrial and scientific problems. More recently, these groups have expanded to several countries, including the Mathematical Problems in Industry Workshops (MPI) in the US and the PIMS Industrial Workshop in Canada. The Consortium for

B. S. Tilley (✉)
Worcester Polytechnic Institute, 100 Institute Road, Worcester, MA 01609, USA
e-mail: tilley@wpi.edu

A. Damlamian et al. (eds.), *Educational Interfaces between Mathematics and Industry*, New ICMI Study Series 16, DOI: 10.1007/978-3-319-02270-3_35,

Mathematics and its Applications (COMAP) has been running the undergraduate Mathematical Contests in Modeling (MCM) for nearly 20 years. The common format for all of these programs is that a new, original problem is presented in a scientific or engineering context, and the contestants have a finite time (within five days) to formulate a mathematical model, solve it, and present the work in the original context of the problem. This format is an engaging (and sometimes enraging) experience to any who have participated. If communicated to undergraduate students in a cogent fashion, such an open-ended experience describes the discipline of applied mathematics well.

Question: *Can a mathematical modeling course be run in this format, and if so, how does the student benefit?*

Integrative experiences as a pedagogical technique have been of interest in engineering education [see Somerville et al. (2005); Zastavker et al. (2006)]. Our Practicum to Industrial Mathematics course begins with students opening that day's the Science Times section of the New York Times on a Tuesday morning and read over the articles. From this selection of articles, groups of students discuss the different aspects of the topic that are interesting to them. By the end of the same class, and with the instructor's input, the student team formulates a mathematical question that, if solved, would provide some insight into the topic itself or some related problem. This course has been implemented at Olin College, Wellesley College, and WPI.

This process has three benefits for the students. The first is that the question is necessarily current: the matter is a current topic in the popular media. The second is that they provide their own motivation for the problem and its solution, and they are responsible for its success. This is their problem, and not one posed by the instructor. The third is that the students understand how to formulate a tractable question based on a practical set of criteria.

The class time between the assignment of the problem and the report due date is spent with teams working on their projects. There are no formal lectures. Each team is guaranteed to have a specific amount of time with the instructor without interruption, typically around 15–20 mins. The students are advised to come prepared with a small list of the questions, prioritized from most to least important in their view.

Our example below summarizes a model of a technique in pain management called patient-controlled analgesia (PCA). The discussion of the final model as a simple three-dimensional system of first-order nonlinear equations may be found in Tilley (2010). However, the choice of the nonlinearity leads to a stable fixed point that is not present in the linear model (all previous modeling attempts from a subsequent literature search at representing pain in this situation appear to be linear). The qualitative results from this model agree with what can be found from clinical studies of this pain management technique. The fact that the student has multiple opportunities to discuss the impact of modeling assumptions in their quantitative description is a significant benefit to this approach.

2 Example: Patient-Controlled Analgesia

There are two principal methods of administering intravenous pain medication. The first involves the administration of the drug according to a regular medication schedule called continuous analgesia (CA). The second gives control to the patient through a process called Patient-Controlled Analgesia, or PCA. The advantages of CA include the ability of the caregiver to limit the concentration of the pain medication given to the patient in addition to better pain management overall (Sucato et al. 2005). However, when the patient is susceptible to side effects, the response time by staff predicts the renewal of the drug supply. With PCA, the patient is allowed to self-administer a dose by simply pushing a button whenever the pain is sufficiently great. The overall dose is limited by the total amount of drug that can be administered over a typical period: if this dose is administered before the period is complete, the button ceases to function. Since the dose is limited, the degree of pain management is suboptimal by design, but complications due to side effects are greatly reduced.

PCA is also used as a pain management treatment for chronic conditions with the use of opiates (such as morphine) as the medication. These conditions are notable since the physical source of the pain may either not have a definite cause or there is no other treatment option to eliminate the pain source. Further, morphine effectiveness varies greatly from patient to patient, which makes it difficult to prescribe a regular dosage schedule that will treat a particular patient's pain without causing undesirable side effects (Hahn et al. (2003)).

Pain measurement is still primarily performed by asking the patient to rate their pain level on a scale from 1 (no pain) to 10 (as much pain imaginable) with no agreement on a uniform metric. One scale that is used for children is the Wong-Baker faces scale with a 0–5 scale, but a 0–10 scale is also used, such as in the visual analog scale (VAS) (see the MD Anderson Cancer Center website http://www.mdanderson.org/pdf/ptedpainscalefaces.pdf, for more information). These metrics are by definition qualitative: they are based on the patient's perception of the pain level they are experiencing, and not a measure of the strength of the pain source (which is called discomfort in the literature). Of interest in models of PCA are the choice of a pain measure 'pang,' defined as the amount of pain a PCA patient experiences such that they would push the button, when the medication is locked, once per second (Jacobs et al. 1985). Again, this unit relies on the patient's interpretation of the pain being experienced, which varies from patient to patient. If professionals must design a medication regimen keyed to patient feedback, perhaps eliminating the one degree of separation and giving control of the medication to the patient could reduce the complexity of the system and improve the overall treatment of the patient's pain.

The modeling of PCA has centered on using linear control systems with a stochastic pain source (Jacobs et al. 1985, 1986, 1995; Liu et al. 1990; Chase et al. 2004; Shieh et al. 2007). Central to these studies is the connection of how this source of pain is perceived by the patient. Liu and Northrop (1990) make the following distinctions of pain. Discomfort arises from the actual source of pain, which is transmitted from the pain location along nerves to the brain. This level of discomfort is called *neurological pain, Q*. However, the medication acts to prevent this pain signal from being perceived by the patient, and the patient's pain response is a measure of *perceived pain, P*. They modeled the perceived pain as a linear combination of the neurological pain and its rate of increase over time. In our study, we wanted to understand the implications of these two modeling assumptions: the metric used in measuring pain levels and how pain is mitigated through PCA. One case of interest is that of chronic pain, during which the patient reports perceiving pain but no source of discomfort can be identified.

In the modeling, we make the assumptions that the medication has the same effect on pain regardless of food intake, sleep, time of day, or mood. Further, we assume the patient will initiate PCA at exactly the same pain level every time. The time for the medication to take effect once it is administered is assumed to be instantaneous, and the experiencing of intense pain over time will increase a patient's sensitivity to pain for some period of time thereafter.

3 Pain Models

We model pain as a system of differential equations. The principal quantities are the amount of *perceived pain p* and the amount of *neurological pain q*. We assume that the perceived pain is a linear combination of the neurological pain and its rate of change involving phenomenological constants. In Liu and Northrop (1990), the corresponding differential equation is forced by the *level of discomfort D*. If $D = 0$, the case corresponding how a patient, and if the rate of change of perceived pain decays based on healing and a second term from the sensitivity to neurological pain, by introducing dimensionless variables *p, q*, and time *t,* it results in the differential equations

$$\frac{\mathrm{d}p}{\mathrm{d}t} = -\mathrm{a}p + \mathrm{b}q, \ldots \frac{\mathrm{d}q}{\mathrm{d}t} = p - q,$$

where a and b are constants. Note that $\mathrm{b} > 0$ implies that the patient is more sensitive to pain if pain has been experienced in the past, while $\mathrm{b} < 0$ implies that the patient develops a tolerance to the previous pain levels.

Note that both of these equations are effectively compartmental models of the underlying biological phenomena which sense pain and transmit the information to the patient. Mathematically, they are appropriate provided that the level of pain is not considered "too large," or deviations of pain around some known constant level remains small. However, no deterministic representation of these processes exists at this time. There is no measure of what level of pain is "too large," nor an understanding of when these models break down.

A linear stability analysis of this system yields growth rates about the steady state $p = q = 0$, which is consistent with the situation in which a patient would become more sensitive to pain the more pain that is experienced results in a positive feedback loop. One model of chronic pain one can introduce a source term as a measure of discomfort and find a local stable solution. However, this model assumes that a physical source of pain is present.

However, the measurement of pain on a linear scale suggests that the patient senses relative changes in pain on a linear scale. However, scales that relate to human sensation such as touch (Richter), sound (decibel), sight (brightness), and smell (European Odor Unit) are proportional to the logarithm of a known quantity. All of these units are logarithmically based on the amplitude of the original stimulus. This suggests that relative changes in pain can be modeled with a logarithmic rule and the system then becomes

$$\frac{\mathrm{d}p}{\mathrm{d}t} = -p\log(p) + \mathrm{b}q, \ldots \frac{\mathrm{d}q}{\mathrm{d}t} = p - q.$$

However with pain measurement, we note that there is a filter between the source of pain and what the patient perceives. Note that this representation is a different compartmental model for the interaction between neurological and perceived pain. It is not the unique representation of how these two quantities are related, but one in which we feel incorporates how human perception is related to changes in external stimuli. The results could be simulated with different nonlinear models, such as a quadratic function of p in place of the log p term.

To model the management of chronic (perceived) pain, we look at the diffusion of a medication *concentration* c within the bloodstream, which is introduced into the patient through the nonlinear source h. In dimensionless form the system becomes

$$\frac{\mathrm{d}p}{\mathrm{d}t} = -\mathrm{a}p\ \log p + \mathrm{b}q - \mathrm{ec}, \ldots \frac{\mathrm{d}q}{\mathrm{d}t} = p - q, \ldots \frac{\mathrm{d}c}{\mathrm{d}t} = -\mathrm{fc} + h(t, p),$$

where the normalized *pain threshold* $p*$ at which the patient will initiate PCA yields the jump in the Heaviside function: $h = 0$, if $0 < t < t*$, $p < p*$, and $h = 1$, if $t > t*$, $p = p*$.

This model allows a mathematical analysis with interesting conclusions. For instance, for certain parameters the patient never receives the equivalent dose for a continuous analgesic treatment, and has heightened discomfort. However, the patient will be less likely to experience side effects from the drug compared to a continuous treatment. This result is found in clinical findings comparing these two pain management schemes for postoperative pain (Sucato et al. 2005, Karci et al. (2004)).

We simulate the model using Matlab's standard numerical integration package ode45. We may plot several examples of the simulation showing the concentration level over time of the simulation, where the transient approaches a periodic-solution where the pain is reduced significantly during the drug infusion and then gradually increases along the nullcline as the concentration of the drug decays in the bloodstream. In some case, the final periodic state has not been achieved.

In summary, we have developed a nonlinear model for patient-controlled analgesia. This model incorporates a pain model where the perceived pain depends on the impact of neurological pain linearly, but its rate of change depends on the perceived pain nonlinearly. This pain model can then exhibit equilibrium pain solutions that may correspond to chronic conditions where no physical cause of the pain is found. The analgesic scheme is found to be suboptimal in that the desired concentration for treatment is never achieved. This agrees qualitatively with clinical results for postoperative pain management as well as other models of PCA. Extensions to this model would include allowing the healing coefficient a to vary in time along with a comparison of the pain response to patient movement or stimuli.

Acknowledgments The author would like to acknowledge support from the National Science Foundation, Grant DUE-0231231.

References

Chase, J.G., Rudge, A.D., G.M. Shaw, Wake, G.C., Lee, D., Hudson, I.L., and Johnston, L. (2004). Modeling and control of the agitation-sedation cycle for critical care patients, *Med Eng Phys*, *26*, 459.

Hahn, T. W., Mogensen, T., Lund, C., Jacobsen, L. S., Hjortsoe, N.-C., Rasmussen, S. N., et al. (2003). Analgesic effect on I.V. paracetamol: Possible ceiling effect of paracetamol in postoperative pain. *Acta Anaesthesiology Scandinavian, 47*, 138.

Jacobs, O. L. R., & Lammer, P. (1986). Modelling post-operative pain and its relief in demand analgesia. *Biomedical Measurement of Information Control, 1*(1), 41.

Jacobs, O. L. R., Bullingham, R. E. S., Lammer, P., McQuay, H. J., O'Sullivan, G. O., & Reasbeck, M. P. (1985). Modelling estimation and control in the relief of post-operative pain. *Automatica, 21*(4), 349.

Jacobs, O. L. R., Liu, Y. P., & McQuay, H. J. (1995). Modeling and estimation for patient control analgesia of chronic pain. *IEEE Transactions on Biomedical Engineering, 42*(5), 477.

Karci, A., Tasdogen, A., Erkin, Y., Aktas, G., & Elar, Z. (2004). The analgesic effect of morphine on postoperative pain in diabetic patients. *Acta Anaesthesiology Scandinavian, 48*, 619.

Liu F.-Y, Northrop, R.B (1990). A New Approach to the Modeling and Control of Postoperative Pain. *IEEE Transaction Biomedical Engineering*, *37*, 1147.

Shieh, J.-S., Chang, L.-W., Wang, M.-S., Wang, Y.-P., Yang, Y.-P., & Sun, W.-Z. (2002). Pain model and fuzzy logic patient controlled analgesia in shock-wave lithotripsy. *Medical and Biological Engineering and Computer, 40*, 182.

Shieh, J.S., Dai, C.Y., Wen, Y.R., Sun, Y.Z., (2007). A novel fuzzy pain demand index derived from patient-controlled analgesia for postoperative pain. *IEEE Transaction Biomedical Engineering*, *54*, 2123.

Somerville, M., Anderson, D., Berbeco, H., Bourne, J. R., Crisman, J., Dabby, D., et al. (2005). The Olin curriculum: Thinking toward the future. *IEEE Transactions on Education, 48*(1), 198.

Sucato, D. J., Duey-Holz, A., Elerson, E., & Safavi, F. (2005). Postoperative analgesia following surgical correction for adolescent idiopathic scoliosis: A comparison of continuous epidural analgesia and patient-controlled analgesia. *Spine*, *30*(2).

Tilley, M. (2010) Using Popular Science in Mathematical Modeling Course. In *Proceedings of the EIMI-2010 Conference* (pp. 501–511). Lisbon: Centro Internacional de Matemática.

Zastavker, Y.V., Crisman, J., Jeunnette, M., & Tilley, B. S. (2006). Kinetic Sculptures: A centerpiece project integrated with mathematics and physics. *International Journal of Engineering Education*, *22*(5).

Part VII
WG Technology Issues

Technology Issues

Helmer Aslaksen and Fadil Santosa

1 Introduction

This article discusses two separate but connected issues (1) Technology in the workplace, (2) The use of technology in education. The unifying thread is modelling and simulation. Mathematics is at the heart of the technologies used in industry. It is essential that we articulate its important role, and how to position technology in the context of mathematical education.

Mathematics has been a crucial tool for understanding complex phenomena and for predicting their behaviour. Models of physical systems are often best described with variables which interact through equations. Activities which abstract physical, biological, and other phenomena into mathematical descriptions are referred to in this report as *Mathematical Modelling*. Through mathematical modelling, complex events can be understood, simulated, and their outcomes, predicted.

One of the earliest successes of mathematical modelling is the calculation of the trajectory of projectiles. Through the use of Newton's law and calculus, trajectories of cannon balls can be predicted with great accuracy. More recently, mathematical modelling has been used to simulate car crashes, and has helped engineers design safer automobiles.

Co-Chairs: Helmer Aslaksen [Norway] & Fadil Santosa [USA].
Working Group Members: France Caron [Canada], Jaime Carvalho e Silva [Portugal], André Garon [Canada], Djordje Kadijevich [Serbia], Rui Gomes Neves [Portugal], Youyu Phillips [USA].

H. Aslaksen (✉)
Department of Teacher Education/Department of Mathematics, University of Oslo, Oslo, Norway
e-mail: helmer.aslaksen@gmail.com

F. Santosa
Institute for Mathematics and its Applications, University of Minnesota,
207 Church Street SE, 408 Lind Hall, Minneapolis, MN 55455, USA
e-mail: santosa@ima.umn.edu

A. Damlamian et al. (eds.), *Educational Interfaces between Mathematics and Industry*, New ICMI Study Series 16, DOI: 10.1007/978-3-319-02270-3_36,

The key enabler in mathematical modelling has been the advent of computing technology. In the past, mathematical methods were developed to compensate for the lack of technology. The concepts of approximation and linearization in calculus made it possible to solve problems that could not be computed directly. However, as in the beginning computing machines and then computers developed, it became possible to solve problems that earlier were impossible to solve, or could only be estimated. For instance, it became possible to address many non-linear problems directly instead of making linear approximations.

Today, the benefit of computing technology has far reaching effects. New mathematical ideas are developed thanks to technology. Indeed new mathematical fields, such as mathematical finance and mathematical biology, are examples of the fruitful combination of computing power and mathematical ideas.

Technology has also provided teachers with tools that help students learn. Technology makes it possible to effectively teach topics that previously were considered out of reach for students. In the past, teachers would have to carefully design the problems so that they became doable and the answers were "nice". Now it is possible to give the students realistic problems. At the same time, topics that in the past were core topics in the syllabus are now not as essential since they can easily be solved by computers.

The goal of this paper is to provide some recommendations based on the findings of the Working Group 2 at the EIMI conference in Lisbon in October 2010. The group was assigned to investigate the issue of technology, both in the modern workplace and in the classroom. There were 2 presentations:

- Tackling the challenges of computational mathematics education of engineers, by France Caron and Andre Garon (reprinted in this book).
- Computational modelling in science, technology, engineering and mathematics education, Rui Gomes Neves, Jorge Carvalho Silva, and Vitor Duarte Teodoro (reprinted in this book).

Both presentations provoked lively discussions among the participants.

In her presentation, Caron made an argument for encouraging students to open up black boxes that they might use in industrial settings. They cite several reasons for this opening of black boxes. The most important among them is that users of black boxes need to know the range of use and limitation of such tools. Given the importance of quantitative analysis on decision making, having confidence in numerical answers produced by such tools is essential. One of the important aspects discussed was legal issues. Garon made a point that engineers, i.e., users of black boxes, often need to certify the claims made by a company. The implications are that the stakes are high and that critical thinking should be encouraged at all levels. There was a chorus of agreement with the ideas presented in their paper, which we reproduce here for the readers.

The second talk was given by Neves who spoke about the centrality of modelling in Science, Technology, Engineering, and Mathematics (STEM) education. He spoke about a software package Modellus, to which he has contributed. The package is designed to engage the students in activities that facilitate learning and reinforce concepts being taught. From the demonstration, it is clear that Modellus does more—it also makes learning fun.

Some of the issues discussed were how such modelling software could be incorporated into the curriculum. Several participants, while acknowledging the value of Modellus, expressed concerned that the syllabus for STEM classes are already quite crowded. What then, needs to be given up in order to have time for the students to interact with packages like Modellus. The paper by Neves and his co-authors are also reproduced in this book.

After the conference, we also received a paper by Qixiao Ye from the School of Mathematics at the Beijing Institute of Technology in China (reprinted in this book). Ye presented a talk at a follow-up meeting to the EIMI Conference in Macau which took place November 3–4, 2011. The paper is entitled "Incorporating the Ideas and Methods of Mathematical Modelling into Calculus Teaching". It presents an initiative to introduce realistic and non-trivial mathematical modelling in calculus classes through modules. The author describes the experience with these modules at several universities and reports on the outcome. Early feedback from the experience is encouraging. Students have become more aware of the usefulness of mathematics and appreciate learning about modelling through the modules.

2 Technology in the Workplace

Although the Discussion Document uses the term 'technology' in its broadest sense, from traditional machinery to modern information technology, from operational research to system simulation, our discussion will focus on computer technology. The task of this group is to look into the issues brought forth by technology and its impact on mathematical education.

Computer technology is pervasive in all aspects of our daily lives. It has become one of the most important tools available to engineers, scientists, and decision makers in industry. Much of computer technology used in industry comes in the form of black boxes. They are often very specialized tools used to solve specific narrow classes of problems. Hidden behind these powerful tools is mathematics. Mathematics comes into these tools in various forms:

- Modelling—first-principle modelling and empirical modelling, modelling from data,
- Numerical computations, including statistical analysis,
- Visualization.

Technology is a means of packaging mathematics with other conceptual and material tools into automatic solutions of problems. As a consequence users are often unaware of the mathematics that is involved.

3 Technology in the Classroom

Technology is playing an important role in mathematical education. Software such as MATLAB, Mathematica, Maple, have been integrated into the curriculum in many university level mathematics classes in many countries. These tools are essential in many ways. They help students learn difficult concepts. They can be used as tools for problem solving. As students trained in this way enter the workforce, these tools have become industry standard.

Aside from mathematical software, software for simulation—such as finite element modelling systems—are introduced into the classroom as tools for analysis of engineering problems. Other tools such as for discrete simulation are used to solve operational research problems. Still there are tools for financial modelling, for ab initio chemistry and the like. They are incorporated at various levels, typically in college classes. In the bellies of these tools lie mathematics, which is a key enabler that is often hidden from the students.

4 Mathematics in Technology

It has become clear to us that the role of mathematics in technology must be articulated to the users of technology. Mathematics, together with other sciences, provides the basis to design and operate hardware and systems that are critical to the health of the World's economy. Without mathematics, communication networks, GPS, electronic trading, and e-commerce will not be possible. Computer models and simulation tools used in engineering design and in making decisions are based on mathematics. They allow users to understand and predict complex phenomena.

On the other hand, we are cognizant of the fact that 'hidden mathematics' is unavoidable. For tools to be used, they must be self-apparent and simple. Much of the mathematics behind these black boxes is highly complicated. But some understanding of the mathematics behind these tools is important as they allow the user to understand the inner workings of a technology. Such understanding better informs the user of limitation of the technology, and provides the platform to develop new capabilities for the technology.

There are reasons why technology is often packaged in the form of black boxes. For one thing, they facilitate knowledge transfer—it is easier to show how a tool is used than to explain the inner workings of a tool. This is particularly true in the case where the black box involves sophisticated mathematics. Also, hiding the technology behind a black box protects intellectual property from being stolen.

At the same time, we are aware that the 'hidden mathematics' in these black boxes can have serious consequences. Like every tool, the result of using a tool depends on the knowledge the user has of the tool. Lack of knowledge often leads to results that cascades into poor decisions. When this happens in engineering, the result could be catastrophic, such as unsafe bridges. For the industry, poor decisions could lead to lost productivity and revenue. The secondary effect, which is equally serious, is that poor decisions from abuse of technology may lead to irrational distrust of science in the public.

5 What Should a Student Learn?

Our vision is that as educators, we should provide students with the basic mathematical knowledge to be informed, sophisticated, and enlightened users of technology used in industry. We should train our students to have critical minds so that they can make autonomous decisions based on information gleaned from the use of technology. They should have the mindset to be interested in the inner workings of a technology that they use, and thus, understand its limitations. Furthermore, they should be willing to find solutions beyond a tool's limitation. An enlightened user of a technology is one who is adaptable to seeking other possible solutions.

Critical to the training of students who end up using technology in their work is teaching them how to model. At the heart of all technology used in industry is mathematical modelling. Modelling, together with computational power, allows students to understand phenomena, to develop intuition about complex phenomena, and to predict their behaviour. The importance of modelling and simulation is that with this tool, one can control and optimize very complicated systems.

There are several competencies which we think are essential:

- Numerical computations and symbolic calculations—Numerical computations are of course at the core of much of mathematics, mathematics education and industrial mathematics. The increase in computing power allows us to solve problems that were previously impossible. Similarly, symbolic calculations have become an important tool for all mathematicians, and have reduced the time spent on topics like integration techniques in calculus classes.
- Visualization—Graphical visualization of data generated by simulation or data gathered through measurements is an important aspect of decision making. Visualization is becoming ever more important.
- Statistics—The ability to use statistical method to make decision is important in industry. Its value is greatest when there are uncertainties and risks involved in the decision.
- Programming—This is an important skill for the students to have. Programming also teaches students to think and organize in logical manner.

In teaching mathematics to the users of technology it is important to teach them the mathematical concepts at the root of algorithms. As educators we must provide insights into the scope and limitations of mathematical methods and their implementation. We also feel strongly that students should be trained to develop intuition of the phenomena that they are modelling with the tools.

When numerical methods are heavily used in a simulation, the user must perform validation and verification steps to ensure that the information created by the simulation can form a reliable basis for decision making. Validation aims to ensure that the equations are solved correctly. Validation tests the extent to which the model accurately represents reality. Thus an introduction to validation and verification during a student's education is recommended.

We encourage the use of appropriate technology in the classroom. Some mathematical concepts are better learned when students are provided with hands on, interactive tools. These tools should provide intuition and insight but they should not be a substitute for fundamentals. They can also provide the motivation for learning. These technologies should enhance but not distract.

Tackling the Challenges of Computational Mathematics Education of Engineers

France Caron and André Garon

We live in a world of black boxes, all of us do, however well we may be educated.

Isaac Asimov, 1967.

1 Introduction

With the presence of powerful simulation tools in today's engineering practice, the challenge in designing an introductory course in computational mathematics is twofold: students must be convinced of the necessity of opening some of the black boxes, and the mathematics education they receive must prove useful in controlling the solving process and use of the tools. The paper provides guidelines and suggestions to help meet these challenges.

As part of an effort to reduce the perceived gap[1] between engineering education and real-world demands on engineers, a growing number of engineering schools are integrating an industrial practicum into their curriculum. This certainly provides an early and valuable encounter with the physical and organisational complexity of the problems engineers are now required to solve, along with the elements of the working environment (resources, teams, rules and regulations) which now shape the engineering practice, but it also tends to create the perception that there now exists a wide array of software tools to do the job, that one of the

[1] This perception has led to the CDIO Initiative, an educational framework developed by engineering schools around the world with input from academics, industry, engineers and students. It aims at providing students with an education stressing engineering fundamentals set in the context of conceiving—designing—implementing—operating real-world systems and products. http://www.cdio.org

F. Caron (✉)
Département de didactique, Université de Montréal, C.P. 6128, Succ. Centre-ville, Montréal, QC H3C 3J7, Canada
e-mail: france.caron@umontreal.ca

A. Garon
Département de génie mécanique, École Polytechnique de Montréal, Montréal, Canada
e-mail: andre.garon@polymtl.ca

A. Damlamian et al. (eds.), *Educational Interfaces between Mathematics and Industry*, New ICMI Study Series 16, DOI: 10.1007/978-3-319-02270-3_37,

main objectives of engineering education should be to master the interface to these tools and that there is little need to know what lies within.[2]

In fact, one may say that engineers live in an increasingly vast and thick world of black boxes: their work relies more and more on simulation software tools which are based on sophisticated models and state-of-the-art numerical algorithms that only experts on these topics fully understand. These tools (e.g. FLUENT, ANSYS, COMSOL, MODFLOW) can typically solve most of the problems engineers will submit to them, but there always comes a time when a problem no longer fits the necessary conditions for the software to provide the solution or requires a careful selection and customisation of the "advanced features" which may look enigmatic to the neophyte.

In large corporations, this may not represent an issue, as there is often a division of modelling and simulation experts who can act as consultants for their generalist colleagues. But in smaller engineering firms, such consulting services may only be found outside the company, at a prohibitive cost—unless a kind university professor is willing to help. From our own experience and perspective, the need for consulting experts in modelling and simulation only seems to be growing, just as the need for statistical consulting services rose significantly when statistical software tools became more accessible and widely used by researchers from all fields (Hodgson 1987).

In this context, the challenge in designing an introductory course in computational mathematics is twofold: students must be convinced of the necessity of opening some of the black boxes, and the mathematics education they receive must prove useful in controlling the solution process and the use of the tools. The following provides suggestions to help meet these challenges. Although these suggestions stem from our computational fluid dynamics background, we believe they are applicable to any field of engineering where computational mathematics is used.

2 Addressing Motivation

2.1 Getting from A to B

When talking about software tools or technology devices, people often compare them to cars, as "just a means for getting from A to B", and this comparison seems to relieve from the obligation of learning about their inner workings. Although one could argue that there is a minimum of mechanics and technology that still needs to be learned for driving a car safely and efficiently, what makes this statement particularly inappropriate for describing the use of specialised software tools for simulations is the difficulty of deciding whether "B" has been reached. After all,

[2] These perceptions can be inferred from students' internship reports.

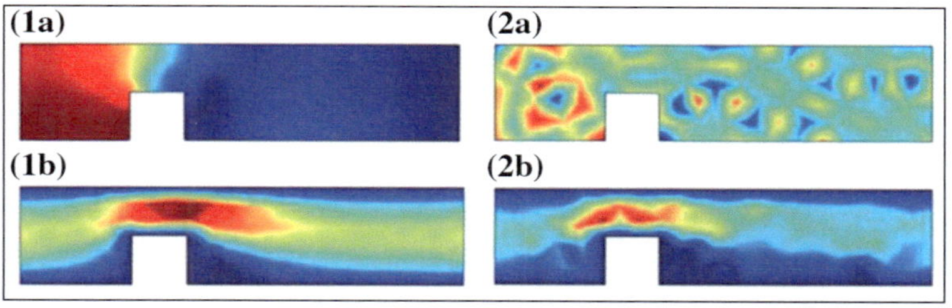

Fig. 1 Distribution of pressure (**a**) and norm of the velocity field (**b**) for two different simulations (1 and 2) of the same situation

this is why we do simulations: to get a better appreciation of where we might end up going. And even when we have the knowledge to determine that an improper "B" has been reached, what may still remain unclear is why we got there in the first place and what we could do differently to get to a more "realistic" destination.

Saying that by no means guarantees buy-in from students. But a couple of pictures may help—especially if they were obtained using commercial software. The graphs above (Fig. 1) represent the flow within a channel with a square obstruction; they were produced with the COMSOL software for the same boundary conditions and the same physical parameters. Which view (1 or 2) is right? Can you explain the difference?

Actually, view 2 was obtained simply by deactivating the default numerical stabilisation method. The choice of the stabilisation method depends on the partial differential equations, their type and order, the interpolation of the variables considered and the physics of the problem (which can be reduced to the Reynolds number in this example). Stabilisation is not always required, and users may not be always aware of the options or default values used by the program to favour robustness over accuracy. By smoothing the solution, stabilisation favours convergence, but this is sometimes done at the cost of hiding the symptoms of an inadequate modelling decision, the choice of a linear interpolation scheme in this case. Such an example provides a clear illustration of the need to get a little deeper in order to assess the validity of a solution.

2.2 *Social Accountability*

News headlines may also provide additional incentive for learning about the inner working of simulation tools and their sensitivity to modelling decisions. The following case could announce the emergence of a new trend in public movement, especially when environment, health and safety are concerned.

In Canada, the choice of a particular landfill site in the Simcoe County (Ontario) was the object of debate for some time and received wide media coverage on a few occasions. The county had been promoting and developing a new landfill site ("Site 41"), with the approval of the Ministry of the Environment and despite the strong opposition from residents in the surrounding communities, concerned with the risk of contaminating the groundwater beneath the site. What is rather exceptional here, but could become more typical in the years to come, is the fact that a group of citizens, through one of its members, asked access to the hydrogeological model and input data which had been prepared and used by the county's external engineering consultant for simulating groundwater flow in the proposed site. The consultant had performed these simulations with open source software (MODFLOW[3]), documented the results into a report submitted to the county, and this report contributed significantly to obtaining approval for the landfill development project. An order, issued in June 2009, from the Information and Privacy Commissioner to the county to provide the requested access gave no results, as the engineering firm refused to provide the county with the model and data, and the county felt in no position to claim them for the following reasons: this information was not part of the agreed upon "deliverables" with the firm, such a request would fall outside the "custom in the trade", and the county did not see any basis upon which it could take legal action against the engineering firm. Although an appeal from the commissioner and the active resistance from citizens finally led the county councillors to vote in September 2009 for a permanent moratorium on Site 41, some elements of the story continue to raise interesting questions.

First, there is the question of ownership of the model and data, where the development perspective and commercial interest may come at odds with the public's right to access to information. On this question, we may witness in the near future some evolution of the legal framework and implications, especially when models are used to help make decisions with potentially significant social impact; it would only make sense from both a scientific perspective and an ethical perspective. But irrespective of whether or not direct access is given to the model, the fact that the software used is publicly or commercially available makes it possible for a third party to come up with different simulation results, based on a different model or simulation parameters. As a consequence, engineering firms that deal with sensitive matters such as the environment, where public awareness and concern are only growing, can now expect a greater pressure to document and justify their assumptions, modelling decisions and solving parameterisation. Relying too much on black boxes and standard procedures could only expose them to the risk of revealing themselves as the blind trying to lead the blind. Without great surprise, this would go against some of the principles of ethics and conduct established by the Canadian Engineering Qualifications Board (2001) with respect to the environment:

---

[3] MODFLOW is open source software that was developed by the US Geological Survey (USGS), which is a science agency within the US Department of the Interior.

> Engineers must recognise and respect the boundaries of their competence and must only undertake the environmental evaluation of those aspects within this competence.

The question now becomes: how can computational mathematics education significantly contribute to developing a competence to undertake sound evaluation of the performance and impact of engineering development? How should we define competence in computational mathematics for engineers?

3 Developing Competence in Computational Mathematics

Computational mathematics typically encompasses both the solving of mathematical problems by computer simulation and the numerical methods which can be used in these simulations. Yet, as we will see in the following, computational mathematics in the engineering practice can hardly make abstraction of the real-world context from which the problem emerged, as it can have a direct incidence on the method to be selected and used.

In an introductory course in computational mathematics, the numerical methods that are taught typically deal with relatively simple mathematical objects and algorithms. Rather than focusing on the mathematical content, we will direct our attention to some of the general ideas and attitudes which we believe should be developed even when solving problems with elementary numerical methods, so as to favour adequate transfer to the use of the more sophisticated numerical methods that are used in solving today's complex engineering problems.

3.1 The Unavoidable Physics of the Situation

To illustrate the necessity of taking into account the real-world context in computational mathematics, we select as example the classical Poisson equation:

$$d^2T/dx^2 = 1.$$

This equation is often used in introductory courses to illustrate the process of discretisation through which a continuous differential equation is transformed into a system of algebraic equations.

Faced with such an equation, the students who have learned to discretise first-, second- and third-order differential operators using forward, backward or central finite difference schemes systematically reproduce this process to any given differential equation, without questioning the assumption of continuity. To avoid such pitfall, our didactical orientation makes us look for a situation where the need for revisiting this assumption will naturally emerge.

The Poisson equation can be seen as a one-dimensional simplified version of the Darcy equation used in the MODFLOW software. A more general version of the Poisson equation could read

$$\mathrm{d}/\mathrm{d}x\,(k\;\mathrm{d}T\,/\mathrm{d}x) = q.$$

In the context of groundwater flow modelling, T would be the pressure head, k would be the hydrogeological diffusivity which can vary in space, and the right-hand side would represent either a source or a sink within the hydrogeological system being studied.

To solve this equation, the natural tendency of students is to rewrite it so as to show explicitly the differential operators on the primary variable T and proceed from there to the application of the familiar finite difference schemes. Their initial transformation thus takes the following form:

$$(\mathrm{d}k/\mathrm{d}x)\,(\mathrm{d}T/\mathrm{d}x) + k\;\mathrm{d}^2T/\;\mathrm{d}x^2 = q.$$

Such transformation is valid only if k is a "sufficiently continuous" function since this indirectly determines the continuity of the primary variable and its derivatives, but this condition is often overlooked by students. Interestingly enough, a typical hydrogeological structure falls outside the domain of validity for this condition, as it is composed of different layers of soil, where k is discontinuous from one layer to the next. Trying to solve the discretised version of the transformed equation on such problem, a student will be faced with a solution that strongly oscillates near the interface between two layers of soil. As the student tries, unsuccessfully, to overcome the problem by refining the discretisation (Fig. 2), he must figure out that the problem must lie with the mathematical model used at the interface between two layers of soil, where the conservation of flux has been lost.

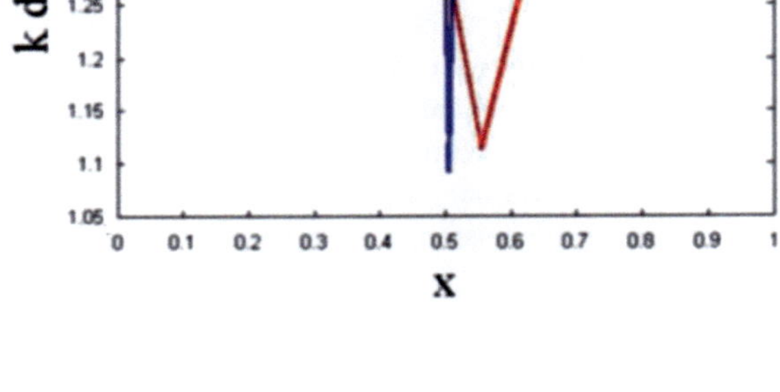

Fig. 2 Effect of grid refinement on the simulated flux

The problem is that a typical numerical introductory course focuses on the differential operators and rather tends to overlook the underlying physical principles, but principles such as conservation laws can and should guide the selection of variables for which the necessary continuity conditions are met, thereby enabling the application of finite difference formula. In this example, the flux $k\ \mathrm{d}T/\mathrm{d}x$ is continuous over the computational domain, whereas $\mathrm{d}T/\mathrm{d}x$ is not.

In designing the numerical model, we must be coherent with the physical problem. More important than the perceived accuracy of the numerical model is its capacity to reflect in its transfer of information the physical principles that govern the situation (e.g. transport, diffusion). This knowledge should determine the appropriate choice of the discretisation schemes, interpolation polynomials and stabilisation methods.

3.2 The Role of Analytical Mathematics

The discretisation targets not only the equations but also the computational geometric domain since the primary variables are solved for a finite number of grid points chosen to maximise the accuracy. The application of the discretisation schemes over this grid results into an algebraic system of equations. In an introductory course, students are often confronted to the difficulty of reasoning the equations in terms of discrete variables; it is as if the time spent in calculus in making h tend to zero acts as obstacle to their capacity or acceptance of applying the finite difference schemes to a set of predetermined points.

This argument, combined with the foreseen mathematical practice of engineers, could militate in favour of earlier exposure to numerical methods in calculus courses for engineers, in the direction proposed by Enelund and Larsson (2006). With this approach, the numerical methods are typically used either to illustrate some of the mathematical concepts being taught or to tackle more complex problems than those that can be solved symbolically. Yet, trying to teach simultaneously analytical and numerical approaches does not come without risks. Depending on the choice of examples and tasks, a student can either develop a false sense of security with regard to the use of the numerical methods or be completely discouraged from their use. But it is mainly the wide domain of applicability of numerical methods that tends to turn students away from analytical solutions, despite the precious information these solutions can provide in appreciating and anticipating the physical properties and solving challenges of the modelled phenomenon.

From our perspective, one of the key challenges in the mathematics education of engineers actually resides in the articulation of the two paradigms (analytical and numerical) to take advantage of their respective pragmatic and epistemic contributions in tailoring the solving process to a given engineering field of application. Rather than aiming for a systematic concurrent exposure of the two approaches in any course of mathematics, we propose that the courses that mainly deal with analytical calculus go deeper into the assumptions upon which some of the analytical methods are based and the properties[4] that they may bring to light.

This rather bold position may be perceived as a way to rehabilitate a more theoretical orientation to the mathematics education of engineers. But it actually comes from the necessity, for both academia and industry, to provide verified and validated engineering results. And, as shown by the Site 41 story, this may be further stressed by growing public awareness.

3.3 Verification and Validation as a Framework for Designing the Computational Mathematics Curriculum

Requirements for verification and validation have been specified by ERCOFTAC[5] (2000) and AIAA[6] (1998). ERCOFTAC defines verification as the "procedure to ensure that the program solves the equations correctly" and validation as the "procedure to test the extent to which the model accurately represents reality". Incorporating verification into computational mathematics education of engineers thus requires to go beyond the sole use of numerical methods as these come with intrinsic uncertainty.

A first step in verification consists in verifying that "the code is capable of achieving correct mathematical solutions to the governing continuum equations in the limit of $\Delta \rightarrow 0$, and the order of convergence is verified at least for well-behaved problems", typically nonlinear problems with an exact closed-form solution (Roache 2008). This requirement clearly points to the necessity of integrating an analytical approach to computational mathematics.

A second step in verification typically makes use of numerical techniques such as Richardson's extrapolation. This technique is usually taught in introductory numerical courses as a recursive means to improve the accuracy of a solution. For instance, its application to the trapezoidal rule leads to the Romberg integration. But what must be made clear to the students is that a sufficiently improved solution is an acceptable surrogate to the closed-form solution, which can be used to derive a reliable error estimator; the latter takes the form of the Grid Convergence Index (GCI) as proposed by Roache (1998). We believe this is where the focus should be when teaching about these techniques: as they allow for error assessment, they can

---

[4] One such property often overlooked is the linearity of the solution which, when it applies, can allow the breaking of a problem into its elementary components and can help reduce the cost of a solution.

[5] This was done by the Special Interest Group on "Quality and Trust in Industrial Computational Fluid Dynamics" within the European Research Community on Flow, Turbulence and Combustion.

[6] American Institute of Aeronautics and Astronautics.

serve as powerful tools to ensure that the engineer remains in control of the numerical solving process. Learning activities should be designed to hone skills and expertise in using these techniques for assessing the quality of the solution. Abstract mathematical concepts and techniques (e.g. vector norm in Sobolev space) naturally reveal their usefulness in such learning context.

In contrast, validation is defined as the "procedure to test the extent to which the model accurately represents reality". This can only be done after a thorough and successful verification process. It is usually accomplished by measuring the difference between simulation results and experimental data. As such, the results produced for meeting the second requirement of verification can be used to provide "discretisation–error free" simulation results, in an attempt to isolate the modelling error.

4 Conclusion

We have provided in this paper some general orientations and more specific suggestions for designing an introductory course in computational mathematics for engineers. Acknowledging the presence of powerful simulation tools in today's engineering practice, we aim at developing in students a sense of control over the solution process and the use of these tools, while nurturing their motivation to go beyond the user interface. In particular, it appears clearly to us that to teach computational mathematics effectively to future engineers, elements of physics specific to their field, mathematical analysis and engineering practice must be used to articulate within the lectures and tasks, the ideas, skills and attitudes expected from an engineer.

Acknowledgments This work was made possible by a Discovery Grant from the National Science and Engineering Research Council of Canada.

References

AIAA (1998). Guide for the Verification and Validation of Computational Fluid Dynamics Simulations (G-077-1998e). AIAA Standards Series. Canadian Engineering Qualifications Board (2001). Guideline on the Environment and Sustainability for all Professional Engineers. Ottawa: Canadian Council of Professional Engineers.

Enelund, M., & Larsson, S. (2006). A computational mathematics education for students of mechanical engineering. *World Transactions on Engineering and Technology Education, 5*(2), 329.

ERCOFTAC Special Interest Group on Quality and Trust in Industrial CFD (2000). Best practice guidelines for industrial computational fluid dynamics. Lausanne, Switzerland: ERCOFTAC Coordination Centre.

Hodgson, B. R. (1987). Symbolic and numerical computation: the computer as a tool in mathematics. In D. C. Johnson & F. Lovis (Eds.), Informatics and the teaching of mathematics—Proceedings of the International Federation for Information Processing (IFIP) TC 3/WG 3.1 Working Conference. North-Holland.

Information and Privacy Commissioner/Ontario (2009). ORDER MO-2449—Appeal MA07-365—County of Simcoe. Toronto: Tribunal Services Department.

Roache, P. J. (1998). *Verification and validation in computational science and engineering*. Albuquerque: Hermosa Publishers.

Roache, P. J. (2008). Validation: Definitions or Descriptions? 3rd Workshop on CFD Uncertainty Analysis. Lisbon.

Integrating Computational Modelling in Science, Technology, Engineering and Mathematics Education

Rui Gomes Neves, Jorge Carvalho Silva and Vítor Duarte Teodoro

1 Introduction

Science, technology, engineering and mathematics (STEM) are evolving structures of knowledge which are symbiotically interconnected. On one hand, science is based on hypotheses and models, leading to theories, which have a strong mathematical character as scientific reasoning, concepts, and laws are represented by mathematical reasoning, entities, and relations. On the other hand, scientific explanations and predictions must be consistent with the results of systematic and reliable experiments, which depend on technological developments as much as these depend on the progress of science and mathematics (see, e.g., Chalmers 1999; Crump 2002; Feynman 1967). The creation of STEM knowledge is a dynamical cognition process, which involves a blend of individual and collective reflexions where modelling occurs with a balance between theoretical, experimental and computational elements (Blum et al. 2007; Neunzert and Siddiqi 2000; Schwartz 2007; Slooten et al. 2006). In this research paradigm, computational modelling plays a key role in the expansion of the STEM cognitive horizon through enhanced calculation, exploration and visualization capabilities.

Although clearly related to real world phenomena, STEM knowledge is built upon abstract and subtle conceptual and methodological frameworks, which have a complex historical evolution. These epistemological and cognitive features make

R. G. Neves (✉) · V. D. Teodoro
Unidade de Investigação Educação e Desenvolvimento (UIED) e, Departamento de Ciências Sociais Aplicadas (DCSA), Faculdade de Ciências e Tecnologia, Universidade Nova de Lisboa, Lisbon, Portugal
e-mail: rgn@fct.unl.pt

V. D. Teodoro
e-mail: vdt@fct.unl.pt

J. C. Silva
Centro de Física e Investigação Tecnológica (CEFITEC) e, Departamento de Física, Faculdade de Ciências e Tecnologia, Universidade Nova de Lisboa, Lisbon, Portugal
e-mail: jcs@fct.unl.pt

A. Damlamian et al. (eds.), *Educational Interfaces between Mathematics and Industry*, New ICMI Study Series 16, DOI: 10.1007/978-3-319-02270-3_38,

STEM difficult fields to learn and to teach. To develop an approach to STEM education which aims to be effective and in phase with the rapid scientific and technological development, it is of crucial importance to promote an early integration of computational modelling in learning environments which reflect the exploratory and interactive nature of modern research (Ogborn 1994). However, even in technologically advanced countries, computers, computational methods and software, as well as exploratory and interactive learning environments, are still not appropriately integrated in most STEM education curricula for the high school and undergraduate university levels. As a consequence, these curricula are generally outdated and most tend to transmit to students a sense of detachment from how science is currently made. These are contributing factors to the development of negative views about the education process and to an increase in student failure.

Physics education is a good example to illustrate this problem. Consider the general physics courses taken by first year university students. These are courses which usually follow a traditional lecture plus laboratory instruction approach and cover a large number of physics topics which students find particularly difficult. Due to a lack of understanding of the necessary fundamental concepts in physics and mathematics, the number of students that fail on the course examinations is usually very high. What is worse is that many of those students that do actually succeed also reveal several weaknesses in their understanding of elementary physics (Halloun and Hestenes 1985; Hestenes 1987; McDermott 1991).

Research in physics education has shown that this situation can be improved when students are involved in the learning activities as scientists are involved in research (Beichner et al. 1999; Mazur 1997; McDermott 1997). This is not a surprising result. Scientific research in physics is an interactive and exploratory process of creation, testing and improvement of mathematical models that describe observable physical phenomena. It is this cognitive process that leads to an inspiring understanding of the rules of the physical universe. As a consequence, physics should be expected to be more successfully taught in interactive and exploratory environments where students are helped by teachers to work as scientists do. In this kind of class environment knowledge performance is better promoted and common sense beliefs as well as incorrect scientific ideas can be more effectively fought.

The scientific research process is supported by a continuously evolving set of analytical, computational and experimental techniques. The same should be true for research inspired learning environments. Consequently, another important aspect of these learning environments is the possibility to balance the role of computational modelling methods and tools. This would set the learning process in phase not only with modern scientific research where computation is as important as theory and experiment, but also with the rapid parallel development of technology.

Several attempts have already been made to introduce computational modelling in research inspired learning environments. The starting emphasis was on professional programming languages such as Fortran (Bork 1967) and Pascal (Redish and Wilson 1993). Although more recently this approach has evolved to Python

(Chabay and Sherwood 2008), it continues to require that students develop a working knowledge of programming, a time-consuming task which hinders the process of learning physics. The same happens with scientific computation software such as Mathematica and Matlab. To avoid overloading students with such programming notions or syntax, and focus the learning process on the relevant physics and mathematics, several computer modelling systems were created, for example, Dynamical Modelling System (Ogborn 1985), Stella (High Performance Systems 1997), Easy Java Simulations (Christian and Esquembre 2007) and Modellus (Neves et al. 2013; Teodoro and Neves 2011).

A proper and balanced integration of computational modelling methods and tools in STEM learning environments is, thus, both a curricular and a technological development problem. In this work, we discuss how Modellus (a freely available software tool created in Java which is able to run in all operating systems, see the software webpage a http://modellus.fct.unl.pt) can be used as a central element of an approach to develop exploratory and interactive computational modelling learning activities relevant for STEM education. These activities can be adopted by high school and university curricula, and used as a valuable instrument for the professional development of teachers. To illustrate, we consider computational modelling with Modellus to teach physics, namely introductory mechanics, and discuss its impact on the student learning process.

2 Modellus: Interactive and Exploratory Computational Modelling for STEM Education

The construction of STEM knowledge requires unambiguously clear declarative, operational and conditional specifications of abstract concepts and of the relations among such concepts. Of crucial importance in the understanding of the resulting models or theories is the interpretation process which involves operational familiarization and connection with the relevant referents in the observable universe (Reif 2008). In education, as in research, computers, computational methods, and software are cognitive artifacts (Teodoro 2005), which may amplify the learning cognition horizon due to more powerful calculation, exploration and visualization capabilities. As a consequence, these artifacts may play a key role in enhancing the operational familiarization and the connection with the real world referents, which must necessarily be involved in the STEM learning processes. To be able to fulfill such a potential key role, computational methods and tools should be used not only to display text, images or simulations but as mathematical modelling tools integrated in learning environments which reflect the exploratory and interactive nature of modern research. In addition, the modelling process should be focused on the meaning of models and avoid learning opacity factors such as too much programming and specific software knowledge.

To meet this educational challenge, it is not enough to simply choose a subset of programming languages and professional computational software. It is necessary to develop computer software systems with computational modelling functionalities, which contribute to nurture the progressive growth of solid STEM cognitive competencies. Among those systems which have already been created, Modellus stands out as a computational modelling tool for STEM education because of the following main advantages: (1) an easy and intuitive creation of mathematical modells using standard mathematical notation; (2) the possibility to create animations with interactive objects that have mathematical properties expressed in the model; (3) the simultaneous exploration of multiple representations such as images, tables, graphs and animations; and (4) the computation and display of mathematical quantities obtained from the analysis of images and graphs.

These are features that allow a deeper cognitive contact of models with the relevant real world referents and a deeper operational exploration of models as objects, which are simultaneously abstract, in the sense that they represent relations between mathematical entities, and concrete, in the sense that they may be directly manipulated in the computer. In a word, Modellus allows a deeper reification of abstract mathematical objects. Because of these characteristics, computational modelling activities built with Modellus can be readily conceived as exploratory and interactive modelling experiments performed by students in collaborative groups or individually. They can also be designed with an emphasis on cognitive conflicts in the understanding of STEM concepts, on the manipulation of multiple representations of mathematical models and on the interplay between the analytical and numerical approaches applied to solve STEM problems.

As a domain general environment for modelling, Modellus can be used to conceive STEM learning activities, which involve the exploration of existing models and the development of new ones (Bliss and Ogborn 1989; Schwartz 2007). As much as possible, such modelling activities should consider realistic problems to maximize the cognitive contact with the real world referents. This is a challenge because more realistic problems are generally associated with more complex analytic solutions, which are beyond the analytic capabilities of high school or first to second year university students. With Modellus and numerical methods, which are conceptually simpler and yet powerful, the interactive exploration of models for more realistic problems can start at an earlier age, allowing students a closer contact with the model referents, an essential cognitive element to appreciate the relevancy and power of models, necessarily a partial idealized representation of their referents.

Clearly, the development of the appropriate computational modelling activities for STEM research inspired learning environments is bound to call for a richer set of modelling functionalities, which are not yet available in Modellus. These events are seeds for technological evolution, which should be accomplished by a Modellus enhancement program. Currently under development and set to appear in forthcoming versions of Modellus is, for example, the following set of new functionalities: spreadsheet, data logging and curve fitting capabilities, advanced animation objects like curves, waves and fields, 3D animations and graphs, creation of a physics engine for motion and collisions, video analysis, and cellular automata models.

The simultaneous development of new functionalities to meet appropriate teaching goals is important because it reduces the learning opacity factor associated to an unnecessary proliferation of tools. However, there is a learning stage where it is advantageous to allow some diversity in the use of computer software tools and complement Modellus with other available tools. Indeed, in a research-based STEM learning environment one of the objectives is to make a progressive introduction to professional STEM computation methods and software. For example, Excel is a general purpose spreadsheet where modelling is focused on the algorithms. In addition, it already allows data analysis from direct data logging. On the other hand, Mathematica and Matlab (or wxMaxima, a similar but freely available tool) have powerful symbolic computation capabilities. Using these different tools to implement the same algorithm is an important step to learn the meaning of the algorithm instead of the syntax of a particular tool. If more realistic simulations are needed, Modellus animations can be complemented, for example, with EJS.

3 Computational Modelling Learning Activities: An Illustrative Example from Rotational Dynamics

Let us consider a computational modelling activity about rigid body rotational dynamics and angular momentum, a topic in general physics which in a course program for first year university students should be introduced after computational modelling activities covering vectors, kinematics and Newton's fundamental laws of motion, including simple numerical and analytical solutions (Neves et al. 2009, 2011). A rigid body is a system of particles whose relative distance does not change with time. When a rigid body rotates around a fixed axis, each one of its particles has a circular motion around the axis, which is characterized by a rotation angle, an angular velocity and an angular acceleration. The kinetic rotation energy is the sum of the kinetic energies of all the particles of the body and is given in terms of the moment of inertia of the body relative to the rotation axis. From a dynamical point of view, the rotational motion of a rigid body around a fixed axis is characterized by two vectors, the angular momentum of the rigid body and the moment of the sum of all the forces acting on the body, the latter also called the net applied torque. Newton's laws of motion imply that the instantaneous rate of change of the angular momentum is equal to the net applied torque.

A real world system, which may be considered as a rotating rigid body is a wind turbine. With Modellus it is possible to model the action of the wind on the rotor blades and analyze their motion using at the same time different representations such as graphs, tables, and object animations. In this model, the fundamental equations of the rotational motion are written in the form of Euler iterations. Students are thus taught to apply this numerical method in a new realistic context, extending the applicability range of knowledge already acquired with a previous analysis of analogous numerical solutions of Newton's equations in translational motion settings (Neves et al. 2009). In this new application, students can determine the angular velocity and the rotation angle knowing the net applied torque, the moment of inertia of the system and the motion initial conditions. The model animation is constructed with three objects: a bar representing the rotor blade, a vector representing the angular momentum and a vector representing the net wind torque. Because the coordinates of the net torque are independent variables and the model is iterative, students can manipulate this vector at will and in real time control the motion of the rotor blade. With this activity students can confirm that the choice of the time step is an important one to obtain a good simulation of the motion and that this is the same as determining a good numerical solution of the equations of motion. While exploring the model, students can determine, for example, what are the values of the angular momentum, the angular velocity and the rotational kinetic energy, 8 s after increasing the net wind torque to 2,000 Nm. The possibility to change the mathematical model and immediately observe this action on the animation, graphs and tables is a powerful cognitive element to enhance the students learning process. Students can also change the model. Introducing a vector to represent the wind force they can explore the effect of the wind direction on the net applied torque and on the motion of the rotor blades.

4 Field Actions, Discussion and Outlook

In this chapter, we have shown how Modellus can be used as a key element of an approach to develop exploratory and interactive computational modelling learning activities for research inspired STEM education. As an example, we have discussed the modelling of the rotational dynamics of a wind turbine. This was one of the activities which was tested on the field when we implemented computational modelling activities in the general physics course offered in 2008 and 2009 to the first year biomedical engineering students at the Faculty of Sciences and Technology of the New Lisbon University (Neves et al. 2008, 2009, 2011). More field tested activities are discussed, e.g., in Neves et al. (2013) and Neves and Teodoro (2012). For other earlier educational applications and evaluation tests of Modellus-based computational modelling activities see, e.g., Araújo et al. (2008), Dorneles et al. (2008) and Teodoro (2002).

The 2009 edition of the general physics course for biomedical engineering involved a total of 115 students. Of these, 59 were taking the course for the first time and only they were enrolled in the computational modelling classes. To build an interactive collaborative learning environment, we organized the students in groups of two or three, one group for each computer in the classroom. During each class, the student teams worked on a computational modelling activity set conceived by us to be an interactive and exploratory learning experience with Modellus, built around a small number of problems in mechanics connected with easily observed real world phenomena. The teams were instructed to analyze and discuss the problems on their own using the physical, mathematical and computational modelling guidelines provided by the activity set documentation, a set of PDF documents with embedded video support. To ensure a good working pace with appropriate conceptual, analytical and computational understanding, the students were continuously monitored and helped during the exploration of the activities. Whenever it was felt necessary, global class discussions were conducted to clarify doubts on concepts, reasoning, or calculations. Online support in class and at home was provided in the context of the Model platform where links to class and homework documentation was provided.

The evaluation procedures associated with the computational modelling activities with Modellus involved both group evaluation and individual evaluation. For each computational modelling class, all student groups had to complete an online test written in the Moodle platform answering the questions of the corresponding activity PDF document. The individual evaluation was based on the student solutions to two sets of homework activities and a final class test with computational modelling problems to be solved with Modellus. All students took a pre-instruction and post-instruction FCI test (Hestenes et al. 1992) which did not count for their final course grade. At the end of the semester, the students answered a questionnaire to evaluate Modellus and the new computational modelling activities of the general physics course. As reflected by the student answers to the 2008 and 2009 questionnaires (Neves et al. 2008, 2011), the activities with Modellus were helpful in the learning process of mathematical models in the context of introductory physics. For students, Modellus was easy to learn and user-friendly. Working in groups of two or three was also acknowledged to be more advantageous than working individually. In addition, the PDF documents with embedded video guidance were considered to be interesting and well designed.

During class and homework communications, it was clear that the computational modelling activities with Modellus were being successful in identifying and resolving several student difficulties in key physical and mathematical concepts of the course. The possibility to explore simultaneously several different representations like graphs, tables and animations was a key success factor, in particular the possibility to have a real time visible correspondence between the animations with interactive objects and the object's mathematical properties defined in the model. During the course, Modellus and simple numerical methods, such as the Euler and Euler-Cromer methods, allowed the introduction and exploration of advanced mathematical concepts such as integration in the context of real world physics problems, prior to the introduction of the corresponding analytic techniques. However, FCI test results for the whole general physics course lead to an average FCI gain of 22 %, a typical traditional instruction performance (Hake 1998). Moreover, students manifested clearly that the new content load was too heavy and that the available time to spend working on the computational modelling activities was insufficient. Improvement actions (Neves et al. 2011) are now being implemented.

Acknowledgments Work supported by Unidade de Investigação Educação e Desenvolvimento (UIED) and Fundação para a Ciência e a Tecnologia (FCT), Programa Compromisso com a Ciência, Ciência 2007.

References

Araújo, I., Veit, E., & Moreira, M. (2008). Physics student's performance using computational modelling activities to improve kinematics graphs interpretation. *Computers & Education, 50*, 1128–1140.

Beichner, R., Bernold, L., Burniston, E., Dail, P., Felder, R., Gastineau, J., et al. (1999). Case study of the physics component of an integrated curriculum. *Physics Education Research, American Journal of Physics Supplement, 67*, 16–24.

Bliss, J., & Ogborn, J. (1989). Tools for exploratory learning. *Journal of Computer Assisted learning, 5*(1), 37–50.

Blum, W., Galbraith, P., Henn, H.-W., & Niss, M. (Eds.). (2007). *Modellng and applications in mathematics education*. New York: Springer.

Bork, A. (1967). *Fortran for physics*. Reading, Massachusetts, USA: Addison-Wesley.

Chabay, R., & Sherwood, B. (2008). Computational physics in the introductory calculus-based course. *American Journal of Physics, 76*, 307–313.

Chalmers, A. (1999). *What is this thing called science?*. London, UK: Open University Press.

Christian, W., & Esquembre, F. (2007). Modelling physics with easy java simulations. *The Physics Teacher, 45*(8), 475–480.

Crump, T. (2002). *A brief history of science, as seen through the development of scientific instruments*. London, UK: Robinson.

Dorneles, F., Araújo, I., & Veit, E. (2008). Computational modelling and simulation activities to help a meaningful learning of electricity basic concepts. Part II: RLC circuits. *Revista Brasileira de Ensino da Física*, 30(3), 3308–1.

Feynman, R. (1967). *The character of physical law*. New York, USA: MIT Press.

Hake, R. (1998). Interactive-engagement versus traditional methods: A six-thousand-student survey of mechanics test data for introductory physics courses. *American Journal of Physics, 66*, 64–74.

Halloun, I., & Hestenes, D. (1985). The initial knowledge state of college physics students. *American Journal of Physics, 53*, 1043–1048.

Hestenes, D. (1987). Toward a modelling theory of physics instruction. *American Journal of Physics, 55*, 440–454.

Hestenes, D., Wells, M., & Swackhamer, G. (1992). Force concept inventory. *The Physics Teacher, 30*, 141–158.

High Performance Systems. (1997). *Stella, version 5*. Hannover, NH: High Performance Systems.

Mazur, E. (1997). *Peer instruction: A user's manual*. New Jersey, USA: Prentice-Hall.

McDermott, L. (1991). Millikan lecture 1990: What we teach and what is learned–closing the gap. *American Journal of Physics, 59*, 301–315.

McDermott, L. (1997). *Physics by inquiry*. New York, USA: Wiley.

Neunzert, H., & Siddiqi, A. (2000). *Topics in industrial mathematics: case studies and related mathematical models*. Dordrecht, Holland: Kluwer Academic Publishers.

Neves, R., Neves, M. C., & Teodoro, V. (2013). Modellus: Interactive computational modelling to improve teaching of physics in the geosciences. *Computers & Geosciences, 56*, 119–126.

Neves, R., Silva, J., & Teodoro, V. (2008). Improving the general physics university course with computational modelling. Poster presented at the 2008 Gordon Research Conference, Physics Research and Education: Computation and Computer-Based Instruction, Bryan University, USA.

Neves, R., Silva, J., & Teodoro, V. (2009). Computational modelling with Modellus: an enhancement vector for the general university physics course. In A. Bilsel & M. Garip (Eds.), *Frontiers in science education research* (pp. 461–470). Famagusta, Cyprus: Eastern Mediterranean University Press.

Neves, R., Silva, J., & Teodoro, V. (2011). Improving learning in science and mathematics with exploratory and interactive computational modelling. In G. Kaiser, W. Blum, R. Borromeo-Ferri & G. Stillman (Eds.), *International Perspectives on the Teaching and Learning of Mathematical Modelling Vol. 1, ICTMA14: Trends in Teaching and Learning of Mathematical Modelling* (pp. 331–341). Dordrecht: Springer.

Neves, R., & Teodoro, V. (2012). Improving science and mathematics education with computational modelling in interactive engagement environments. In T. Simos, G. Psihoyios, C. Tsitouras and Z. Anastassi (Eds.), *Numerical Analysis and Applied Mathematics 2012, AIP Conference Proceedings* 1479, 1806–1809.

Ogborn, J. (1985). *Dynamic modelling system*. Harlow, Essex, UK: Longman.

Ogborn, J. (1994). Modelling clay for computers. In B. Jennison & J. Ogborn (Eds.), *Wonder and delight, Essays in science education in honour of the life and work of Eric Rogers 1902–1990* (pp. 103–114). Bristol: Institute of Physics Publishing.

Redish, E., & Wilson, J. (1993). Student programming in the introductory physics course: M.U.P.P.E.T. *American Journal of Physics, 61*, 222–232.

Reif, F. (2008). *Applying cognitive science to education: Thinking and learning in scientific and other complex domains*. Cambridge, USA: MIT Press.

Schwartz, J. (2007). Models, simulations, and exploratory environments: A tentative taxonomy. In R. A. Lesh, E. Hamilton, & J. J. Kaput (Eds.), *Foundations for the future in mathematics education* (pp. 161–172). Mahwah, NJ: Lawrence Erlbaum Associates.

Slooten, O., van den Berg, E., & Ellermeijer, T. (Eds.). (2006). *Proceedings of the International Group on Research on Physics Education (GIREP) 2006 conference: modelling in physics and physics education*. Amsterdam, Holland: European Physical Society.

Teodoro, V. (2002). Modellus: learning physics with mathematical modelling. PhD Thesis. Lisboa, Portugal: Faculdade de Ciências e Tecnologia, Universidade Nova de Lisboa.

Teodoro, V. (2005). Cognitive artefacts, technology and physics education. *Interactive Educational Multimedia, 11*, 173–189.

Teodoro, V., & Neves, R. (2011). Mathematical modelling in science and mathematics education. *Computer Physics Communications*, *182*, 8–10.

Incorporating the Ideas and Methods of Mathematical Modeling into Calculus Teaching

Qixiao Ye

1 Introduction

Of what use is mathematics? How scientists and engineers solve real world problems through mathematical modeling (MM)? These are questions asked by freshmen students. In helping them to get deeper understanding of these problems, especially the main ideas and steps of MM at their early mathematics studying in the universities and colleges, we design and develop several modules on MM which can be integrated into calculus teaching. Two of them are related to the loan problem and the shape design of beverage cans (pull-tab cans). Since 2009, class tests of these two modules have been implemented in calculus teaching for the freshman students in the autumn semester at Guizhou University, Beijing University of Chemical Technology, Teachers' College of Beijing Union University, and Beijing Institute of Technology. Each class test includes 2 teaching-hour (50 min for one teaching hours) lecture, questionnaire, and quiz; we also leave several take-home exercises to the students.

Experience tells us that the more solid basis in mathematics a university student has the more chance for success will go to him (her) no matter what kind of job he/she will be working for. MM is a "bridge" for using mathematics to solve real world problems, often MM is critical in solving some kind problems.

With the rapid and healthy development of the China Undergraduate Mathematical Contest in Modeling (CUMCM), we increasingly realize that we should incorporate some contents into the regular teaching of main math courses in universities and colleges to benefit more students.

From 2002 to 2005, we executed a reform project named "Reform Experiment on incorporating ideas and methods of MM into Main Math Courses in Universities" (Project leader: Tatsien Li). More than 20 universities and higher vocational colleges participated in this project. Since then quite a lot texts in which

---

Q. Ye
School of Mathematics, Beijing Institute of Technology, Beijing, China
e-mail: yeqx@bit.edu.cn

A. Damlamian et al. (eds.), *Educational Interfaces between Mathematics and Industry*, New ICMI Study Series 16, DOI: 10.1007/978-3-319-02270-3_39,

many math models included were published and many teachers convey the ideas and methods of MM through teaching MM examples in their own classes. But there were some disadvantages, namely, less feedbacks from students and not easy to assess the students' achievements.

From 2009 to 2011, I joined the project named "Reform and Innovation in the MM and Mathematical Experimentation (ME) Courses" (Project leader: Yongji Tan) from the Centre for research and development of College mathematics teaching in colleges and universities. Since then I begin to implement our own reform practice.

2 Features of our Project

2.1. We begin our reform practice in the early calculus teaching, namely in autumn semester;

2.2. We design and develop several modules on MM which can be integrated into calculus teaching (2 teaching-hours (50 min for one teaching hours) for each lecture; by module we mean:

> A unit of education or instruction with a relatively low student-to-teacher ratio, in which a single topic or a small section of a broad topic is studied for a given period of time. Quote from *the American Heritage Dictionary of the English Language*, 4th Edition, Houghton Mifflin Company, Boston, New York, 2000, p. 1131.

2.3. Implement of each module includes lecture, questionnaire, and quiz; we also leave several take-home exercises to the students. Questionnaire, quiz, and exercises are all printed in advance; it is easy to make statistics. We will also design a tracking surveys and appropriate modeling activities to know how many students become studying math much hard.

2.4. Our expectation is not high. We hope most of the students will get deeper understanding about math is very useful and how scientists and engineers using mathematics to solve the real world problems through MM, etc., and to remember the key steps of MM, and a small part of students will interested in taking MM and ME courses, even participate in CUMCM later. We also want students to understanding the function of MM, which are useful, important, sometimes critical, but its function is also limited, namely, MM is not universal and dominate, cannot solve every aspect of the real world problems. Just like Turing have said:

> 1. *A model of the embryo. Morphogens—In this section a mathematical model of the growing embryo will be described. This model will be a simplification and an idealization, and consequently a falsification. It is to be hoped that the features retained for discussion are those of greatest importance in the present state of knowledge.—The Chemical Basis of Morphogenesis*, by A. M. Turing, F.R.S. (Fellow of the Royal Society), *University of Manchester*, Philosophical Transactions of the Royal Society of London, Series B (Biological Sciences), v.237 (1952), 37–72.

2.5. Since 2009, class tests of these two modules have been implemented in calculus teaching for the freshman students in the autumn semester at Guizhou University, Beijing University of Chemical Technology, Teachers' College of Beijing Union University, and Beijing Institute of Technology.

3 Module 1: Loan Problem

First lecture: Brief introduction of what is MM and its key steps, in short, four key steps are: reasonable assumptions, building of a mathematical model, solving the model, and interpretation and validation of the results; then we posed a real world problem to teach the whole MM process through a discrete model of the loan problem, at last let students to fill out the questionnaires. You can find a similar and simpler problem on a newly published book: Christiane Rousseau, Yvan Saint Aubin, *Mathematics and Technology*, Springer, 2008, Chap. 5, Savings and Loans.

Second lecture: first 15 min for a quiz, then teach a continuous model of the loan problem, and its connection with the discrete model, emphasizing the importance of Taylor formula, etc.

Discrete model of the loan problem

We first show students a small electric dictionary. In its menu, there is an item named loan computation, opening it you will see

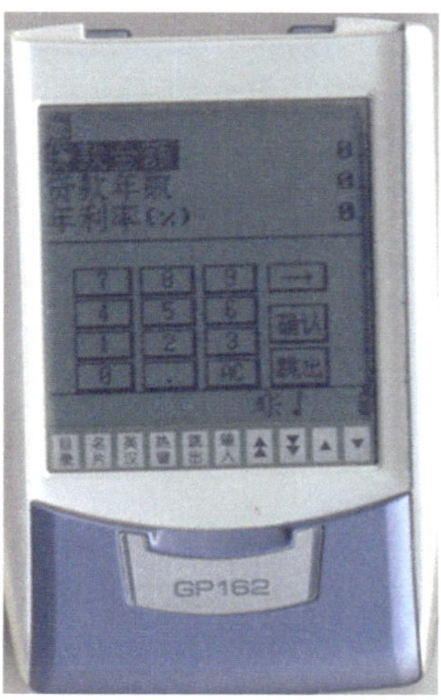

It requires you to give three inputs. If your inputs are

Loan amount or initial balance: 200,000
Amortization period: 20 years
Annual interest rate (%) 6.39 % = 0.0639
(Monthly interest rate = 6.39/12 % = 0.5325 %)

Then you get the output

Monthly payment: 1478.22
Total payment 354,773.41
Total Interest 154,773.41

Now, the question is: How the electric dictionary computes this result (Using what math model)?

Assumptions: Equal monthly payment, your outstanding balance should be zero at the end of 20 years.

Building mathematical model:

Let

Loan amount be $A_0(= 200{,}000)$,
Amortization period be $N(= 240$ months),
Annual interest rate be $R = 0.0639$,
Monthly interest rate be $r = R/12 = 0.005325$.
Monthly payment is x.

Mathematical model

$$\begin{cases} A_n = A_{n-1}(1+r) - x & n = 1,2,3,\ldots,N \\ A_N = 0 \end{cases}$$

Solving mathematical model:

We only need the summation formula of a geometric sequence

$$S_n = 1 + q + q^2 + \cdots + q^{n-1} = \frac{1-q^n}{1-q}, \quad q > 0, q \neq 1$$

From

$$0 = A_0(1+r)^n - x\frac{(1+r)^n - 1}{r}$$

We get the solution

$$x = \frac{A_0 r(1+r)^N}{(1+r)^N - 1}$$

Then using a calculator or some kind math software to compute, get the same result

$$x = 1478.22$$

Interpretation and validation

You can use different data and calculator or math software to check and convince that the electric dictionary use exactly this math model.

Questionnaires, quiz, home work, and their statistics

Questionnaire at Guizhou University 2009-12-17

About 80 students from school of science majoring in engineering Seventy two of them filled out the following questionnaires

Questions	Yes	No
Do you know that there is a "competition on applying mathematical knowledge for middle school students (CAMK)" in Beijing, Shanghai etc. cities?	15	57
If you ever participated in CAMK?	9	63
If you ever heard the term "mathematical model"?	55	27
If you ever heard the term "mathematical modeling"?	52	20
Do you know that there is a "China undergraduate mathematical contest in modeling (CUMCM)" in our country?	34	38
Do you know that there is a "mathematical contest in modeling (MCM)" in the USA?	5	67
Before this lecture, did you ever participate in activities related to mathematical modeling?	4	68
After this lecture, do you know more about mathematical modeling?	63	9
Your evaluation for this lecture, critics, and suggestion? Suggestions about topics and teaching methods when teaching Mathematical modeling or applications?	43 answer the question most are positive and encouraging 27 waive	

Questions	Yes	No	Waive
Do you know that there is a "competition on applying mathematical knowledge for middle school students (CAMK)" in Beijing, Shanghai etc. cities?	25	55	
If you ever participated in CAMK?	11	68	
If you ever heard the term "mathematical model"?	75	5	
If you ever heard the term "mathematical modeling"?	74	6	
Do you know that there is a "China undergraduate mathematical contest in modeling (CUMCM)" in our country?	67	13	
Do you know that there is a "mathematical contest in modeling (MCM)" in USA?	20	60	
Before this lecture, did you ever participate in activities related to mathematical modeling?	13	67	
After this lecture, do you know more about mathematical modeling?	69	2	9
The key steps of mathematical modeling are:	71 Basically right		9
Your evaluation for this lecture, critics and suggestion? Suggestions about topics and teaching methods when teaching mathematical modeling or applications?	40 Almost all the responses are positive and encouraging		40

Questionnaire at Beijing Institute of Technology (BIT) 2011-10-16

Eighty nine students all majoring in mathematics in their mathematical analysis class 80 of them filled out the following questionnaires

One student from Guizhou University said: "I was very shocked by today's lecture. I feel once more the huge influence of mathematics, as well as mathematics applies the economic efficiency which brings in the real life. This lecture will give me great influence now and in future".

Quiz problem

Strictly follow four key steps: "assumptions, building the math model, solving the math model, and interpretation and validation the results" to answer the following problems.

Someone would like to borrow some amount of money for buying an apartment. He is sure that the monthly payment about 3,000 yuan is no problem in 10 years. He also knows the yearly interest rate is 6 %.

1. Build a math model for how much money he needs to borrow.
2. Numerically to compute the approximate loan amount he needs to borrow.

$$\left(\text{Hint}: (1.005)^{120} = 1.8194\right)$$

Statistics of the quiz:

At Guizhou University: Grading standards: 3 + 2

Statistics:

5 = 3 + 2	5− = 3 + 1.5	4 = 3 + 1	3 = 3 + 0	2 = 2 + 0	1 = 1 + 0
6	2	2	6	6	4

For first question

Get 3	Get 2	Get 1
16	6	4

For second question

Get 2	Get 1	Get 0
8	1	16

At Beijing Institute of Technology:

Grading standards: assumption + interpretation = 1, Building and solving model = 4

Score distribution

0	1	1.5	2	2.5	3	3.5	4	4.5	5
7	1	0	8	2	4	2	48	9	0

Take home exercises

You can choice any one of the following three problems and present them in a week.

Strictly follow four key steps: "assumptions, building the math model, solving the math model, and interpretation and validation the results" to answer the following problems.

1. The following is an advertisement of Citibank in Hong Kong

What's the monthly interest rate? In order to find out this rate you have to solve what kind of problem? Can you compute it by hand?

2. Smith borrows $60,000 from a bank. Yearly interest rate is 12 %; amortization period is 25 years, equally monthly payment. His monthly payment is about $632.00 (Why?) And he learned that another loan company said they ban help Smith to shorten the amortization period up to 2 years, but he has to pay 3 monthly payment, $632 × 2 = $1896. Smith compute his 2-year total payment is $632 × 24 = $15168, which is much larger than $1896. So he thinks it is a good bargain. Please give suggestion to smith. (This problem is adapted from a similar American problem.)
3. Same lending interest rate, different total payment, why?

Please carefully read a report in the Jinling evening news in Nanjing on 1998-12-30, and answer the following questions:

1. How the Construction bank of Jiangsu Province computes their monthly payment according to their equal monthly loan amount and interest method (first pay the interest than pay part of loan amount)?
2. What is the so-called "equal loan amount, equal interest payment method" of the Construction bank of Hangzhou city?
3. Analysis the advantage and disadvantage of these two methods?

Statistics at BIT

Number of student who present their home work

#1	#2	#3	#1 + #2	#1 + #2 + #3
65	6	1	3	5

Continuous model of the loan problem

Math model

$$\begin{cases} \dfrac{dA(t)}{dt} = rA(t) - x \quad t > 0 \\ A(0) = A_0 \end{cases}$$

Its solution

$$A(t) = \left(A_0 - \frac{x}{r}\right)e^{rt} + \frac{x}{r}$$

Taylor formula with Lagrange remainder

$$f(x) = f(x_0) + f'(x_0)(x - x_0) + \frac{f''(\xi)}{2!}(x - x_0)^2$$

$$e^r = 1 + r + \frac{r^2 e^{\xi}}{2!}$$

$$A(k) = \left(A_0 - \frac{x}{r}\right)(1+r)^k + \frac{x}{r} = A_0(1+r)^k - \frac{x}{r}\left((1+r)^k - 1\right)$$

$$k = N, \quad A(N) = 0, \quad x = \frac{A_0 r(1+r)^N}{(1+r)^N - 1}$$

It is the same solution formula of the discrete model.

Take home exercise

In the discrete model, If A_0, r, N are known, than equally monthly payment is $x = \frac{A_0 r(1+r)^N}{(1+r)^N - 1}$.

In the continuous model, If A_0, r, N are known, than equally monthly payment is $x_c = \frac{A_0 r e^{rN}}{e^{rN}-1}$.

Prove that $x_c < x$.

4 Module 2: The Shape Design of Beverage Cans (Pull-Tab Cans)

Only give lecture at Guizhou University on 2009-12-18 Contents of the 2 lectures

Brief history

A story is told that the product that has given the world its best-known taste was born in Atlanta, Georgia, on May 8, 1886. Dr. John Smith Pemberton, a local pharmacist, produced the syrup for Coca-Cola®, and carried a jug of the new product down the street to Jacobs' Pharmacy, where it was sampled, pronounced "excellent" and placed on sale for five cents a glass as a soda fountain drink. Carbonated water was teamed with the new syrup to produce a drink that was at once "Delicious and refreshing," a theme that continues to echo today wherever Coca-Cola is enjoyed.

The first Coca Cola cans was made in 1955 for shipment to Japan and the Pacific for consumption by military personnel. The introduction of the '202' aluminium can, with a reduced end diameter, reduced cost by a mere one-tenth of a penny. However, considering the Coca-Cola Enterprises Ltd. Sell around two million cans a year you can work out that these savings become significant.

John Halliwell, Barry Lambert, Dave Webster *Revise for Product Design: Graphics with Materials Technology* (Advanced Design and Technology for Edexcel) **Heinemann Educational Publishers, 2004, p. 85**.

Take home exercise—*A Space Shuttle water Container*

Consider the space shuttle and an astronaut's water container that is stored within the shuttle's wall. The water container is formed as a sphere surmounted by a cone (like an ice cream cone), the base of which is equal to the radius of the sphere (see the figure). If the radius of the sphere is restricted to exactly 6 ft and a surface area of 450 ft^2 is all that is allowed in the design, find the dimensions x_1 and x_2 such that the volume of the container is a maximum.

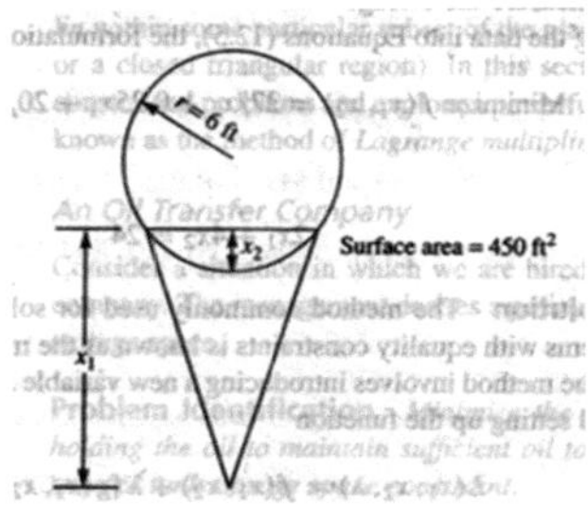

5 Conclusions

1. Students welcome the two lectures. Especially, the freshman class university students are willing to know extremely anxiously why they have to learn mathematics better. They need real guidance; this is precisely our teacher's responsibility. Of course, we also need to do a lot of follow-up work.
2. To implement our project will help young teacher to develop their own teaching style.
3. We have difficulty, especially many teachers are seeking stability and fearing change, and they are being afraid of paying too much and getting too little.

Acknowledgments I would like to sincerely thank Professor Dawei Shen of School of Mathematics of Beijing Institute of Technology, who invite me to give lectures in his Mathematical Analysis class, and has carefully done the quiz grading and many statistics. I would also like to thank Professor Wei Wei of School of Science of Guizhou University for her inviting me to give lectures at one of their calculus class.

Part VIII
WG the Mathematics-Industry Interface

The Mathematics–Industry Interface

Jofré Alejandro and Lutz-Westphal Brigitte

The Working Group had been given the mandate of mainly looking into questions from Sects. 2 and 6 of the Discussion Document, i.e., questions like:

- How can mathematics, especially industrial mathematics, be made more visible to the public at large?
- How can mathematics be made more appealing and exciting to students and the professionals in industry?
- How can mathematics serve a progressive rather than a restrictive role in education and training for the workplace?
- What is the best way to teach analytical skills to various groups of students?
- To what extent is it necessary or desirable to describe the inner workings of black boxes?
- What are the social implications of not explaining the inner working of black boxes?

In the discussions of the Working Group, the following major trends were most often mentioned and used to explain the current development. Mathematics is

---

Co-Chairs: Alejandro Jofré [Chile] & Brigitte Lutz-Westphal [Germany].Working Group Members: Jonas Bergman Ärlebäck [Sweden], Matsuzaki Akio [Japan], Burkhard Alpers [Germany], Silvia Barbiero [Italy], France Caron [Canada], Johan Deprez [France], Peter Frejd [Sweden], André Garon [Canada], Vince Geiger [Australia], Edwige Godlewski [France], Avenilde Romo Vasquez [Mexico], Gloria Stillman [Australia], Jinxing Xie [China]—Report written with the help of Rudolf Sträßer.

J. Alejandro (✉)
Centro Modelamiento Matematico and Departamento Ingenieria Matematica, Universidad de Chile, Santiago, Chile
e-mail: ajofre@dim.uchile.cl

L.-W. Brigitte
Didaktik der Mathematik, Fachbereich Mathematik und Informatik,
Freie Universität Berlin, Königin-Luise-Str. 24-26, 14195 Berlin, Germany
e-mail: brigitte.lutz-westphal@math.fu-berlin.de

A. Damlamian et al. (eds.), *Educational Interfaces between Mathematics and Industry*, New ICMI Study Series 16, DOI: 10.1007/978-3-319-02270-3_40,

applied to new territories, illustrated by a mathematization of many areas in sciences like biology, engineering, medicine, computer sciences, and social sciences. In turn and inside Mathematics itself, new interdisciplinary and intensive areas have emerged, like biomathematics, bioinformatics, behavioral economics, signal and image analysis applied to medicine, natural resources exploration and astronomy; cryptography; climate change and contaminant diffusion as well as intelligent transportation, financial and risk analysis, robotics, data mining, among others. In most cases, these developments are accompanied by an increase in technological power and requirements, especially an intensive use of computers, which usually requires an enormously increasing amount of data, which in turn is pushing the current state of art in data analysis to the limit.

In spite of these substantial contributions to the modern society based on old and new mathematical developments, we do not observe a valorization of mathematics in industry, governments, and society at similar level. In the Working Group, we made explicit at least two important reasons:

1. The way mathematics is introduced in elementary and secondary schools as well as in the training curricula of high school teachers;
2. The existence of many black boxes within industry and mathematics education.

With respect to the first issue, it is not an easy task, because today, mathematics education is not well connected with the "real world" of problems, models, and requirements, the language of applications and their complexity and variability. At the level of Elementary Schools, the introduction of the basic concepts in mathematics should be connected to easy real-world problems, which can be solved or better understood through these new concepts. In high schools and for the training of teacher in mathematics, building, solving, and implementing mathematical models in connection with practical problems require basic concepts and algorithms coming from optimization, numerical analysis, applied probability, differential equations, statistics, discrete mathematics, and also laboratory format courses on specific applied areas including learning basic software for modeling and solving, such as Matlab, Octave, Scilab, Mathematica, Maple, Statistics software, simulation software among others and most important: practice, i.e., learning by doing models.

In the same vein, we could ask more specifically:

- Which aspects of industry driven mathematics can improve secondary education?
- Which methods can be used to analyze industrial processes containing mathematical tools?
- What kind of input do we need from industry or mathematicians to be able to teach mathematical modeling in an appropriate way?
- There are some classical topics in high school education, which have a long tradition, but how can we find new challenging modeling activities including modern mathematical methods?

Concerning the second issue, one way of opening black boxes could be the creation of educational interfaces through:

- Training current and future workers with different backgrounds in mathematics introducing the key concepts, models and algorithms behind a new technology and/or a software,
- Making visible mathematical concepts, models, algorithms, simulations, visualization and analysis tools usually encapsulated in software and equipments used by the industry at different levels,
- Consulting activities for the industry developed by mathematicians, engineers, economists or consulting companies.

In conclusion, the Working Group came up with some recommendations:

- How should modeling be taught? As with technology, if modeling is introduced only as a means to assist in the learning of mathematics, without revisiting the content to be taught or the nature and form of assessment, then some tensions are bound to emerge. Integration of modeling in the teaching of mathematics will gain in significance and coherence if it is also considered as a learning outcome, as an educational goal, in the development of a mathematical practice that also aims at addressing applications. This may entail significant curricular implications, in the way courses are articulated, in the respective content that they should address, in the learning activities that they should promote, in the role that is awarded to technology, and in the nature and form of the tasks that will be used for assessment.
- Every undergraduate program in mathematics should at least organize a lab course (for instance using matlab, scilab…) with a final project, when possible linked with some instructor coming from industry or a PhD/Postdoc in industry.
- In undergraduate mathematics, it is important to offer interaction with the industry through short-term practices.
- Every graduate program in industrial mathematics should involve an internship period in industry.
- Summer internships in industry and modeling weeks (like ECMI, for Undergraduate/Graduate level) should be promoted.
- Part of the program in modeling should be evaluated and given some credits (for example ECTS in the European credit transfer system).
- Non-academic internships or other on-site problem-solving experiences should be incorporated into degree requirements.
- Professors willing to include (and improve their) teaching of mathematical modeling should be assisted by a wider dissemination of resources like e-learning tools (such as those from COMAP). Textbooks are not enough for this purpose.

In order to develop educational interfaces between Industry and Mathematics, a meta-study could be helpful, which pulls together all the Mathematics in Industry projects together with their findings. More ethnographic studies in the style of Hoyles and Wake are needed to strengthen the nexus between research, industry, and classrooms. Finally, Centres of Excellence in Mathematics well connected with industry, which have been proliferating recently, should promote training activities with high school teachers as well as those training future teachers.

Part IX
Selected Papers Linked to More than One Working Group

Inappropriate Use of Spreadsheets in the Finance Industry

Djordje M. Kadijevich

To my son Aleksandar

1 Introduction

Spreadsheets are in use over a quarter of a century and they are nowadays major tools in the finance industry (Croll 2005). Most even the majority of financial documents are found in the form of spreadsheets. A recent survey in December 2007 evidences that Microsoft Excel is used in nearly half of banks in Serbia for recording and monitoring operational risks (www.ubs-asb.com/Portals/0/Vesti/40/Rezultati_Upitnika_2.pdf). The fact that a skilled work with spreadsheets and other business accounting software is vital for a person working in the finance industry is evidenced, for example, in the requirements for many positions in commercial banks.

Spreadsheet models, as other computers models, have a dual role in using mathematics in the workplace: they hide mathematics that is built in those models, but also, through using the software, can help users understand the built-in mathematics (Straesser 2000). The use of technology to help employees understand (for them) hidden mathematical models has been, for example, examined in project *Techno-mathematical literacies in the workplace*, which found that "calculation and basic arithmetic are less important than a conceptual grasp of how, for example, process improvement works, how graphs and spreadsheets may highlight relationships, and how systematic data may be used with powerful, predictive tools to control and improve processes." (Hoyles et al. 2007, p. 3) However, the assumption that the models to be understood with technology are error-free and does not always apply. This implies that technology in general and spreadsheets in particular should be also used to test models and data applied (e.g., when using outcomes of what-if analyses and solutions of different scenarios). This relatively neglected afford enables critical evaluations of the results obtained with the software, which is according to the Discussion Document of ICMI/ICIAM

D. M. Kadijevich (✉)
Megatrend University and Mathematical Institute of the Serbian Academy of Sciences and Arts, Belgrade, Serbia
e-mail: djkadijevic@megatrend.edu.rs

A. Damlamian et al. (eds.), *Educational Interfaces between Mathematics and Industry*, New ICMI Study Series 16, DOI: 10.1007/978-3-319-02270-3_41,

Study "Educational Interfaces between Mathematics and Industry" relevant to both industry and education. This opens an important question for the teaching and learning of industry related practice regarding the use of spreadsheets models and the appropriateness of those models that is also acknowledged by this document in a general sense.

The chapter examines how spreadsheets are used in the finance industry. This question was answered by using several research papers located through a detailed search of various electronic databases and the Internet. We also used some data gathered in interviews with seven experienced bankers. One of them worked at a big software firm that has implemented Business Intelligence (BI) solutions in many banks in Serbia and abroad. Other banker worked in the Association for Serbian Banks (www.ubs-asb.com), which is a large finance association in Serbia dedicated to improving the business by proposing, among other things, methodological solutions on various financial issues (e.g., operational risk) to be applied in the practice (possibly previously brought into accord with approaches developed at central banks or bank groups and the National Bank of Serbia). Three of the remaining five bankers were members of top management in their banks.

This examination showed that spreadsheets may be used insufficiently, uncritically, and erroneously. The next section exemplifies such a use of spreadsheets, whereas the closing section briefly deals with some implications for vocational education that would improve the matters.

2 Inappropriate Practice

2.1 Insufficient Use

Although a bank may use spreadsheets even for 80 % of all its documents (revealed by one interview), spreadsheets may be used insufficiently. Consider, for example, bank branch productivity usually expressed as cost/income ratio (should be less than 80 %, for example). A race among banks to expand into different regions is now taking place in Serbia. Even in small towns, within a circle of few hundred meters in their downtowns, one can find a constant presence of 10 or so banks represented by their branches. Despite the facts that their branch productivity is probably low (having in mind such a number of banks represented, town size, and a lack of qualified employees),[1] analyses concerning the branch productivity (or other relevant measures), supported by spreadsheets or other software, seem missing at most banks. This astonishing fact was revealed in one of the interviews mentioned above. The banker clarified that, when he asked colleagues from banks about these analyses, they usually replied that some measures were

[1] In the first nine months in 2009, for example, just twenty of thirty-four banks in Serbia have operated profitably (www.nbs.rs/export/internet/latinica/55/55_4/kvartalni_izvestaj_III_09.pdf).

available at the central bank or bank group headquarters but that they had not yet so far received instructions what to do in that respect.

Although banks can introduce their own new indicators, it seems that they rarely do so. It may primarily be because this requires formalized IT solutions and these do cost. However, as Johnston et al. (2009) underline, the finance industry needs new indicators (e.g., growth rate in credit that is the prime empirical predictor of credit default), which would provide early warnings of (potential) financial problems. In general, business executives should be aware that indicators in use may be just those that are easier (easiest) to measure, rather than those that are the most informative (Hahn and Hoerl 1998).

Even when mathematics used in models is quite simple and spreadsheets thus very easy to build, the results generated by them may be questionable. Consider, for example, the profitability of child savings displayed at Fig. 1 (data are fictional). As it stands, this service is clearly not productive because this bank needs to invest 1.56 € to get 1 €. This outcome may be questionable, however, because, as one interviewed banker underlined, banks usually do not have reliable and complete data concerning expenses for some of their services, especially for some of its organizational units. This is an instance of a general business problem. It is widely known that production or service total expenses can be divided into fixed and variable ones, but, as the author's colleagues (an economist) highlighted several years ago, it is not easy task for firm to come up with reliable and complete data regarding these expenses. This opens an opportunity to use spreadsheets to test the validity of the data applied. Processing costs may, for example, be estimated by two spreadsheet models before the relevant number is determined. Also, data for one year may be compared with data for another year knowing that some ratios (e.g., material costs/processing costs) would not vary much by year. Finally, using different scenarios for two kinds of costs in question may indicate problematic data. It should be kept in mind that when some test does not detect problematic data, it may not mean that the data are reliable, but rather that this test gave no information on the issue in question (O'Beirne 2009). This implies that the quality of data should not be examined by a single test, which brings us to the issue of critical use of spreadsheets.

Fig. 1 Determining profitability of child savings (example suggested by one interviewed banker)

	A	B	C
1	Profitability of child savings		
2	**Expenses**		
3	Material costs (saving books and money-boxes)	9,000 €	
4	Marketing costs	7,000 €	
5	Processing costs	10,000 €	
6	Interest paid on savings	24,000 €	
7	Total	50,000 €	=SUM(B3:B6)
8	**Revenue**		
9	Interest received on loans	32,000 €	
10	**Profitability** (Rev/Exp)	**0.64**	=B9/B7
11	**For revenue of 1 €, expenses are**	**1.56 €**	

2.1.1 Uncritical Use

Mathematics, and therefore software that makes use of it, may uncritically be applied in modeling. In the case of finance, the sophistication of mathematics used for financial models has increased considerably in the last few decades, whereas the common sense of situations being modled dropped. This situation is nicely illustrated in Wilmott (2000) where a distribution of rate of return (a simple tool) could help Proctor and Gamble (P&G) to analyze the outcome of a financial deal with Banker Trust in a proper context, which eventually caused P&G to lose about $165 million.

An uncritical use of software was revealed by the interviews because, as one banker sad, central bank did not explain procedures to be used just asked regional and branch offices to apply them. Another banker said that to his knowledge, despite its importance, critical use of applied financial models had been a neglected issue at the Association (he also added that some bank(s) might take care about this issue). The expert from the above-mentioned software company remarked that mostly because of budget limitations, just one model was usually implemented per financial issue.

We can, for example, observe an uncritical use of financial models in Serbia in the case of credit risk. Serbian banks seems to use just one model (revealed by the interviews) and it is usually a credit portfolio model called CreditMetrics implemented in Credit Manager software, or by using Microsoft Excel or Matlab tools (Jazić 2007). As there is no best method for the assessment of credit risk, several methods should be combined (Kalapodas and Thomson 2006). Such an approach to other risks and other important financial issues would promote a critical use of financial models that seems largely missing at present.

As one interviewed banker said, main reasons for uncritical use of spreadsheets may be found in the fact that employees typically work with pre-defined templates to be filled with financial data as well as they do so in very demanding working conditions. Undoubtedly, such workplace features promptly suppresses employees' creativity if any. Other interviewed banker estimated that about 30 % of employees may be interested to improve the business matters concerning business and mathematics, but just for themselves to ease their daily work or understand it (better). He added that just 5 % of all employees may have gut to propose some improvements to the board of directors, taking a risk to disclose unwanted issues and eventually even loose their (present) jobs.

As psychologically, employees (or people in general) are resistant to changes, implementing new business procedures, even when they ease financial matters, does not go in a simple way. As an example, consider the work with pivot tables and pivot charts.[2] These business intelligence tools promptly and usually effortlessly enable multidimensional views of large quantities of data, enabling answers to questions such as what region has the highest credit risk and in what area type

---

[2] NumberGo is a free BI tool for multidimensional data views (www.numbergo.com).

(see Fig. 2; data are fictional). As the interviewed banker from the software company (the e-banker for short) highlighted, typically just one or two of all managers and analysts that in one bank received training to use pivot tables and charts regularly do so. It may be that, for most users, pivot tables are not effective and easy-to-use interface because they requires, among other things, the understanding of the multidimensionality of data being manipulated (O'Donnell 2005). Also, as the e-banker remarked, the work with pivot tables and graphs differs from the work with spreadsheets and, with employees' frequent resistance to changes, changes are denied and the opportunity to critically examine the data by using several pivot summary representations (that may reveal problematic data) is not used. Thus, in order to attain a critical use of spreadsheets, (some of) employees should use, compare, and reconcile the outcomes not only of different models of financial issues, but also of several representations of financial data.

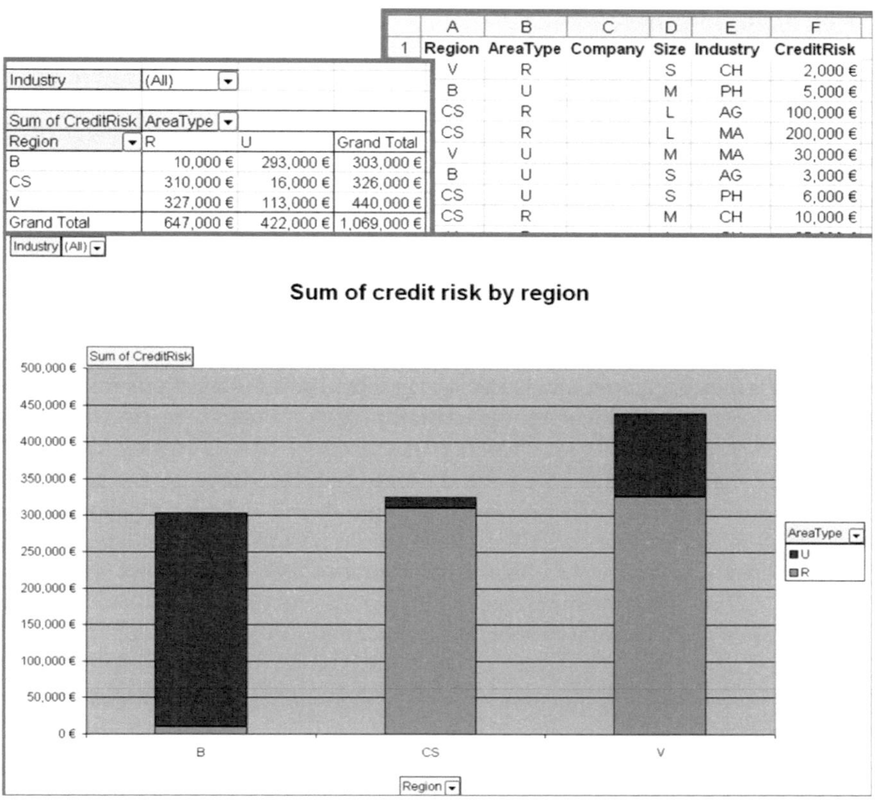

Fig. 2 Credit risk: original data, pivot table, and pivot chart

2.1.2 Erroneous Use

There are more errors in operational spreadsheets then one would expect and these errors may lead to inadequate or wrong decisions causing (considerable) financial losses. This surprising issue is well documented by, for example, EuSpRIG (European Spreadsheets Risk Interest Group), whose Internet presentation contains many case studies, which reveal that errors are usually found in formulas (see www.eusprig.org/horror-stories.htm). A detailed analysis evidenced that, on average, about 1 % of all formulas in operational spreadsheets contain errors (Powell et al. 2009a). By auditing 25 operational spreadsheets, Powell et al. (2009b) found 117 errors: while 70 errors did not have impact of the output, as many as 27 had an impact of at least $100,000 (with seven being $10,000,000 or more). Among these spreadsheets some were used in a large financial firm that calculated tax liabilities measured in the billions of dollars. As the three authors underlined, "these (tax liability) spreadsheets were astonishingly complex, difficult to understand, difficult to work with, and error-prone." (p. 131) A recent report from PricewaterhouseCoopers revealed that "95 % of the excel spreadsheets examined had errors out of which 80 % had significant monetary errors" (Spreadsheet mistakes 2009, p. 1). According to Croll (2009), uncontrolled use of spreadsheets in credit derivatives marketplace even played a substation part in the destruction of capital of the global financial system in the period 2007–2009. Surprisingly, a recent research from Deloitte showed that almost 60 % of companies do not evaluate and protect spreadsheets that are important to their business (Spreadsheet management 2009).

In general, most spreadsheet errors can be attributed to the use of wrong data, model, or function attribute (Powell et al. 2008). They are usually result of (1) applying chaotic spreadsheet design, (2) using numbers in formulas, (3) calculating similar results in different ways, (4) using formulas in a row or a column that change their structure, and (5) using complex formulas. Reasons that prevent their developers to build better spreadsheets may primarily be found in time pressure, unstructured design, changing specification, lack of testing, as well as lack of relevant knowledge and skills (Powell et al. 2009b).

In classifying credit risk, for example, some banks in Serbia assume that an increase in exchange rate of EUR to RSD would be followed by a similar, even higher increase in borrower's salary received in RSD. This means that if the borrower's 200 EUR installment could be monthly covered by 30 % of his/her salary in 2009 (a 30 % limit is applied for most loans), the same would happen in 2010. Let us assume a fixed salary of 66,000 RSD in 2009 and 2010 (it is slightly above 660 EUR at that time, but a 10–20 % reduction might occur as the result of the global financial crisis as did happen in some private firms). Depending on the increase in the exchange rate (e.g., 10 % per year), it may happen that this 200 € in a year time will consume 35 % of the salary or more, compromising the borrower's financial credibility. This undesirable outcome could not be anticipated by using general data a year ago. Of course, credit risk category of each loan is, if need be, updated few times each year, but it is a post-action not a pre-one.

However, this individual approach to credit risk may only be relevant to a small number of larger, long-term loans, not justifying (on the basis of an informal or formal estimation, if any) its application to all loans of various types.

Concrete data regarding financial errors due to the use of spreadsheets or other software in some Serbian banks could not be collected, and, although they do occasionally occur as the interviews revealed, these errors cannot be, for understandable reasons, publicly available. However, it can be said that BI solutions may be questionable if based upon poor data generated by banks themselves, but this outcome banks usually overlook relating errors to the software used. Also, an opportunity to work on financial documents in Microsoft Excel (generated by BI solutions or other employees) by using additional (usually simple) calculations opens a space for introducing errors in these documents.

Although having large and complex spreadsheets that are error-free is an unattainable goal, their developers should avoid errors that can be located and corrected, especially those that may cause considerable financial losses. It may be attained with, for example, self-checking built into a spreadsheet (see O'Beirne 2009, for such checks). If spreadsheet is error free, examining the same financial issue in different ways must produce the same result. This does not happen on Fig. 3 because the last attribute of function VLOOKUP (omitted at present) must be FALSE (supporting exact match with an arbitrary ordered credit risk categories). The error in this hypothetical example was relatively easy to locate. Its source was suggested by an MS Excel add-in for spreadsheet auditing called *Spreadsheet Professional*

C10 fx =A10*VLOOKUP(B10,$A$2:$B$5, 2)

	A	B	C	D
1	Credit risk category	Part of loan amount to secure credit risk	No of loans by category	Total amount to secure credit risk by category
2	A	1.00%	2	193.46 €
3	B	5.00%	2	568.02 €
4	G	25.00%	1	6,250.00 €
5	D	100.00%	1	4,280.33 €
6		Total	6	11,291.81 €
7				
8	Amount of loan to return	Credit risk category	Amount to cover credit risk	**Total amount to cover credit risk** (amount frosen for bank use; when a loan is returned this total is reduced for the credit risk of its loan that is bank free to use; the amount updated quarterly).
9	25,000.00 €	G	6,250.00 €	
10	4,280.33 €	D	214.02 €	
11	2,000.00 €	B	100.00 €	
12	12,645.75 €	A	126.46 €	
13	6,700.00 €	A	67.00 €	
14	9,360.38 €	B	468.02 €	
15			7,225.49 €	

Fig. 3 Securing credit risk according to borrower's financial credibility

(trial copy available at www.spreadsheetinnovations.com). Although this tool reported that no general test errors were discovered, it pointed out that, for VLOOKUP function, the range being searched had to be in the ascending order or errors might occur.

3 Closing Remarks

Financial industry cannot be imagined without the widespread use of spreadsheets. Because of that, financial institutions should cultivate a critical approach to spreadsheet models and their data (including software tools for auditing too), which questions their appropriateness and correctness as well as compares and reconciles the outcomes of different models and data views. Such an approach, as this chapter shows, seems missing in most financial institutions at present. As many people who create and use spreadsheets are self-taught (Croll 2005), this approach calls for the development of an appropriately designed pre-service and in-service vocational courses at secondary and tertiary levels, which would also enhance the management of spreadsheet environment that is today poor in the majority of companies (Spreadsheet management 2009).

According to Hoyles et al. (2007), a good learning design promoting techno-mathematical literacies in the workplace calls for authenticy (examine actual workplace events), visibility (make invisible relationships visible) and complexity (reflect real situations in alternative ways). Inspired with these requirements, promoting these literacies in the context of this chapter should require: (1) understanding data and, if needed, questioning and improving their quality, (2) understanding (some of) relationships in models and, if needed, questioning and improving these relationships, and (3) understanding models and, if needed, questioning and improving them.

In the courses above-mentioned students may develop various deterministic and nondeterministic spreadsheet models involving optimisations. However, the development and application of simple deterministic spreadsheet models without optimizations should be widely practiced first. Even in this kind of modeling students may make many errors in selecting, initializing, and relating variables (Kadijevich 2009, 2013). These errors need to be carefully pedagogically treated, which would promote a critical approach to spreadsheet models and ease the work with more complex models latter.

Acknowledgments The author thanks to seven bankers who participated in the interviews that revealed important data about some features of the finance industry in Serbia with respect to general issues and the use of spreadsheets. This contribution resulted from the author's work on projects OI-174012 and III 44006 funded by the Serbian Ministry of Education, Science and Technological Development.

References

Croll, G. J. (2005). The Importance and criticality of spreadsheets in the city of London. EuSpRIG—European Spreadsheets Risk Interest Group. www.eusprig.org.

Croll, G. J. (2009). Spreadsheets and the financial collapse. Paper presented at the 10th annual conference of the European Spreadsheet Risks Interest Group, Paris, France, 2–3 July 2009. http://arxiv.org/ftp/arxiv/papers/0908/0908.4420.pdf.

Hahn, G., & Hoerl, R. (1998). Key challenges for statisticians in business and industry. *Technometrics, 40*(3), 195–200.

Hoyles, C., Noss, R., Bakker, A., & Kent, P. (2007). TLRP Research Briefing 27—Techno-mathematical Literacies in the Workplace: A Critical Skills Gap. London: Teaching and Learning Research Programme. www.tlrp.org/pub/documents/Hoyles%20RB%2027%20FINAL.pdf.

Jazić, V. (2007). Credit risk management (in Serbian). *Finansije*, LXII(1–6), 113–146. www.mfin.gov.rs/download/pdf/casopis_finansije/casopis_finansije_1-6-2007.pdf.

Johnston, R. B., Psalida, E., de Imus, P., Gobat, J., Goswami, M., Mudler, C. & Vazquez, F. (2009). Addressing Information Gaps (IMF Staff Position Note). Washinghton, DC: International Monetary Fund. www.imf.org/external/pubs/ft/spn/2009/spn0906.pdf.

Kadijevich, D. (2009). Simple spreadsheet modeling by first-year business undergraduate students: Difficulties in the transition from real world problem statement to mathematical model. In M. Blomhøj & S. Carreira (Eds.), *Mathematical applications and modeling in the teaching and learning of mathematics: Proceedings from Topic Group 21 at the 11th International Congress on Mathematical Education, Mexico, July 6–13, 2008* (pp. 241–248). Roskilde University, Denmark: Department of Science, Systems and Models (IMFUFA tekst nr. 461/2009).

Kadijevich, D. (2013). Learning about spreadsheets. In D. Kadijevich, C. Angeli & C. Schulte (Eds.), *Improving computer science education* (pp. 19–33). New York: Routledge.

Kalapodas, E., & Thomson, M. E. (2006). Credit risk assessment: a challenge for financial institutions. *IMA Journal of Management Mathematics, 17*, 25–46.

O'Donnell, P. (2005). The problem with pivot tables. *Business Intelligence Journal, 10*(1), 25–31. http://tdwi.org/research/2005/04/business-intelligence-journal-vol-10-no-1.aspx.

O'Beirne, P. (2009). Checks and controls in spreadsheets. Paper presented at the 10th annual conference of the European Spreadsheet Risks Interest Group, Paris, France, 2–3 July 2009. Internet: http://arxiv.org/ftp/arxiv/papers/0908/0908.1186.pdf.

Powell, S. G., Baker, K. R., & Lawson, B. (2008). A critical review of the literature on spreadsheet errors. *Decision Support Systems, 46*(1), 128–138.

Powell, S. G., Baker, K. R., & Lawson, B. (2009a). Errors in operational spreadsheets. *Journal of Organizational and End User Computing, 1*(3), 4–36. http://mba.tuck.dartmouth.edu/spreadsheet/product_pubs_files/Errors.pdf.

Powell, S. G., Baker, K. R., & Lawson, B. (2009b). Impact of errors in operational spreadsheets. *Decision Support Systems, 47*(2), 126–132. http://mba.tuck.dartmouth.edu/spreadsheet/product_pubs_files/Impact.pdf.

Spreadsheet management: Not what you figured (2009). Deloitte. www.deloitte.com/assets/Dcom-UnitedStates/Local%20Assets/Documents/AERS/us_risk_spreadsheet_mgt_022509%20(2).pdf.

Spreadsheet mistakes (2009). PricewaterhouseCoopers Cyprus. www.pwc.com.cy/en_CY/cy/challenges/assets/pwc-cy-spreadsheet-mistakes-2.pdf.

Straesser, R. (2000). Mathematical means and models from vocational perspectives: A German perspective. In A. Bessot & J. Ridgway (Eds.), *Education for Mathematics in the Workplace* (pp. 65–80). Dordrecht: Kluwer Academic Publishers.

Wilmott, P. (2000). The use, misuse and abuse of mathematics in finance. *Philosophical Transactions: Mathematical, Physical and Engineering Sciences, 358,* 63–73. www.wilmottwiki.com/wiki/uploads/e/e1/Millennium.PDF.

MITACS Accelerate: A Case Study of a Successful Industrial Research Internship Program

Sarah Petersen and Marsh Rebeccah

1 Introduction

Since its inception in 1999, MITACS (www.mitacs.ca) has been funding mathematical sciences research projects across Canada. MITACS has played a leadership role in linking businesses, government, and not-for-profits with over 50 of Canada's universities to develop cutting-edge tools to support the growth of our knowledge-based economy.

A core objective of all MITACS programs is to support the development of the up- and coming generation of Canadian researchers. In 2003, MITACS introduced its Graduate Research Internship Program, with 18 internships in the mathematical sciences in its 1st year. The program, rebranded MITACS Accelerate (www.mitacsaccelerate.ca), grew rapidly and was expanded to all disciplines nationwide in 2008, while retaining a substantial presence in the mathematical sciences. The program supported 608 internships across all disciplines in 2008–2009 and is set to deliver nearly 1,300 internships in 2009–2010.

MITACS Accelerate is directly transforming the way Canada recruits, trains and retains a highly skilled workforce. In addition, by proactively matching the needs of the public and private sectors with university expertise, MITACS Accelerate is driving knowledge and technology transfer between traditionally isolated sectors.

S. Petersen · M. Rebeccah (✉)
MITACS, Inc, Mitacs Inc, Suite 301 Technology Enterprise Facility 6190 Agronomy Road, Vancouver, BC V6T 1Z3, Canada
e-mail: rmarsh@mitacs.ca

S. Petersen
e-mail: spetersen@mitacs.ca

A. Damlamian et al. (eds.), *Educational Interfaces between Mathematics and Industry*, New ICMI Study Series 16, DOI: 10.1007/978-3-319-02270-3_42,

2 Program Strategy

2.1 Objectives

Headed by MITACS as a collaborative effort of Canadian federal and provincial governments, industry, universities, and other organizations, MITACS Accelerate aims to dramatically increase Canadian research intensity through the creation of hundreds, soon to be thousands, of internships. The objectives of the program are to:

- Enhance the training experience of graduate students and postdoctoral fellows by providing opportunities to work on industrially relevant research challenges with impact on Canada's social and economic sectors;
- Provide industry with access to state-of-the-art research;
- Promote technology transfer and knowledge exchange to improve Canada's position at the leading edge of technology innovation and commercialization;
- Introduce and build lasting connections between businesses and academic researchers;
- Increase industrial R&D spending; and
- Through government matching of these contributions, increase funding for university research.

2.2 Challenges

Envisioning and implementing the program to reach these objectives required a major effort to:

- Convince Canadian companies of the benefits of research expertise in addressing business issues;
- Encourage the academic community to consider the needs of the industrial sector in their research programs;
- Persuade federal and provincial governments to include the program in their growth strategies and budgets; and
- Recruit and place significant numbers of trainees in industrially relevant and academically sound research projects.

2.3 Unique Approach

MITACS was able to address the needs of all stakeholders (governments, industry, and academia) by developing a program unlike any other at Canadian universities with:

- A strong research focus (different from co-op or work terms) that results in open, peer-reviewed research (different from consulting contracts);
- Clear deliverables for the industrial partner (different from most research grants); and
- An emphasis on the expertise of the academic supervisor as well as that of the intern (different from scholarship programs which focus exclusively on the merits of the student).

3 Details of the Internship Program

A MITACS Accelerate internship is a collaborative research project that links a graduate student or postdoctoral fellow, their academic supervisor, and a non-academic partner. Internships with for-profit companies form the bulk of the program, with a small amount of funding available for internships with not-for-profit or local government partners. Internships are undertaken in a wide range of areas including manufacturing, technical innovation, business processes, IT, design, and many more.

Each Accelerate internship is 4 months long. The intern gains experience outside the university environment, spending half of the internship working on site with the partner organization to understand the research problem, collect data, interact with employees, etc. The balance of the intern's time is spent at their university, collaborating with their supervisor and other researchers and accessing specialized research resources as needed.

Interns are encouraged to publish their research results, working with the industrial partner to identify aspects of the project which contribute to the research expertise in the field without compromising the company's intellectual property (IP). MITACS takes no stake in IP generated through internships; it is shared between the university and company according to their own negotiated agreements.

Internships may be combined into larger 8- or 12-month projects, which can be applied for through a single proposal. Interns gain proposal-writing experience during the application process and receive valuable feedback on their projects through the peer review process.

3.1 Funding

The industrial partner contributes $7,500 per 4-month internship, and this amount is matched by the federal Industrial Research and Development Internships (IRDI) program (www.nce.gc.ca/irdi_e.htm), provincial grants, universities, and other sources. A research grant of $15,000 is provided to the academic supervisor, of which the intern receives at least $10,000 as a stipend. The remainder of the grant can be used toward the stipend or for research-related costs such as equipment, travel, and conference attendance.

3.2 Flexibility

While the core features of an internship—a research project and time spent on site with the partner organization—are non-negotiable, there is a great deal of flexibility built into the program to enable participation by a broad cross-section of companies and interns. Options include:

- The Intern Travel Subsidy Program provides matching funds over and above the amount of the internship grant, enabling a partner organization to be matched with academic expertise from another geographic region.
- Interns may apply for multiple 4-month internships on one proposal.
- Multiple interns may collaborate on a group of internships related to a larger research project.
- While we recommend that the intern write their own proposal as part of the learning experience, supervisors may apply for internships in order to gain funding to attract a student.
- There are no application deadlines. Proposals are reviewed as they are received, and proposals which pass peer review are approved (subject to availability of funds).
- Internships for postdoctoral fellows are eligible for double-funding ($30,000 per 4-month internship) to provide more competitive salaries.

3.3 Benefits to Stakeholders

MITACS Accelerate provides unique benefits to all three participant groups:

INTERNS: Internships provide exposure to industrially relevant applications of graduate-level research. They are an opportunity to connect with potential employers and to gain business skills in a non-academic setting, as well as to gain additional funding for graduate school.

FACULTY: MITACS Accelerate provides opportunities to apply research expertise to realworld business problems and to build long-term collaborations with companies.

PARTNER ORGANIZATIONS: The program is a low-investment way for companies to tap into research expertise at Canadian universities. Successful internships can lead to long-term collaborations with academia and further access to early-stage research results. Internships may also be a valuable recruitment tool for future employees.

The low investment required for a single internship ensures that the program is accessible to start-up companies and small and medium enterprises, providing them with access to research expertise they may not have in-house, and increasing their competitiveness in the global marketplace.

4 Success Factors

MITACS has identified two primary factors that have led to the ongoing success of the Accelerate program: a business development team and rigorous peer review.

4.1 Building Connections

The success of MITACS Accelerate hinges on building connections between academia and industry. Professors are often expected to take on this task, but in a large and sparsely populated country such as Canada, a company with a research challenge may be thousands of kilometres away from applicable university research expertise. Making these connections is an overwhelming task for most faculty members, resulting in a perceived lack of interest in collaborating with industry.

It is our experience that the small number of joint research projects is less indicative of interest and more of frustration on both sides; organizations often have difficulty finding appropriate academic talent, and professors often cannot identify partners who would benefit from their research expertise.

The solution to this problem lies at the core of the success of MITACS Accelerate: its business development (BD) team. The BD team consists of highly talented individuals with a mix of academic and business experience, who are familiar with the talent available in Canadian universities. This team is skilled at matching industrial opportunities and academic expertise, facilitating initial interaction between parties, and providing feedback on proposal development. BD staff manage expectations on both sides and help applicants frame proposals for mutually beneficial research projects.

Each member of the BD team has responsibility for a geographic region, an industry section, and a number of academic disciplines, ensuring coverage of the entire country. The team approach allows opportunities to be shared and matches made across sectors. To maximize the program's reach, the BD team seeks out companies with little or no history of university collaboration and academic disciplines traditionally distant from industrial interaction.

In addition, the BD team actively seeks internship opportunities in areas of high priority to our federal and provincial government funders. For example, Ontario provincial priorities include clean and sustainable technologies, health science and advanced health technologies, digital media, information and communication technologies, finance, advanced manufacturing, and the automotive sector. First Nations and northern communities are also priority areas across the country.

MITACS also relies on the Accelerate Consortium, which consists of 17 leading Canadian research networks that partner with MITACS to identify and deliver internship opportunities.

4.2 Peer Review

The objective of the process is to maintain a high standard of research quality for internships funded by the program and to provide valuable feedback to students and their supervisors from expert reviewers. Peer review is an essential part of maintaining the academic relevance of the internship program and ensuring that research faculty view internships as part of their student's research training.

Internal reviews of large proposals are conducted by the Accelerate Research Review Committee (RRC). The RRC is a committee of the MITACS Board of Directors mandated to ensure the research excellence of the internship program through a fair and transparent evaluation process.

Arms-length external reviews of all proposals are provided by the MITACS College of Reviewers (COR), a body of over 350 faculty from Canada and around the world. Reviews may be solicited from other sources if appropriate expertise is not available on the COR.

5 Outcomes

- The continued growth of the program and positive feedback from participants are indicators of its success. Exit surveys[1] completed since 2007 show the following outcomes for MITACS Accelerate:
- 87 % of interns were of the opinion that there might be future employment opportunities with their internship partner company
- 98 % of interns felt the internship experience would assist them in securing future career opportunities
- 99 % of interns would recommend the MITACS Accelerate Program to other potential interns
- 92 % of professors anticipated future collaborative projects with the industry partner
- 98 % of professors felt the internship was beneficial to the intern's research program

[1] The results comprise responses from 133 interns, 86 professors, and 86 industry partners

- 99 % of professors would recommend the MITACS Accelerate Program to other university professors
- 96 % of industry respondents indicated that they will utilize the techniques, tools and research advances developed during the internship
- 97 % of industry partners would recommend the MITACS Accelerate Program to other organizations
- 95 % of industry respondents indicated that the opportunity to collaborate with university researchers was beneficial to their organizations.

6 Future Directions

Building on the success of the standard internship model described above, MITACS is exploring several new options to expand the program. One possibility under discussion at several universities is the integration of internships with graduate programs, especially in master's degree programs in fields with a strong applied focus, such as health informatics.

"Development internships'', focusing on the commercialization of research results, are also under consideration for a pilot in 2010.

"Internship clusters" consist of six or more related internships that form a coherent research program. The cluster model was developed in early 2009 in response to growing interest from both academia and industry for larger collaborative research projects. A single research proposal covering the whole project streamlines the application and review process. Clusters provide a higher matching ratio, allowing more student stipends per industry dollar, and provide flexibility in budgeting research costs across all internships in the cluster.

Finally, the MITACS international program connects Canadian mathematicians with scientists in similar organizations around the world, and discussions are underway about the possibility of international internships for graduate students.

Acknowledgments MITACS Accelerate is funded by the federal government of Canada through the Industrial Research and Development Internship program, and through individual funding agreements in each province. We gratefully acknowledge the support of all of our funders.

We would like to thank Tracie Williams for the survey results, and all the MITACS staff whose hard work has made this program a success.

A Meta-Analysis by Mathematics Teachers of the GIFT Program Using Success Case Methodology

Richard Millman, Meltem Alemdar and Bonnie Harris

The ICMI-ICIAM Discussion Document (2009), Educational Interfaces between Mathematics and Industry, talks about "… the intimate connections between mathematics and industry" and, for example, then says that" … mathematics is said to be used almost everywhere. However, these uses are not generally visible except to specialists." The GIFT program is one whose goal is to bridge this gap through substantive projects, which bring practicing middle and high school mathematics teachers into industrial projects and then have them integrate their new experiences back into their classroom. Because of time constraints, teachers without firsthand knowledge cannot provide mathematical situations, which involve "real life problems." The last sentence of 1.1 in the Discussion Document is especially in agreement with our approach; to wit, "In other words, learners should be equipped for flexibility in an ever-changing work and life environment, globally and locally." From the GIFT experience, the flexibility should change the teachers' outlook and, ultimately, the outlook of their students. The last four of the bullets of the "What are the aims of the Study?" and the last two of "Why is there a need for this Study?" fit well with the goals of the GIFT program. The issues of Sections 8 (Curriculum and Syllabus issues) and 9 (teacher training) of the discussion document fit well with regard to the goals of the GIFT project. The present work lays the groundwork for future projects motivated by the discussion document. The results of the present meta-analysis will ultimately be used as a basis for further analysis in a long-term project using a variety of methods for both math and science teachers.

R. Millman (✉) · M. Alemdar · B. Harris
Georgia Institute of Technology, 760 Spring Street, N.W., Suite 102, Atlanta, GA 30308, USA
e-mail: richard.millman@ceismc.gatech.edu

M. Alemdar
e-mail: meltem.alemdar@ceismc.gatech.edu

B. Harris
e-mail: bonnie.harris@ceismc.gatech.edu

A. Damlamian et al. (eds.), *Educational Interfaces between Mathematics and Industry*, New ICMI Study Series 16, DOI: 10.1007/978-3-319-02270-3_43,

In a commitment to providing firsthand connections between classroom activities and real-world applications, the Georgia Intern-Fellowships for Teachers (GIFT) initiated in 1991 by the Georgia Institute of Technology is a collaborative effort between industry and education. GIFT provides mathematics, science, and technology teachers in grades 6 through 12 (students between 12- and 17-years old) "real life" experiences in the applications of those disciplines. Over time, GIFT has placed 1,515 teachers into summer internship positions of 4–7 weeklong in corporate and university research laboratory settings. From the beginning, participants in GIFT benefited from internships provided by long term industry partners such as UPS, Georgia Power, Cisco, and EMS Technologies; and more recently by partners Nordson Corporation, Solvay Pharmaceuticals, CIBA Vision, Gwinnett Hospital System, RFS Pharma, Optima Chemicals, PCC Airfoils, and Stiefel Laboratories, and General Electric. We know anecdotally that these internships have contributed to teachers having increased content knowledge and enhanced teaching practices based on evidence based experiences and now move towards an in-depth analysis. Programs like GIFT are sometimes referred to as "externship" programs.

GIFT is designed with the following goals:

- Provide industry mentors an efficient method of identifying and selecting teachers interested in participating in internships;
- Quickly orient teachers to industry work environments, and mentors to K-12 workplace culture;
- Provide participants (teachers and mentors) support throughout the summer by assigning small groups of teachers to a master-teacher facilitator,
- Assist teachers with creating an "Action Plan" for implementing summer experiences into the classroom or more generally applying the GIFT experience in the classroom;
- Provide support for Action Plan implementation in the classroom through visits by GIFT staff;
- Foster the development of an extended professional community of learners; and
- Encourage extended partnerships for communication and collaboration between teachers and industry mentors and pass that approach on to the students of the GIFT teachers.

To participate in GIFT, sponsors from industry or a university submit an online survey, which includes a position description describing the nature of the summer work, a list of the skills required of the teacher, and a letter of intent for participation. The teachers complete an online application that includes information about their background, courses they have taught, their technology skills, and their geographical preference for work locations. GIFT uses information from the sponsor and teacher databases to coordinate the matching of skills with the preferences of both. Sponsors are given access to applications of teachers who meet their job requirements. Sponsors then interview prospective applicants and select a teacher to hire for the summer. Approximately 150 teachers from in Georgia, U.S. apply to the program each year, with a current average placement of 80 teachers per summer. Once the sponsorships are arranged, each GIFT teacher works with two people. One is a mentor from the industry whom we call an (industry) mentor or sponsor and the second is a facilitator. A facilitator is generally an experienced teacher who has served as a GIFT intern in previous years or a college professor with significant grades sixth through twelfth experience. Facilitators provide guidance in the development of interns' Action Plans, assess the Action Plan, and make recommendation on the allocation of state granted Professional Learning Units (PLUs), a requirement for teachers in the State of Georgia.

1 Rationale for the Study

Research suggests that the quality of the teaching workforce is the single most important factor in predicting student achievement (Darling-Hammond and Ball 1997). "Quality" has many dimensions, however. Effective teachers must have a solid knowledge of academic content, a high mastery of different pedagogical techniques, an understanding of student developmental issues and different ways of learning, and a strong sense of professionalism. Teachers also must have a satisfactory answer to the inevitable question by students—"When am I ever going to use this"? Other than student learning or developmental issues, industrial workplace environments are in the unique position of being able to help teachers develop their strengths in most of these categories through summer internships. When teamed with facilitators, industry mentors can provide motivated teachers summer experiences that show the uses of math skills in industry, that increase the teacher's content knowledge, and that provide new teaching strategies.

These experiences also provide teachers with first-hand knowledge about how industrial scientists actually approach problems, how they design experiments, how they interpret data, how they communicate orally and in writing, and how they come to and implement workplace solutions. And, in perhaps the most powerful effect of all, the teachers' sense of professionalism from these experiences has a continuing influence on them. In that regard, the GIFT program provides teachers an opportunity to connect classroom activities to real-world applications and vice versa.

The point of this evaluation research is to see, using Success Case Methodology, whether the goals of the program that we see informally being achieved are, in fact, supported by data. In addition, we will use any "unanticipated consequences" from this analysis to improve the GIFT program regardless of teacher instructional specialty as long as the insights are transportable to other disciplines. The purpose of this study is to document the success cases of GIFT mathematics teachers in industrial workplace environment. We identified 23 mathematics teachers, each of whom worked in industrial workplace environment, to construct individual cases. This approach will help us to uncover patterns and develop themes across cases (Yin 1994). The case studies document the prior experiences, knowledge, and beliefs these mathematics teachers brought to the program, as well as how those factors interacted with their learning from the program and from their own teaching experiences.

It is hoped that this study will broaden the awareness of mathematics teachers and other educators with regard to industrial work place environment and needs with respect to education. In addition, we hope to explore the relationship between mathematics teachers and industrial workplace environment, and the impact of GIFT program on industrial workplace environment.

2 Study Design

The Success Case Evaluation Method (SCM). Brinkerhoff (2003) originally developed the Success Case Method (SCM) to evaluate the impact of interventions on business industry goals. It is a simple process that combines analysis of outstanding groups with case study and story-telling. The primary goal of the model is to assess how well an organizational intervention is working by focusing on extreme (that is, both "success" and "unsuccess") groups (Coryn et al. 2009). It is a way of exploring whether and how well an initiative is working. Furthermore, it is designed to identify the contextual factors that differentiate successful from unsuccessful cases. The stories are supported with evidence to confirm their reality. According to Brinkerhoff (2003); *A success story is not considered valid and reportable until we are convinced that we have enough compelling evidence that the story would 'stand up in court' … if pressed we could prove it beyond a reasonable doubt .*

The core questions of the SCM approach are:

- What is really happening?
- What results are being achieved?
- What is the value of the results?
- How can it be improved?

Although SCM has been used to evaluate training initiatives and new work methods, it has also been used in educational setting to determine the reasons that influence the academic achievement of minority students (Coryn et al. 2007). More recently, SCM has been proposed as an alternative approach to reexamine causal relationships when more scientifically rigorous designs are not practical and not feasible (Brinkerhoff 2003; Shriven 2006a). SCM is a five-step procedure:

1. Focus and plan the SCM;
2. Create an impact model;
3. Survey all program recipients to identify success and nonsuccess cases;
4. Interview a random sample of success and nonsuccess cases and document their stories;
5. Document findings, conclusions, and recommendations.

According to Coryn et al. (2007), in step one, the researcher needs to determine the focus of the SCM study that can be used for both formative and summative reasons. In step two, the cases are identified as high ("success cases"), moderate (average cases) and low ("unsuccess cases"). The survey method usually is used to identify cases. We are now in the process of constructing the required survey. In step four, these identified cases are used to create a sampling strata that represents both success and unsuccess cases. Therefore, SCM method is an analysis of extreme or outlier cases where independent evidence is sought to support claim of success or failure. Next, the underlying reasons for success and unsuccess case are investigated using a semi-structured interview method that searches for an explanation from a random sample of extreme cases. In final step, the SCM findings, conclusions, and explanations are put together across all cases. Furthermore, the final report is usually presented as a meta-analysis of success stories.

Why SCM method? In order to learn from GIFT's program impact and to explore the impact of leadership improvement on industrial work place environment, an evaluation methodology was needed. SCM is a valid method to evaluate the GIFT program because it allows the researcher to assess the past efforts in a particular program (Coryn et al. 2009). It is particularly useful as an evaluation method for GIFT because it allows for an in-depth exploration of (1) what impact it is having on mathematics teachers and (2) what are the specific barriers that exist in the industrial workplace environment that GIFT mathematics teachers encountered. The result of using the SCM is that it can be used to understand and evaluate the effectiveness and impact of GIFT mathematics teachers. Applying SCM to historical data will provide insight for the evaluation through other methods of the GIFT program.

The Developed Impact Model. At the core of the SCM approach, the following Impact Model will be followed:*Capability* → *Critical Actions* → *Key Results* → *Organizational Goals*

Capability: What new or improved capabilities were acquired as a result of participating in GIFT?

Critical Actions: What the participants did with the new or improved capability?

Key Results: What did the mathematics teachers achieve?

Organizational Goals: How did the results achieved affect goals of the GIFT program and industrial work environment?

With the SCM method, we will be able to assess the GIFT program with these four categories. In other words, we will assess to what extent the GIFT intervention program developed improved capabilities, and then resulted in improved actions in industrial workplace environment, and whether those actions helped the industrial work environment. Furthermore, we will investigate how those achieved results affected the overall program goals. We will gather information to support this impact model through a combination of survey(s) and interview(s) as well as data verification(s). This research will be also a case study of "stories" as these are compelling ways to demonstrate improved leadership.

The participants for this study will consist of 23 mathematics teachers who did a total of 26 internships and were in the program at least one of the summers of 2007, 2008, and 2009. Demographically, 14 are high school teachers and 9 are middle school teachers. There were African American (16), Caucasian (6), and one Asian-American in the sample. In addition, 17 of them are female and 6 male. Their numbers represent 14 suburban, 6 urban, and 3 rural school districts. To assess the program impact, we will interview the teachers about their experience in industrial working environment, and ask them to complete a survey about their experiences with the program.

This proposed study will examine 23 mathematics teacher cases (here, case refers to per internship experience) to determine what impact GIFT is having in the industrial workplace environment. Furthermore, we will identify the barriers to the work. It is possible that we may reconstruct our own version of an Impact Model based on the information that will be found in the cases while we are analyzing each of the 23 cases. An SCM study results, typically, in two immediate ways. One is "in-depth stories of documented business impact that can be disseminated to a variety of audiences with the company," and a second "Knowledge of factors that enhance or impede the impact of training on business results" (Brinkerhoff and Dressler 2002). According to Brinkerhoff (2003), in some cases, the researchers view the logic differently from what had originally been written, and they can change some components of the model for consistency across the models. It is important to emphasize that this method allows us to evaluate many cases across the state, and not just cases that are considered successful.

References

Brinkerhoff, R. O. (2003). *The success case method*. San Francisco: Berrett Koehler Publisher.

Brinkerhoff, R. O., & Dressler, D. E. (2002). Using evaluation to build organizational performance and learning capability: A strategy and a method. *Performance Improvement, 41*(6), 14–21.

Coryn, C. S., Schroter, D. C., & Hanssen, C. E. (2009). Adding a time-series design element to the success case method to improve methodological rigor: An application for nonprofit program evaluation. *American Journal of Evaluation, 30*(1), 80–92.

Coryn, C. S., Schroter, D. C., & Hanssen, C. E. (2007). *A study of successful schools for Hawaiians: Identifying that which matters*. Kalamazoo: Western Michigan University. The Evaluation Center.

Darling-Hammond, L., & Ball, D. L. (1997). *Teaching for high standards: What policymakers need to know and be able to do*. Paper prepared for the National Education Goals Panel, Washington, D.C.

ICMI-ICIAM Discussion Document (2009). Educational interfaces between mathematics and industry.

Shriven, M. (2006a). *Can we infer causation from cross-sectional data?* Retrieved 2 Oct 2009, from http://www7.nationalacademies.org/bota/1School-Level%20Data_Michael%20Scriven-Paper.pdf.

Yin, R. (1994). *Case study research: Design and methods* (2nd ed.) Thousand Oaks: Sage.

Cultivating an Interface Through Collaborative Research Between Engineers in Nippon Steel & Sumitomo Metal and Mathematicians in University

Junichi Nakagawa and Masahiro Yamamoto

1 Introduction

Nippon Steel & Sumitomo Metal is the second largest producer of crude steel in the world, currently manufacturing all types of steel products at ten different steelworks and supplying economical, high-quality, innovative steel products. Our software and hardware resources amassed through steel operations offer great potential for applications to urban development such as efficient energy utilization, power generation, waste recycling, water treatment, transportation, regional developments, and computer systems. We are reinforcing our manufacturing skills and continuing to refine our technologies.

The steel-making process requires control of a diverse range of phenomena: combustion, chemical reactions, granulation of raw materials, fluid dynamics of molten steel, heat transfer, plastic deformation, crystallization, and so forth, which involve mathematical applications for solving and modeling such problems.

Nippon Steel & Sumitomo Metal has collaborated internationally with various mathematicians for decades and resolved industrial problems by enhancing practical insights with mathematical reasoning. Engineers in Nippon Steel & Sumitomo Metal have learnt how to view the phenomena only by the rules of pure logic. On the other hand, mathematicians in universities have learnt how to link mathematics with the physical reality of the phenomena. As a result, the collaborative research is playing a major role in mathematical innovation to broaden applications of the diverse range of mathematics in both academia and industry.

J. Nakagawa (✉)
Advanced Technology Research Laboratories, Energy and Environment Research, Technical Development Bureau, Nippon Steel & Sumitomo Metal Corporation, Shintomi 20-1, Futtsu, Chiba 293-8511, Japan
e-mail: nakagawa.junichi@nsc.co.jp

M. Yamamoto
Graduate School of Mathematical Sciences, The University of Tokyo, 3-8-1 Komaba, Meguro, Tokyo 153, Japan
e-mail: myama@ms.u-tokyo.ac.jp

A. Damlamian et al. (eds.), *Educational Interfaces between Mathematics and Industry*, New ICMI Study Series 16, DOI: 10.1007/978-3-319-02270-3_44,

2 Collaboration Style

Figure 1 shows our style of collaboration with engineers and mathematicians in the case of Nippon Steel & Sumitomo Metal and the University of Tokyo. We formed international task force teams made up of faculty members, post-doctoral fellows, and doctor course students. Team members are selected flexibly to create a task force according to the characteristics of the task. Our collaboration is composed of six indispensable phases.

The first is "intuition and expertise" from industry. Intuition and expertise can be gained exclusively by insight based on observation of phenomena in the manufacturing process. The insight should be enhanced by mathematical reasoning. The second is "communication." Communication is bilateral translations: the translation of phenomena to mathematics and the translation of mathematics to phenomena. Engineers in industry need to understand real problems on-site, express them in the language of physics, and offer possible model equations to mathematicians. Mathematicians explore the underlying mathematics to the model equations. This forum for communication through the interpretation of phenomena is extremely important in order that engineers and mathematicians may reach a common understanding of the nature of the problem and the mathematical components. The third is "logical path." This corresponds to the extraction of mathematical principles from phenomena. Better communication can create a more logical path. The fourth is "analysis of data." This means reasonable and quantitative interpretation of observations carried out on-site. This enables us to extract the essence of phenomena. The fifth is "manufacturing theory." This means the integration of logical paths from viewpoints of operation and economic rationality on site. The last is "activation to mathematics." Motivation for mathematicians has launched new mathematical research fields.

Engineers in industry, have been eager to free themselves from restrictions in our conventional thinking by making full use of mathematical reasoning that is free from specific industrial fields, through wider borderless collaborations. We have examined various conjectures by mathematicians and gained better practical solutions and further utilized analysis results. By repeating such phases of collaboration many times, we are able to pursue economic rationality, and mathematicians are able to find new results and describe them as theorem for future wider uses. It is important that mathematicians work not only for mathematical interests but also for the economic rationality through teamwork with engineers from a long-term point of view.

The cultivating interface between mathematics and industry has come into being as a forum for communication with the mathematicians mentioned above. Communication between the team members who are engineers in industry, and faculty members, post-doctoral fellows, and doctor course students in university mathematical departments, has enhanced their communication skills day by day. As a result, several new themes have been launched.

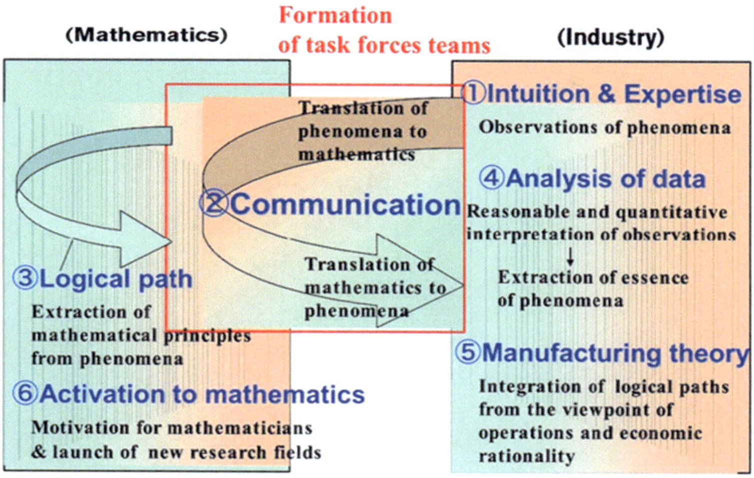

Fig. 1 Collaboration with engineers and mathematicians in the case of Nippon Steel & Sumitomo Metal and the University of Tokyo

3 Example of Interdisciplinary Collaboration

Figure 2 shows a challenge faced by Dr. Yuko Hatano. She is an associate professor affiliated with University of Tsukuba whose major is Risk Engineering, and she had already collaborated with Nippon Steel & Sumitomo Metal on another subject.

The objective is to predict the progress of soil contamination. It is concerned with mass diffusion in a porous medium such as soil and numerical simulations using traditional advection diffusion equations often fail to predict observation results of a real phenomenon observed in the field or laboratory tests. For instance, there are cases where actually the concentration is beyond the environmental standard as shown in Fig. 3, even when a simulation by traditional diffusion equation indicates that the concentration of the pollutant is below the relevant environmental standard and the danger of soil pollution is unlikely. Diffusion not following the prediction based on such a simulation is called anomalous diffusion, in contrast to the traditional diffusion equations, and is often observed in different manners with various substances in the soil or atmosphere in the real environment.

The above issues frequently occurred when using coarser grids than voids and numerically simulating a soil system where voids are distributed unevenly between particles. This type of problem will not occur when the grid spacing is smaller than the voids between soil particles, for instance, about 0.1 mm. However, since several kilometers or more is a normal scale for environmental studies, in view of computer load the use of such a fine grid for a three-dimensional case is extremely difficult, and is practically unsuitable for on-line field analysis. Moreover, whereas

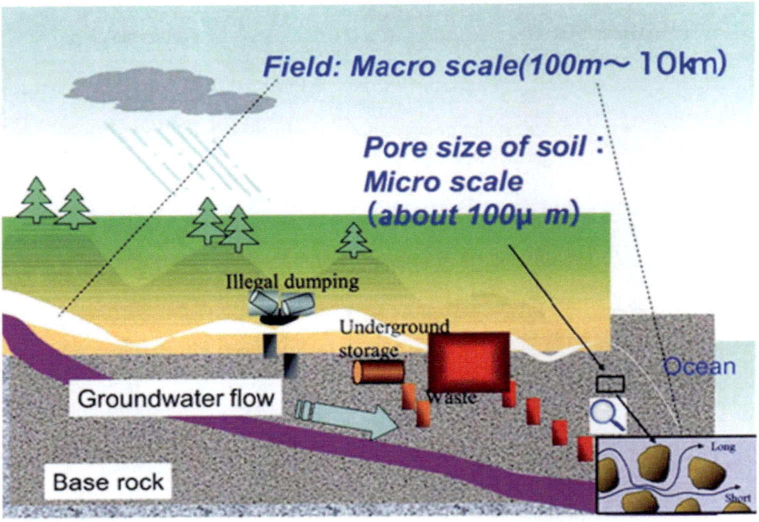

Fig. 2 Prediction of soil contamination in large scale and long term

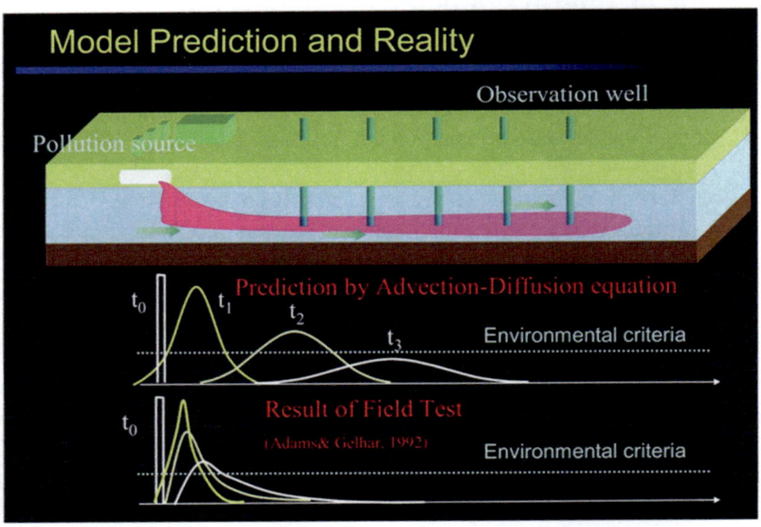

Fig. 3 Comparison between model prediction and results of field tests

a model test covers a timescale of as short as minutes to days, the prediction of a real environmental problem must deal with a timescale as large as a few years to tens of years.

Although we have to treat widely varied sizes of data obtained through physical and numerical tests based upon different scales of space and time, the scaling law allows us to combine those data together in accordance with principles of phenomena.

Large-scale numerical simulation is the principal method for the dynamic analysis of substances in any environmental medium: air, water, or soil. Many detailed chemical and bio-chemical reactions are incorporated in the program codes for environmental simulation, and as a result, simulation programs seem to be becoming increasingly complicated these days. While a great number of numerical simulations are conducted on environmental issues, it is often difficult to tell whether each of such simulation results is valid, which fact is most serious for the problems.

Therefore the present study aims at dynamic prediction of environmental phenomena not totally depending on conventional numerical simulations but also employing mathematical methods such as scaling law. Toward this end, it is desirable to create a new field of environmental study involving mathematicians.

4 Launch of New Research Field in Mathematics

A stochastic method employing random walk in consideration of the distribution of the waiting time of particles is used for describing mass transfer in soil. The stochastic method has been effective when applied to the small space dealt with in laboratory tests, but the limitation on the number of particles is a bottleneck due to the limit of computer capacity, and thus the method cannot respond effectively to more pragmatic requirements of calculation in a larger volume of space.

On the other hand, some fields of physics and engineering employ numerical simulation based on a diffusion equation that includes a fractional-order derivative in time. While the concept of a fractional-order derivative can be traced back to as long ago as Leibniz (see Podlubny 1999), a theory of fractional-order partial differential equation has not yet been established well for applications to such numerical simulations, and the application of such a method has been limited to some special cases of lower spatial dimensions or the whole spaces. It is reported in the literature (Sokolov et al. 2002) that, according to the scaling law to the effect that the root mean square of the displacement of particles is in proportion to time raised to the αth power (t^{α}), the stochastic method using the random walk mentioned earlier is closely related to the Fokker–Planck equation, which leads to a fractional-order derivative:

$$(\partial/\partial t)^{\alpha}u(x, t) = \nabla \cdot (k\nabla u(x, t)) - \mu \cdot \nabla u(x, t),$$

where $u(x, t)$ k and μ are the probability density function of particles, their diffusion coefficient, and mobility acting on them, respectively. It is expected that a scaling law combines stochastic methods such as the random-walk model for anomalous diffusion with the theory of partial differential equation including a fractional-order derivative to form a new field of research for mathematical concept and methodology. In (Cheng et al. 2009), we discuss a related topic with such a theory.

Besides the above, Hatano et al. found that a formula empirically derived from two short-term atmospheric pollution cases (emission of inert gas Kr–85 from a nuclear plant in U.S.A. and the data of aerosol collected by an international team on global warming in the Arctic Ocean region) can describe the behavior of the pollutant of a long-term atmospheric pollution case (the accident of the Chernobyl Nuclear Power Plant) reasonably well (Hatano and Hatano 1997; Hatano and Hatano 1999). The formula is also written as a scaling law, but it is not yet been fully clarified why the formula has such a form.

Thus, through the collaboration of mathematicians and engineers from both academic and industrial fields, the present study establishes the fundamental logical structure that lies behind the scaling law observed in the behavior of pollutants in different environmental media such as soil and atmosphere, and thus clarifies the essence of scaling law.

5 Future Plan

In industrial practice, a reduced-scale model is constructed to analyze a phenomenon that takes place in real-size equipment, significant physical values for the phenomenon in question are described by dimensionless numbers, and the dimensionless numbers obtained from the model analysis are made to match with those of real-size equipment. This matching operation secures the similarity of the dynamic physical values between the model and real-size equipment. This similarity refers also to the scaling law. It has been found from the above viewpoint of scaling law that, in addition to the physical parameters such as time and length which have been conventionally used for scaling up, the fractional powers in the differentiation of time and space are essential. This means that mathematics is expected to present a new "angle of view" for the scaling law that deals with inhomogeneous media. Practically, environmental analysis deals with a scale of several kilometers or more in size. In this relation, establishment of scaling laws including an a priori choice of an exponent will make it possible to appropriately use results obtained through reduced-scale tests and clarify a real phenomenon across a large space.

By establishing scaling laws and developing mathematical methods based on them, we can significantly reduce costs for producing high-quality products as well as energy consumption and CO_2 emission by improving production efficiency in various problems of manufacturing industries such as monitoring of sintering processes, reactions in a blast furnace, and other metallurgical reactions in steelmaking processes as shown in Fig. 4.

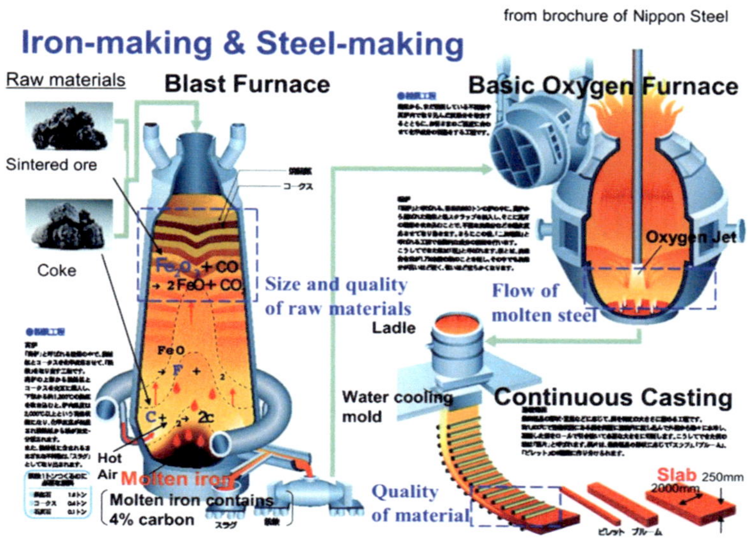

Fig. 4 Example of inhomogeneous media in iron making and steel making processes

Scaling laws and mathematical methods are applicable also to a wide variety of fields such as chemical engineering, mechanical engineering, geotechnical engineering, biotechnology, etc., and therefore, the establishment of such scaling laws is expected to be useful in remarkably accelerating the development of science and technology through the solution of important industrial problems.

Furthermore, the concept of scaling law combining micro and macroscopic aspects is closely related to that of multi-scale modeling, the application of which is rapidly expanding in material science, chemistry, and other widely varied fields. The present study is expected to lead to proposals of new mathematical concepts and methodologies for multi-scale modeling, bringing about new problem recognition and methodology to mathematics.

6 Suggestions Regarding Future Research of Activities in the Field of EIMI

"Mathematics for phenomena" will be the key for combining mathematics with industrial technology. Mathematical science can be understood as mathematics for phenomena; it is aimed at extracting fundamental principles behind different natural phenomena and engineering problems, and crystallizing them into mathematical structures.

Beyond conventional numerical operation of physical model equations, a methodology based on the principles and rules of mathematics makes it possible to construct mathematical models that describe the essence of a phenomenon selectively. Such mathematical models serve as important basis for understanding and controlling a phenomenon. When a mathematical model describes the essence of a phenomenon as simply and comprehensibly as possible (a minimum necessary model), it becomes easier for engineers and researchers from a variety of technical fields to study, and it becomes easier to conceive ideas that can lead to innovations.

In order to construct such a minimum necessary mathematical model that describes the essence of a phenomenon efficiently, a framework is required for the joint work of mathematicians and engineers from academic and industrial fields where they can thoroughly discuss subject phenomena and define suitable targets and milestones for different study stages. In addition, it is indispensable to mutually confirm work progress. At present, however, applied mathematics in Japan, compared with other developed countries, seems to lack such teamwork experience that helps to combine a phenomenon with mathematical methodology. In order to solve a problem as promptly as required in industry, it is too late to begin studying methodology after posing of the problem. It is necessary to continue to improve the skill to combine a phenomenon with mathematical methodology for its prompt application, and in this respect, each individual must improve their qualification to be "the right person" who can meet the above conditions and the role.

It is desirable that both mathematics and industry foster people capable of working jointly with each other from the viewpoint of "mathematics for phenomena" through academic-industrial collaboration. Towards this end, it is necessary to create a new framework independent of the structure of present industry and academic organizations. We must reinterpret and reconstruct the fundamental concept of manufacturing based on field practice, which constitutes the competitive edge in developed countries, from the standpoint of mathematical methodology while learning about interdisciplinary collaboration from abroad. By so doing, we will be able to command the most advanced industrial technology of the world.

References

Cheng, J., Nakagawa, J., Yamamoto, M., & Yamazaki, T. (2009). Uniqueness in an inverse problem for a one-dimensional fractional diffusion equation. *Inverse Problems, 25*, 115002.

Hatano, Y., & Hatano, N. (1997). Fractal fluctuation of aerosol migration near Chernobyl. *Atmospheric Environment, 31*, 2297–2303.

Hatano, Y., & Hatano, N. (1999). Aeolian migration of radioactive dust in Chernobyl. *Zeitschrift für Geomorphologie, 116*, 45–58.

Podlubny, I. (1999). *Fractional differential equations*. San Diego: Academic Press.

Sokolov, I. M., Klafter J., & Blumen, A. (2002). Fractional kinetics. *Physics Today, 55*, 48–54.

An Introduction to CUMCM: China/Contemporary Undergraduate Mathematical Contest in Modeling

Jinxing Xie

China Undergraduate Mathematical Contest in Modeling (CUMCM) is a national event held annually in China since 1992. The first character "C" in CUMCM is now redefined as "Contemporary" for participants from outside China Mainland. During the three-day contest, teams of up to three undergraduate students investigate, model, and submit a solution to one of two simulated real-world problems in engineering, management, etc. The aim of the contest is to expose students to the real-world challenges inherent to mathematical modeling and applications, and provide educational (creativity, challenge, etc.) experience unique to problem bare learning. The contest has been influenced by and has also significantly influenced the teaching of mathematical modeling and applications in China. In this paper, we briefly introduce the aims and scope, organization and achievement of CUMCM. In particular, we explain why CUMCM is so successful in China. We also talk about some problems and difficulties for the contest we currently face.

The tertiary education in China has a very short history compared with the long history of China and the long history of the tertiary education in Europe. In fact, the first university in China was not established until about 100 years ago, and before 1950s there are only about 100 universities in China, each of them with very few students. After 1950s, the number of universities increased dramatically and now there are more than 2,000 universities in China. However, before 1980s, the mathematical education system in nearly all China universities follows the (former) Soviet Union style and focuses on teaching the students "mathematical knowledge and skills" other than the ability of using mathematics to solve real-world problems.

In the last 30 years, as reform-and-open policy was carried out in China, more and more universities recognized this kind of teaching style should be changed and a more close relationship between mathematics and industry should be emphasized. In particular, mathematical modeling courses and related activities are

J. Xie (✉)
Department of Mathematical Sciences, Tsinghua University, Beijing 100084, China
e-mail: jxie@math.tsinghua.edu.cn

A. Damlamian et al. (eds.), *Educational Interfaces between Mathematics and Industry*, New ICMI Study Series 16, DOI: 10.1007/978-3-319-02270-3_45,

highlighted as the breakthrough of reforming mathematical education in China universities (Xiao 2000; Jiang et al. 2007; Xie 2010). The reason behind is that mathematical modeling is not only the first step to apply mathematics in industry, but also very difficult for the modeler and original innovation is needed. The key to mathematical modeling teaching process is to create an environment to arouse students' desire to learn and develop their ability of self-study, and to enhance their application and innovation ability. In order to improve the students' quality in mathematics, the emphasis is put on the students' ability of acquiring new knowledge and the processes of problem solving, rather than only knowledge and skills in pure mathematics. Therefore, mathematical modeling is gradually becoming the best bonding point to enhance students' mathematical knowledge and application ability.

In order to promote the teaching of mathematical modeling courses and providing the students with more chances to doing mathematical modeling, a contest named CUMCM was created in China in 1992. In my own opinion, CUMCM is the most important event in China university mathematical education in the last 30 years, which significantly changed, and will continue to change the contents and forms of mathematical education in China universities. This paper summarizes the history and current status of CUMCM and its influence in China universities.

1 Aims, Scope, and History

CUMCM is a national annual contest in China for undergraduates. The first character "C" in CUMCM is now redefined as "Contemporary" for participants from outside China Mainland. The aim of the contest is to give students an opportunity for practicing the whole mathematical modeling process; to improve students' understanding of mathematics, especially mathematical modeling; to enhance students' motivation for learning and applying mathematics; and to cultivate students' overall competency (creativity/innovation, collaboration and practice competencies).

The contest rules are completely different from traditional mathematical contests such as Olympic mathematics competition. The students should participate in teams other than individually, with each team consisting of up to three undergraduate students from the same institution. Each team's task is to investigate and develop models to solve one of two given contest problems, which simulate real-world problems in engineering, management, etc. The contest lasts for three days (72 h), usually starting at 8:00 a.m on the second or third Friday in September and ending at 8:00 a.m on Monday of the following week. During the contest, teams are permitted to reference any materials they wish (including any software, references or data source they can find in library, Internet, etc.), but they must cite all sources. Failure to credit a source will result in a team being disqualified from the competition. Team members may not seek help from or discuss the problem with their advisor (which is only optional) or anyone else, except other members of the same team. That is to say, inputs of any form from anyone other than the team members are strictly forbidden. Before the ending time of the contest, based on their own research work, each team should submit a solution paper to the contest organizer, which is the only evidence used for rank the teams.

This unique feature of the contest is first initiated in Unites States other than in China. In fact, the Consortium for Mathematics and its Applications (COMAP) first organized the Mathematical Contest in Modeling (MCM) in 1985. In 1999, COMAP start to organize a similar contest named ICM (Interdisciplinary Contest in Modeling). Teams from China participated in MCM every year since 1989 and ICM since 1999. In recent years, the majority of the participants (more than 80 % of the MCM teams and 90 % of the ICM teams) are from China (see Table 1).

CUMCM can be seen as a Chinese copy of MCM/ICM. Recognizing the MCM is beneficial to the students and helpful to the mathematics education reform in

Table 1 The statistics on Chinese students participating in MCM/ICM and CUMCM

Years	MCM in USA		ICM in USA		CUMCM in China	
	Number of all teams[a]	Number of Chinese teams[a] (%)	Number of all teams[a]	Number of Chinese teams[a] (%)	Number of institutions	Number of teams
1989	211	4(1.9)[b]				
1990	235	6(2.6)				
1991	260	21(8.1)				
1992	292	26(8.9)			74	314
1993	259	40(15.4)			101	420
1994	315	84(26.7)			196	867
1995	320	84(26.3)			259	1,234
1996	393	115(29.3)			337	1,683
1997	409	107(26.2)			373	1,874
1998	472	138(29.2)			400	2,103
1999	479	155(32.4)	[c]	[c]	460	2,657
2000	495	169(34.1)	[c]	[c]	517	3,210
2001	496	198(39.9)	83	38(45.8)	529	3,887
2002	525	216(41.1)	106	54(50.9)	572	4,448
2003	492	218(44.3)	146	83(56.8)	637	5,406
2004	600	297(49.5)	143	102(71.3)	724	6,881
2005	644	389(60.4)	164	126(76.8)	795	8,492
2006	748	466(62.3)	224	194(86.6)	857	9,898
2007	949	627(66.1)	273	234(85.7)	969	11,742
2008	1,162	847(72.9)	380	357(93.9)	1,023	12,846
2009	1,675	1,282(76.5)	374	340(90.9)	1,135	15,042
2010	2,254	1,854(82.3)	356	331(93.0)	1,196	17,317
2011	2,775	2,357(84.9)	735	683(92.9)	1,251	19,490

Note [a] The data are collected from http://www.comap.com

[b] The bracketed numbers represent the proportion of Chinese teams in all teams participated

[c] For the MCM and ICM contests of 1999 and 2000, ICM teams are included in the MCM teams in this table

universities, the China Society for Industrial and Applied Mathematics (CSIAM) began to organize CUMCM in 1992. CUMCM is co-organized by CSIAM and the Ministry of Education of China since 1994, and from 1999, the contest has been divided into two levels (categories)—University Level for four-year university students, and College Level for two-or-three-year college students. Because of the very challenging nature of the contest, it attracts the most competitive students in China in an ever-increasing manner. Currently, CUMCM has become the most widespread extra curricular scientific activity for undergraduates in China. In 2011, there was participation by 19,490 teams from 1,251 institutions in the contest, representing almost all of the most prominent institutions and more than 50 % of all institutions in China. It is also interesting that more than 80 % of the participants are engineering, economics, management, and even humanities majors, other than mathematics majors one might expect. Table 1 gives the statistics on Chinese students participating in the contests MCM, ICM, and CUMCM. More details about CUMCM can be found in Li (2011), or from the web site http://en.mcm.edu.cn (in English) or http://cn.mcm.edu.cn (in Chinese).

In order to celebrate its 10th, 15th, and 20th anniversaries, CUMCM organized university students' summer camps on mathematical modeling in 2001, 2006 and 2011 respectively. Besides, motivated by CUMCM, many similar contests, with possibly smaller sizes and dedicated scopes, are also organized by various organizers in China.

2 Organizing and Judging System

There is a CUMCM National Organizing Committee (NOC), which is set up by the Ministry of Education of China and CSIAM. In most of the provinces or regions in China, there is a CUMCM Local Organizing Committee (LOC). Currently, 28 of the total 34 provinces (or regions) in China have established their own LOC. Currently, the registration fee for each team is 200 Yuan (RMB), but no registration fee is needed for teams from outside China Mainland. Some industrial companies (e.g. China Higher Education Press, MathWorks China Ltd, etc.) provide financial support to the contest.

The contest problems are usually simplified versions of real world problems. Each year, the contest organizer (i.e., NOC) sends "call for problems" to professionals (professors, engineer, manager, etc.) to ask them to submit their problem suggestions as candidates. Currently NOC can usually collect more than 40 suggested problems each year. NOC will choose from them the most suitable ones and then finalize them. When choosing a problem for the modeling contest, it is important to make balance between two aspects: (1) the problem should be challenging and attractive to the students, usually with no solutions can be easily found from anywhere (that's not easy in current Internet age); (2) to solve the problem, the students do not need to use very complicated knowledge and skills in mathematics, since most of the participants do not major in mathematics and they only learn standard mathematical courses in universities. A list of all the contest problems for CUMCM 1992–2011 is provided in Appendix for your reference. The details of the problems (in English) can be freely downloaded from http://en.mcm.edu.cn. As you can notice, these problems are from very different industrial backgrounds, but all of them are interesting and attractive to students.

Teams register, obtain contest materials, and download the problems and data at the prescribed time through the CUMCM website. During the three-day contest, teams of University (College) Level can choose any one from the two contest problems of *A* and *B* (*C* and *D*). There is no need for all the participants to get together during the contest, since the students finish their tasks on their own campus. Each team should submit a solution paper to the corresponding LOC before the contest deadline. After the contest, LOCs will starts judging to rank the submissions by the contestants. About top 10 % of all entries will be submitted to NOC for second round evaluation, and the others will also be ranked and perhaps, awarded prizes at the regional level. After the second round evaluation by NOC, only about top 1 % of the total solution papers will be awarded with the national-level first prize, and the figure for the second prize are about 6 %. Finally, each year about 15 outstanding papers will be selected and published in the journal *Engineering Mathematics*, which is one of the official journals of CSIAM.

The solution papers (including appendices such as computer programs) are the only materials used for rank the teams. Although there are no absolutely correct or incorrect answers for the contest problems, there are basically four criteria which can be followed to evaluate and rank the solution papers: the reasonability of the model assumptions, the creativity/innovation of the model, the correctness of the solutions, and the readability of the presentation.

3 The Training and Its Influence

Most students who register in the contest can get some guidance from their advisors before the contest. They usually have some kind of training on how to participating in CUMCM from their mathematical modeling courses, related mini-courses, or seminars. Some students prepare for the contest independently through studying materials related to mathematical modeling.

The most important achievement of CUMCM is that the contest has successfully promoted mathematical modeling courses and related activities in China. As CUMCM getting more and more popular, the mathematical modeling courses are also getting more and more popular. Before 1990s, only about 30 top universities offered the mathematical modeling course to students in mathematics majors; while currently, almost all students in all majors can get some training in mathematical modeling. Conversely, the popularization of mathematical modeling courses has enhanced the quality of the contest.

Teachers of mathematical modeling courses are the key to the success of CUMCM since they not only teach the students mathematical modeling skills but also usually serve as the contest teams' advisors. In order to properly prepare the university professors as contest advisors and good teachers in mathematical modeling courses, in the last 20 years NOC have cooperated with CSIAM and several universities to organize short-term training seminars. NOC also organizes a national conference titled China Conference on the Teaching of Mathematical Modeling and Applications (CCTMMA) every two years to as an educational forum of exchange where teachers share information and discuss how to mentor the students effectively and how to prepare and teach a high quality mathematical modeling course. These activities have had great impact on the mathematics education reform, and have enhanced the teaching quality of courses related to the mathematics in most universities and colleges. Currently, mathematical modeling courses are offered in about one thousand universities in different forms, which are more than half of all the universities in China. The contest also boosts the publication of many innovative textbooks on mathematical modeling and mathematical experiments. In the last decade, more than 110 Chinese textbooks on mathematical modeling were published, which are suitable to be used in courses for universities of different levels and students from different majors. In these textbooks, various industry problems are modeled as mathematical problems which can be solved with the students' knowledge learnt from their fundamental mathematics courses. Most of the university students, no matter which majors they enrolls in, can get some training in mathematical modeling. Furthermore, more than 200 universities have their own students' societies on mathematical modeling, and the societies are very active in the campuses for organizing extra-curriculum mathematical modeling activities by the students themselves.

Why the contest CUMCM is so successful in China? I think there are at least two reasons. The first reason is that the contest is a real challenge to its participants and thus is very welcomed by the students. The special and unique experience students have got during the contest, e.g., solving problems from real industry, cooperative teamwork and 72-h hardworking, improved students' competency in a lot of aspects, enriched extracurricular lives, etc., is greatly helpful to tap their innovative potential and strengthen their cooperative spirit. The contestants conclude their experience with one sentence "Once participated, benefit for life long." The whole contest process consists of three stages, namely, the training and preparation before the contest, the hard work during the three-day contest, and the summing up of students' own experience and doing further work on the contest problems after the contest. Through these stages students' creativity and overall ability are greatly improved. Indeed, most of the winners of CUMCM have done very well in their successive courses and projects before their graduation. This naturally leads to the second reason why the contest is so attractive to students—the scientific and industry communities are getting to know more and more about CUMCM, and they are glad to accept the students who have the experience of the contest when they go to graduate schools or find jobs after their graduation. Therefore, the participants, especially the winners of the contest, have a much brighter future in their life (good job, good salary, more chance, etc.). An example is that one advertisement of the IBM China Research Lab in 2009 announced that in order to get some of its research positions, "award in highly regarded mathematical modeling contest is a plus."

4 Difficulties and Prospects

CUMCM does encounter some difficulties as the participation continues to grow. The most important task facing NOC is about how to improve the whole quality of the contest. First of all, the good contest problems are vital to the success of the contest. Contributing a good modeling problem, which is both a meaningful real-world problem and also a solvable problem by most teams within three days, is a real challenging task to the organizers. Another difficulty the organizer faces is how to ensure the equity and fairness of the contest. Since the contest lasts for three days and it is essentially a completely open contest, it is not easy to supervise if some teams violate the contest rules. As a matter of fact, some teams do violate the contest rules, for instance, looking for help from teachers or other persons outside the team even on the Internet. The organizer emphasizes the very importance of self-discipline. A firm policy is observed by NOC, namely, once we have the evidence of violating the contest rules by some teams, these teams will be disqualified from the competition.

5 Conclusion

It now becomes an important part of contemporary industrial and applied mathematics and an absolutely necessary step for connecting mathematics and applications. In order to promote a close relation between mathematics and the world outside mathematics (other sciences, industry and high tech, social and human life etc.), mathematical modeling plays a crucial role. For university students, it should emphasize and organize their formation and training on mathematical modeling. Based on this thought, a contest in mathematical modeling named CUMCM is organized in China to promote mathematical modeling activities. According to our practice, the contest provides a good educational interface between mathematics and industry. CUMCM and the related teaching activities in mathematical modeling and applications are very successful in China universities. They play an important role in reforming the contents and forms of mathematical education. It will be helpful to investigate whether this experience can be extended to other countries.

Appendix

PROBLEM TITLES FOR CUMCM 1992–2011 (For details, see http://en.mcm.edu.cn)

(92A) Analysis on Fertilization Effect. (92B) Decomposition of Some Experimental Data. (93A) Frequency Design of Intermodulation Signal. (93B) Sorting Football Teams. (94A) Designing a Road in a Mountain Area. (94B) Packing Locks into Boxes. (95A) Air Traffic Control. (95B) Operations Management of Cranes and Furnace. (96A) Optimal Strategy for Fishing. (96B) Water Saving Procedure for Washing Machine. (97A) Designing Parameters of the Component Parts. (97B) Sectional Cuttings. (98A) Returns and Risks of Investment. (98B) Inspection Tours in a Disaster Area. (99A) Management of an Automated Lathe. (99B) Layout of Exploratory Wells. (99C) Heaping up the Gangue. (99D) Layout of Exploratory Wells (similar to 99B). (00A) Classification of DNA sequences. (00B) Ordering and Transportation of Steel Pipe. (00C) Flying over the North Pole Area. (00D) Hole Exploration. (01A) 3D Rebuilding for Blood Vessel. (01B) Buses Scheduling. (01C) Fund Plan. (01D) Buses Scheduling (similar to 01B). (02A) Optimal Design for Light Source of Headlights. (02B) Mathematics in Lottery Ticket. (02C) Optimal Design for Light Source of Headlights (similar to 02A). (02D) Match Schedule Arrangement. (03A) Transmission of SARS. (03B) Truck Planning for an Opencast Iron Mine. (03C) Transmission of SARS (similar to 03A). (03D) Speedily Crossing the Yangtze River. (04A) Planning Temporary Mini Supermarkets for the Olympic Games. (04B) Transmission Congestion Management of Electricity Market. (04C) Drinking and Driving. (04D) Recruiting Government Officers. (05A) Water Quality Evaluation and Prediction of the Yangtze River. (05B) Online DVD Rental Business. (05C) Evaluation of Rain Forecasting Methods. (05D) Online DVD Rental Business (similar to CUMCM-05B). (06A) Resource Allocation of a Publishing House. (06B) The Assessment and Prediction of the AIDS Treatments. (06C) Optimal design of the shape and dimension of beverage cans. (06D) Inspecting and Controlling of Gas and Coal Dust. (07A) Prediction of China Population Growth. (07B) Public Transportation Routes Selection Problem. (07C) How Many Benefits Does A Package Deal Offer? (07D) Scheduling for physical performance tests. (08A) Digital Camera Positioning. (08B) On the Tuition in Higher Education. (08C) Search in an area. (08D) Analysis and Evaluation of the NBA Schedule. (09A) Analysis on control method of brake test bench. (09B) Assignment of Hospital Beds. (09C) Tracking satellites or spaceships. (09D) Conference Preparations. (10A) Position Identification and Capacity Table Calibration of Oil Tanks. (10B) A quantitative evaluation on the influence of Expo 2010 Shanghai. (10C) Planning the oil pipeline. (10D) Evaluation for four designs of student dormitory. (11A) Heavy Metals Contamination in Urban Topsoil. (11B) Setting and Scheduling of Traffic and Patrol Police Service Platforms. (11C) System Reform of Retired Enterprises Employees. (11D) Natural Casing Bundling Problem.

References

Jiang, Q., Xie, J., & Ye, Q. (2007). An introduction to CUMCM. In C. Haines, P. Galbraith, W. Blum, & S. Khan (Eds.), *Mathematical modeling—ICTMA12: Education, engineering and economics* (pp. 168–175). Chichester, UK: Horwood Publishing.

Li, D. (Ed.). (2011). *China undergraduate mathematical contest in modeling* (4th ed.). Beijing: Higher Education Press. (in Chinese).

Xiao, S. (2000). *Research report on reforms of higher mathematics (for non-mathematical specialties)*. Beijing: Higher Education Press. (in Chinese).

Xie, J. (2010). Mathematical modeling courses and related activities in China universities. In A. Araújo, et al. (Eds.), *Proceedings of EIMI 2010 conference (educational interfaces between mathematics and industry)* (pp. 597–606). COMAP: MA.

Part X
Conclusion

Conclusion on Educational Interfaces Between Mathematics and Industry

Alain Damlamian, José Francisco Rodrigues and Rudolf Sträßer

This conclusion identifies major results and open questions related to the industrial and societal use of mathematics. The main issues raised in this final section are the understanding of mathematics, the difficulties of communication between different communities like the workplace and education, differences in time lines, goals and ways to learn, the roles of boundary objects and black boxes within the interaction in and between these communities and the role of modelling activities in the educational interfaces between mathematics and industry.

1 Introduction

In this conclusive section of the Study Book, we try to identify some major issues, which could be seen as "results" of this joint ICMI-ICIAM study. It is simply impossible to mirror the variety and wealth of the contributions to this study. We do not intend to underestimate the details and insights present in the individual contributions of participants of the study and in this Study Book. Nevertheless, some issues stand out and can be identified as overarching topics of the study. This is what this Conclusion tries to identify.

A. Damlamian (✉)
Université Paris-Est Créteil Val de Marne, Créteil Cedex, Val-de-Marne, France
e-mail: damla@u-pec.fr

J. F. Rodrigues
Universidade de Lisboa/CMAF-FCUL, Lisboa, Portugal
e-mail: rodrigue@fc.ul.pt

R. Sträßer
FB 12 Mathe, Inst. Didaktik der Mathematik, Universität Gießen, Gießen, Germany
e-mail: rudolf.straesser@uni-giessen.de

A. Damlamian et al. (eds.), *Educational Interfaces between Mathematics and Industry*, New ICMI Study Series 16, DOI: 10.1007/978-3-319-02270-3_46,

2 The EIMI-Study: Major Issues

2.1 Mathematics: Common, but Different

When talking about mathematics at the study Conference, it was obvious that the two communities have different ways of linking with mathematics. "On average", "industrial" mathematics tends to be advanced and clearly linked to mathematical sub-domains, while "educational" mathematics tends to be more elementary (in some cases even simple Arithmetic) and to be constructed in a learning process. However, in both cases, it seems that mathematics is not ready at hand for the majority of people involved, but has to be identified, sometimes newly constructed (for learners in education) or developed (for use in industry). For the sake of mathematics and for the sake of the exchanges during the conference, it seemed to be helpful to think and speak of "one" mathematics—even if it may be necessary to distinguish different types of "mathematics" for political and/or research purposes. In reality, the Study Conference and the work that followed can be envisaged as the beginning of a co-operation between those two communities.

2.2 Two Communities: Language and Communication

The whole EIMI-Study, and especially the Study Conference, was a major challenge to bring together two (if not three) communities: industrial mathematicians (from academia and industry) and mathematics educators (from primary to tertiary and university level). University mathematicians—coming from traditional and application-oriented institutions—sometimes established bridges between those two "species" of mathematicians. The conference was a place to make them talk and listen to each other—and this was often done in excellent ways.

Whenever different communities meet, they tend to use two different "jargons"—and a communication problem may occur. Sometimes even within each community and within the conference, we could hear different jargons, different research approaches and different paradigms. One way to overcome this problem is to find common metaphors to bridge the gaps. In some instances, the Study Conference contributed to overcoming this communication problem. Hopefully, the EIMI project as a whole can build bridges between the different environments.

2.3 Different Time Lines, Different Goals, Different Ways to Learn

As for details, the contributions in the proceedings and the communications during the Study Conference showed that both communities (Industry and Education) have different time lines. Within industry, the time line is normally shorter than within education. At best, industry wants results in the short run and tends to have clearer goals and purposes of action and clearly defined criteria of success. In education, "soft" goals prevail; there is often a lack of evaluation and of explicit, clear success criteria. In general education in particular, the goal is long-term development of skills and knowledge. The reason for teaching and learning mathematics can be very general and unspecific—like training of rigour, argumentation and logic—together with offering a "correct" image of mathematics to the learner.

Often the traditional teaching is done in institutions with a clear division of labour and clear explicit role models for the actors. Some people (often called "teachers") act as experts who know what is to be delivered and learned, while others (often called "students" or "pupils") act as novices, who have to individually digest what is taught, who have to learn as individuals. From the practice in industry and in some, mostly technical universities, one can see a different framework for the training in industrial mathematics, for industrial mathematics education: somehow following the "community-of-practice" approach (e.g. described by Lave 1988 or Wenger 1998), modelling weeks in general education as well as internships and industrial workshops seem to blur the traditional separation of those who know already and those who have to learn. Instead, people from university and industry—in a joint, social effort—hope to approach problems unsolved to date with the help of mathematics. In general education, the mathematics teacher does not know the answer to a problem set in a modelling day (or week) and has to join her/his students in finding out the constraints of a situation—and hopefully a solution.

Differences in doing mathematics in the various environments also imply different ways to evaluate success. Usually in education, a person will be assessed as an individual—with individual marks as indicators of individual skills and competencies. The 'unit of assessment' is the individual. Normally and especially in examination situations, co-operation is not welcome and will be negatively sanctioned. In an industrial working environment, the situation is often quite different: co-operation is a pre-requisite for success, which can only be achieved in a joint, concerted manner. In many cases, especially in advanced industrial environments, the success of a working group is more important than the individual display of professional skills and competencies.

2.4 Boundary Objects and Black Boxes

During the Study Conference, one particular issue repeatedly came up in presentations and discussions: The question of "black and/or white boxes", of "packaging" mathematics with other conceptual and material tools into (hopefully) automatic solutions to problems, which has the consequence of hiding the mathematics from the immediate view of the users. This packaging can be anything from a fast food cash register, where the keys show only pictures of the items instead of numbers, to the search algorithm in Google™ (cf. the Discussion Document). As long as the user does not know or understand the ingredients of the package, s/he is working with a black box. Getting to understand the inner functioning of the package may change this into a grey and eventually a white box. Using black boxes is an important strategy often employed in industry for the control of the production and distribution process.

Hiding a process in a black box can be a marketing strategy for industry to secure superiority over competitors in the market. In education, black boxes are a challenge for the learning process—even if some learning processes definitely rely on black boxes remaining black, for example when sophisticated software is used to solve advanced problems. Consequently, both communities (education and industry) have to cope with black boxes. In contrast to the intention of industry to keep boxes black for competition purposes, education may be interested in opening black boxes—possibly with the help of technology and simulation to help the learning and understanding process. Here, the two communities may have conflicting interests, but the question whether and which black box should be opened, de-greyed and maybe even made transparent has to be answered.

The plenary of Celia Hoyles at the conference showed a somewhat different approach. Being interested in crossing the boundaries of different communities of practice, a first step could be the identification of boundary objects, which can serve as indicators of limits, but also as a way to cross boundaries. The development of technologically enhanced boundary objects specifically designed to foster understanding of the workplace situation was a strategy employed in projects Hoyles described in her plenary. These specifically designed boundary objects proved helpful for use in the workplace and in vocational education. The concept of 'boundary crossing' fundamental to the work in Working Group 1 is another metaphor to cope with the intricacies of understanding the workplace and making use of workplace situations for the teaching and learning of mathematics.

If one is interested in well-informed and critically minded citizenry, creating boundary objects for understanding the situation and opening black boxes seems to be a necessity. Even if it is impossible to open all black boxes in a technology dominated society, it is desirable that the principles and the process of opening black boxes be taught in the mathematics classrooms of secondary education.

2.5 Modelling

Another major issue was "Modelling". For industry it is the gateway into the use of mathematics. In contrast to education, using a mathematical model in industry is normally linked to a purpose outside of mathematics. For industry, the extra-mathematical part of the famous modelling circle is the reason for using mathematics at all. Generally speaking, it is only if mathematics offers additional insights into an extra-mathematical question, that industry will make use of mathematics. For education, modelling with the help of mathematics is often an aim in itself for classroom activities and curricula and it is definitely an important competence to be acquired. Nevertheless Education should not forget about the importance of the situation to be modelled and the assumptions and interests linked to a certain modellisation. In fact, as observed in the post-conference Macau workshop by Li Ta-tsien, if mathematical modelling is the most important interface between mathematics and industry one can say that a balanced mathematical modelling education is the most important educational interface between mathematics and industry. If one is again interested in well-informed and critically minded citizens, this extra-mathematical aspect of modelling with the help of mathematics is also an unavoidable part of mathematics education.

3 Conclusive Remarks

In the students' training, a balance has to be reached between understanding the theory, understanding the "industry" (or of a particular "industry") and developing practical experience. During the course of the Study, a wide variety of activities susceptible to be used to develop and strengthen links between education and industry came into sight. These activities include modelling weeks in general education, where realistic extra-mathematical problems are analysed with the help of mathematics. During vocational education in colleges and academic institutions, special courses on industrial mathematics are offered, sometimes with mathematics teachers working part-time or full-time in industry. Based inside industry, workshops on (sometimes still unsolved) industrial problems are run at the initiative of industry, which also often offers industrial internships for students to learn more about the industrial use of mathematics. In most cases, this is a "win–win-situation" for both partners, as the students get a more authentic picture of mathematics, while industry may profit from mathematical knowledge not necessarily available in the enterprise. A special target group for activities run by industry can be teachers, who are more than often only acquainted with pure mathematics and not used to the opportunities and constraints of industrial and applied mathematics. All these joint activities also reduce the communication problems mentioned above.

In contrast to these positive developments, a significant fact, which was known before beginning the study, was made more obvious during the course of the Study: there are no research institutions specialising in the subject of educational interfaces between mathematics and industry or industrial mathematics as such. Research on the use of mathematics in industry (taken in the wide sense indicated in the beginning of the paper) has just started in various institutions around the world and in several academic disciplines (such as Didactics of Mathematics, Applied Mathematics or Sociology of Work). The contributions in this study were all submitted by individuals who work in institutions, which do not have the analysis of industrial mathematics as their main mission. Actually, there are very few institutions with this mission. Consequently, research into industrial mathematics and its teaching and learning is frequently a personal, if not private activity, which often cannot count on institutional support. Progress in the field is slow, research methods are imported from other disciplines—they may sometimes be inappropriate. Conceptual, theoretical foundations are insecure, even if education and industry start to analyse the situation with the help of professional organisations like ICIAM and ICMI.

As long as there are no institutes to care for education in industrial mathematics, initiatives like the EIMI-Study are of utmost necessity. The co-operation between diverse societies—the likes of ICMI and ICIAM (as well as the ICIAM member societies)—must compensate for the lack of activities and institutions in the field. If mathematics wants to remain more than playful thinking in the academic ivory tower and as long as industry wants to profit from the steadiness and creativity of university mathematicians, the continuous exchange of ideas between these societies and the sometimes difficult, but continuous co-operation between these societies is necessary.

References

Araújo, A., Fernandes, A., Azevedo, A., & Rodrigues, J. F. (Eds.). (2010). *Educational interfaces between mathematics and industry. Proceedings of EIMI 2010 Lisbon Conference*. Portugal: Centro International de Matemática. Available online at :http://www.cim.pt/files/proceedings_eimi_2010.pdf.

Damlamian, A., & Sträßer, R. (2009). ICMI study 20: Educational interfaces between mathematics and industry. (Discussion Document of the EIMI Study 20) in *L'Enseignement mathématique*, *55*, 197–209, also published in *Zentralblatt für Didaktik der Mathematik*, *41*(4), 525–533. (Discussion Document of the EIMI Study. Also published in L'Enseignement mathématique. 55, 197–209, on behalf of the International Program Committee).

Lave, J. (1988). *Cognition in practice*. Cambridge, Mass: Cambridge University Press.

Organisation for Economic Co-operation and Development. (2008). Global Science Forum. Report on Mathematics in Industry. http://www.oecd.org/dataoecd/47/1/41019441.pdf.

Wenger, E. (1998). *Communities of practice: Learning, meaning and identity*. Cambridge: Cambridge University Press.

Index

A. Damlamian et al. (eds.), *Educational Interfaces between Mathematics and Industry*, New ICMI Study Series 16, DOI: 10.1007/978-3-319-02270-3,

M

MIX
Papier aus verantwortungsvollen Quellen
Paper from responsible sources
FSC® C105338

If you have any concerns about our products,
you can contact us on
ProductSafety@springernature.com

In case Publisher is established outside the EU,
the EU authorized representative is:
Springer Nature Customer Service Center GmbH
Europaplatz 3, 69115 Heidelberg, Germany

Printed by Libri Plureos GmbH
in Hamburg, Germany